Big Data in Radiation Oncology

Imaging in Medical Diagnosis and Therapy

Series Editors
Andrew Karellas
Bruce R. Thomadsen

Stereotactic Radiosurgery and Stereotactic Body Radiation Therapy
Stanley H. Benedict, David J. Schlesinger, Steven J. Goetsch, Brian D. Kavanagh

Physics of PET and SPECT Imaging
Magnus Dahlbom

Tomosynthesis Imaging
Ingrid Reiser, Stephen Glick

Beam's Eye View Imaging in Radiation Oncology
Ross I. Berbeco, Ph.D.

Principles and Practice of Image-Guided Radiation Therapy of Lung Cancer
Jing Cai, Joe Y. Chang, Fang-Fang Yin

Radiochromic Film: Role and Applications in Radiation Dosimetry
Indra J. Das

Clinical 3D Dosimetry in Modern Radiation Therapy
Ben Mijnheer

Hybrid Imaging in Cardiovascular Medicine
Yi-Hwa Liu, Albert J. Sinusas

Observer Performance Methods for Diagnostic Imaging: Foundations,
Modeling, and Applications with R-Based Examples
Dev P. Chakraborty

Ultrasound Imaging and Therapy
Aaron Fenster, James C. Lacefield

Dose, Benefit, and Risk in Medical Imaging
Lawrence T. Dauer, Bae P. Chu, Pat B. Zanzonico

Big Data in Radiation Oncology
Jun Deng, Lei Xing

For more information about this series, please visit:

https://www.crcpress.com/Series-in-Optics-and-Optoelectronics/book-series/TFOPTICSOPT

Big Data in Radiation Oncology

Edited by

Jun Deng
Lei Xing

CRC Press
Taylor & Francis Group
Boca Raton London New York

CRC Press is an imprint of the
Taylor & Francis Group, an **informa** business

CRC Press
Taylor & Francis Group
6000 Broken Sound Parkway NW, Suite 300
Boca Raton, FL 33487-2742

First issued in paperback 2020

ISBN-13: 978-1-138-63343-8 (hbk)
ISBN-13: 978-0-367-78015-9 (pbk)

Library of Congress Cataloging-in-Publication Data

Names: Deng, Jun (Professor of therapeutic radiology), editor. | Xing, Lei, editor.
Title: Big data in radiation oncology / [edited by] Jun Deng, Lei Xing.
Other titles: Imaging in medical diagnosis and therapy ; 30.
Description: Boca Raton : Taylor & Francis, 2018. | Series: Imaging in medical diagnosis and therapy ; 30
Identifiers: LCCN 2018040966 | ISBN 9781138633438 (hardback : alk. paper)
Subjects: | MESH: Radiation Oncology | Data Mining--methods
Classification: LCC RC270.3.R33 | NLM WN 21 | DDC 616.99/40757--dc23
LC record available at https://lccn.loc.gov/2018040966

Visit the Taylor & Francis Web site at
http://www.taylorandfrancis.com

and the CRC Press Web site at
http://www.crcpress.com

To my wife, Jie, and my children, Daniel and Grace,
Thank you for your love, support, and inspiration.

Jun

In loving memory of my father who passed away from
rectal cancer. His spirit lives with me.

Lei

Contents

Series preface   ix
Preface   xi
Acknowledgments   xiii
Editors   xv
Contributors   xvii

1. Big data in radiation oncology: Opportunities and challenges   1
   Jean-Emmanuel Bibault

2. Data standardization and informatics in radiation oncology   13
   Charles S. Mayo

3. Storage and databases for big data   23
   Tomas Skripcak, Uwe Just, Ida Schönfeld, Esther G.C. Troost, and Mechthild Krause

4. Machine learning for radiation oncology   41
   Yi Luo and Issam El Naqa

5. Cloud computing for big data   61
   Sepideh Almasi and Guillem Pratx

6. Big data statistical methods for radiation oncology   79
   Yu Jiang, Vojtech Huser, and Shuangge Ma

7. From model-driven to knowledge- and data-based treatment planning   97
   Morteza Mardani, Yong Yang, Yinyi Ye, Stephen Boyd, and Lei Xing

8. Using big data to improve safety and quality in radiation oncology   111
   Eric Ford, Alan Kalet, and Mark Phillips

9. Tracking organ doses for patient safety in radiation therapy   123
   Wazir Muhammad, Ying Liang, Gregory R. Hart, Bradley J. Nartowt, David A. Roffman, and Jun Deng

10. Big data and comparative effectiveness research in radiation oncology   145
    Sunil W. Dutta, Daniel M. Trifiletti, and Timothy N. Showalter

11. Cancer registry and big data exchange   153
    Zhenwei Shi, Leonard Wee, and Andre Dekker

12. Clinical and cultural challenges of big data in radiation oncology   181
    Brandon Dyer, Shyam Rao, Yi Rong, Chris Sherman, Mildred Cho, Cort Buchholz, and Stanley Benedict

13. Radiogenomics   201
    Barry S. Rosenstein, Gaurav Pandey, Corey W. Speers, Jung Hun Oh, Catharine M.L. West, and Charles S. Mayo

14. Radiomics and quantitative imaging   219
    Dennis Mackin and Laurence E. Court

15. Radiotherapy outcomes modeling in the big data era 241
Joseph O. Deasy, Aditya P. Apte, Maria Thor, Jeho Jeong, Aditi Iyer, Jung Hun Oh, and Andrew Jackson

16. Multi-parameterized models for early cancer detection and prevention 265
Gregory R. Hart, David A. Roffman, Ying Liang, Bradley J. Nartowt, Wazir Muhammad, and Jun Deng

Index 283

Series preface

Since their inception over a century ago, advances in the science and technology of medical imaging and radiation therapy are more profound and rapid than ever before. Further, the disciplines are increasingly cross-linked as imaging methods become more widely used to plan, guide, monitor, and assess treatments in radiation therapy. Today, the technologies of medical imaging and radiation therapy are so complex and computer-driven that it is difficult for the people (physicians and technologists) responsible for their clinical use to know exactly what is happening at the point of care, when a patient is being examined or treated. The people best equipped to understand the technologies and their applications are medical physicists, and these individuals are assuming greater responsibilities in the clinical arena to ensure that what is intended for the patient is actually delivered in a safe and effective manner.

The growing responsibilities of medical physicists in the clinical arenas of medical imaging and radiation therapy are not without their challenges, however. Most medical physicists are knowledgeable in either radiation therapy or medical imaging, and expert in one or a small number of areas within their disciplines. They sustain their expertise in these areas by reading scientific articles and attending scientific talks at meetings. In contrast, their responsibilities increasingly extend beyond their specific areas of expertise. To meet these responsibilities, medical physicists periodically must refresh their knowledge of advances in medical imaging or radiation therapy, and they must be prepared to function at the intersection of these two fields. How to accomplish these objectives is a challenge.

At the 2007 annual meeting of the American Association of Physicists in Medicine in Minneapolis, this challenge was the topic of conversation during a lunch hosted by Taylor & Francis Publishers and involving a group of senior medical physicists (Arthur L. Boyer, Joseph O. Deasy, C.-M. Charlie Ma, Todd A. Pawlicki, Ervin B. Podgorsak, Elke Reitzel, Anthony B. Wolbarst, and Ellen D. Yorke). The conclusion of this discussion was that a book series should be launched under the Taylor & Francis banner, with each volume in the series addressing a rapidly advancing area of medical imaging or radiation therapy of importance to medical physicists. The aim would be for each volume to provide medical physicists with the information needed to understand technologies driving a rapid advance and their applications for safe and effective delivery of patient care.

Each volume in the series is edited by one or more individuals with recognized expertise in the technological area encompassed by the book. The editors are responsible for selecting the authors of individual chapters and ensuring that the chapters are comprehensive and intelligible to someone without such expertise. The enthusiasm of volume editors and chapter authors has been gratifying and reinforces the conclusion of the Minneapolis luncheon that this series of books addresses a major need of medical physicists.

This series "*Imaging in Medical Diagnosis and Therapy*" would not have been possible without the encouragement and support of the series manager, Lou Chosen, Executive Editor at Taylor & Francis. The editors and authors, and most of all I, are indebted to his steady guidance of the entire project.

William R. Hendee
Founding Series Editor

Preface

We initially discussed the possibility of publishing a book on big data in radiation oncology while organizing a symposium on the topic in the 2015 ASTRO annual meeting at San Antonio, Texas. After chatting with Lou Han of the Taylor & Francis Group, it became apparent that this was an undertaking that would benefit the community of radiation oncology and cancer research. We were thrilled to receive highly constructive and encouraging comments from two anonymous reviewers, to whom we are grateful, about our book proposal submitted to Taylor & Francis in late 2016. Here we are—in a period of just a little over two years, we were able to bring our idea to print.

The tremendous possibilities that big data can bring to cancer research and management have triggered a flood of activities in the development and clinical applications of the technology. Particularly, with the support of machine learning algorithms and accelerated computation, the field is taking off with tremendous momentum. We strongly believe that data science will dramatically change the landscape of cancer research and clinical practice in the near future.

This book is intended for radiation oncologists, radiation physicists, radiation dosimetrists, data scientists, biostatisticians, health practitioners, and government, insurance, and industrial stakeholders. The book is organized into four main groups: Basics, Techniques, Applications, and Outlooks. Some of the most basic principles and concepts of big data are introduced in the Basics. Following that, techniques used to process and analyze big data in radiation oncology are discussed in some details. Then some clinical applications of big data in radiation oncology are presented with great details. Finally, future perspectives and insights are offered into the use of big data in radiation oncology in terms of cancer prevention, detection, prognosis, and management.

Compared to a handful of similar books, the major features of this book include: (1) a comprehensive review of the clinical applications of big data in radiation oncology; (2) specially designed content for a wide range of readership; and (3) valuable insights into future prospects of big data in radiation oncology from experts in the field.

Being the first of its kind in this much talked-about topic, by no means did we set out to nor could we cover all the related topics in this book. However, we hope that this book will lay the foundations to many future works and hopefully inspire others to get involved in big data analytics.

Acknowledgments

It was rewarding to undertake this book project. When we began this project, we knew that it would take a huge amount of time and efforts to complete. However, we grossly underestimated the amount of support we would need from our colleagues in medical physics, radiation oncology, and beyond. It was the totality of our efforts and the support we received that made this book a reality.

While the ultimate responsibility for the content of this book is ours, we acknowledge with gratitude the generous help from the lead authors and co-authors of all the chapters, including Drs. Sepideh Almasi, Aditya P. Apte, Stanley Benedict, Jean-Emmanuel Bibault, Stephen Boyd, Cort Buchholz, Mildred Cho, Laurence E. Court, Joseph O. Deasy, Andre Dekker, Sunil W. Dutta, Brandon Dyer, Issam El Naqa, Eric Ford, Gregory R. Hart, Vojtech Huser, Aditi Iyer, Andrew Jackson, Jeho Jeong, Yu Jiang, Uwe Just, Alan Kalet, Mechthild Krause, Ying Liang, Yi Luo, Shuangge Ma, Dennis Mackin, Morteza Mardani, Charles S. Mayo, Wazir Muhammad, Bradley J. Nartowt, Jung Hun Oh, Gaurav Pandey, Mark Phillips, Guillem Pratx, Shyam Rao, David A. Roffman, Yi Rong, Barry S. Rosenstein, Ida Schönfeld, Chris Sherman, Zhenwei Shi, Timothy N. Showalter, Tomas Skripcak, Corey W. Speers, Maria Thor, Daniel M. Trifiletti, Esther G.C. Troost, Leonard Wee, Catharine M.L. West, Yong Yang, and Yinyi Ye.

We would also like to thank the people at the Taylor & Francis Group, particularly Lou Han, for continuous and prompt support during this long journey. We can't remember how many times we have approached Lou for thoughtful advice, suggestions, or last-minute help, for which we are deeply indebted. We are also very grateful to Angela Graven, our Project Manager from Lumina Datamatics, who efficiently managed the typesetting and proofreading of all the chapters in the book.

On a more personal level, we would like to thank our families for their gracious love, unwavering support and encouragement that empowered us to complete this book project.

Editors

Jun Deng, PhD, is a professor at the Department of Therapeutic Radiology of Yale University School of Medicine and an American Board of Radiology board-certified medical physicist at Yale New Haven Hospital. He obtained his PhD from the University of Virginia in 1998 and finished his postdoctoral fellowship at the Department of Radiation Oncology of Stanford University in 2001. Dr. Deng joined Yale University's Department of Therapeutic Radiology as a faculty physicist in 2001. He serves on the editorial boards of numerous peer-reviewed journals and has served on study sections of the NIH, DOD, ASTRO, and RSNA since 2005 and as a scientific reviewer for the European Science Foundation and the Dutch Cancer Society since 2015. He has received numerous honors and awards such as Fellow of Institute of Physics in 2004, AAPM Medical Physics Travel Grant in 2008, ASTRO IGRT Symposium Travel Grant in 2009, AAPM-IPEM Medical Physics Travel Grant in 2011, and Fellow of AAPM in 2013. At Yale, his research has focused on big data, machine learning, artificial intelligence, and medical imaging for early cancer detection and prevention. In 2013, his group developed CT Gently®, the world's first iPhone App that can be used to estimate organ doses and associated cancer risks from CT and CBCT scans. Recently, funded by an NIH R01 grant, his group has been developing a personal organ dose archive (PODA) system for personalized tracking of radiation doses in order to improve patient safety in radiation therapy.

Lei Xing, PhD, is the director of Medical Physics Division and the Jacob Haimson Professor of Medical Physics in the Departments of Radiation Oncology and Electrical Engineering (by courtesy) at Stanford University, Stanford, California. His research has been focused on artificial intelligence in medicine, biomedical data science, medical imaging, inverse treatment planning, image-guided interventions, nanomedicine, and molecular imaging. Dr. Xing is on the editorial boards of a number of journals in medical physics and imaging and is a recipient of numerous awards. He is a fellow of American Association of Physicists in Medicine (AAPM) and American Institute for Medical and Biological Engineering (AIMBE).

Contributors

Sepideh Almasi
Department of Radiation Oncology
Stanford University
Stanford, California

Aditya P. Apte
Department of Medical Physics
Memorial Sloan Kettering Cancer Center
New York City, New York

Stanley Benedict
Department of Radiation Oncology
UC Davis Cancer Center
Sacramento, California

Jean-Emmanuel Bibault
Radiation Oncology Department
Georges Pompidou European Hospital
Assistance Publique—Hôpitaux de Paris
and
INSERM UMR 1138 Team 22:
    Information Sciences to support
    Personalized Medicine
Paris Descartes University
Sorbonne Paris Cité
Paris, France

Stephen Boyd
Department of Electrical Engineering
Stanford University
Stanford, California

Cort Buchholz
SingleMind Consulting
Lake Oswego, Oregon

Mildred Cho
Stanford Center for Biomedical Ethics
Stanford University School of Medicine
Stanford, California

Laurence E. Court
Department of Radiation Physics
The University of Texas MD Anderson Cancer
    Center
Houston, Texas

Joseph O. Deasy
Department of Medical Physics
Memorial Sloan Kettering Cancer Center
New York City, New York

Andre Dekker
Department of Radiation Oncology (MAASTRO
    Clinic)
GROW—School for Oncology and Development
    Biology
Maastricht University Medical Centre
Maastricht, the Netherlands

Jun Deng
Department of Therapeutic Radiology
Yale University School of Medicine
New Haven, Connecticut

Sunil W. Dutta
Department of Radiation Oncology
University of Virginia
Charlottesville, Virginia

Brandon Dyer
Department of Radiation Oncology
UC Davis Cancer Center
Sacramento, California

Eric Ford
Department of Radiation Oncology
University of Washington
Seattle, Washington

Gregory R. Hart
Department of Therapeutic Radiology
Yale University School of Medicine
New Haven, Connecticut

Vojtech Huser
Laboratory of Informatics Development
NIH Clinical Center
Washington, District of Columbia

Aditi Iyer
Department of Medical Physics
Memorial Sloan Kettering Cancer Center
New York City, New York

Andrew Jackson
Department of Medical Physics
Memorial Sloan Kettering Cancer Center
New York City, New York

Jeho Jeong
Department of Medical Physics
Memorial Sloan Kettering Cancer Center
New York City, New York

Yu Jiang
School of Public Health
University of Memphis
Memphis, Tennessee

Uwe Just
Department of Radiotherapy and Radiation
    Oncology
Technische Universität Dresden
Dresden, Germany

Alan Kalet
Department of Radiation Oncology
University of Washington
Seattle, Washington

Mechthild Krause
Department of Radiotherapy and Radiation
    Oncology
Technische Universität Dresden
Dresden, Germany

Ying Liang
Department of Therapeutic Radiology
Yale University School of Medicine
New Haven, Connecticut

Yi Luo
Department of Radiation Oncology, Physics
    Division
University of Michigan
Ann Arbor, Michigan

Shuangge Ma
Department of Biostatistics
Yale University
New Haven, Connecticut

Dennis Mackin
Department of Radiation Physics
The University of Texas MD Anderson Cancer
    Center
Houston, Texas

Morteza Mardani
Department of Radiation Oncology
and
Department of Electrical Engineering
Stanford University
Stanford, California

Charles S. Mayo
Department of Radiation Oncology
University of Michigan
Ann Arbor, Michigan

Wazir Muhammad
Department of Therapeutic Radiology
Yale University School of Medicine
New Haven, Connecticut

Issam El Naqa
Department of Radiation Oncology, Physics
    Division
University of Michigan
Ann Arbor, Michigan

Bradley J. Nartowt
Department of Therapeutic Radiology
Yale University School of Medicine
New Haven, Connecticut

Jung Hun Oh
Department of Medical Physics
Memorial Sloan Kettering Cancer Center
New York City, New York

Gaurav Pandey
Department of Genetics and Genomic
    Sciences
Icahn Institute of Genomics and Multiscale
    Biology
Icahn School of Medicine at Mount Sinai
New York City, New York

Mark Phillips
Department of Radiation Oncology
University of Washington
Seattle, Washington

Guillem Pratx
Department of Radiation Oncology
Stanford University
Stanford, California

Shyam Rao
Department of Radiation Oncology
UC Davis Cancer Center
Sacramento, California

David A. Roffman
Department of Therapeutic Radiology
Yale University School of Medicine
New Haven, Connecticut

Yi Rong
Department of Radiation Oncology
UC Davis Cancer Center
Sacramento, California

Barry S. Rosenstein
Department of Radiation Oncology
and
Department of Genetics and Genomic Sciences
Icahn Institute of Genomics and Multiscale Biology
Icahn School of Medicine at Mount Sinai
New York City, New York

Ida Schönfeld
Department of Radiotherapy and Radiation
   Oncology
Technische Universität Dresden
Dresden, Germany

Chris Sherman
Third Door Media
Boulder, Colorado

Zhenwei Shi
Department of Radiation Oncology (MAASTRO
   Clinic)
GROW—School for Oncology and Development
   Biology
Maastricht University Medical Centre
Maastricht, the Netherlands

Timothy N. Showalter
Department of Radiation Oncology
University of Virginia
Charlottesville, Virginia

Tomas Skripcak
Department of Radiotherapy and Radiation
   Oncology
Technische Universität Dresden
Dresden, Germany

Corey W. Speers
Department of Radiation Oncology
University of Michigan
Ann Arbor, Michigan

Maria Thor
Department of Medical Physics
Memorial Sloan Kettering Cancer Center
New York City, New York

Daniel M. Trifiletti
Department of Radiation Oncology
University of Virginia
Charlottesville, Virginia

Esther G.C. Troost
Department of Radiotherapy and Radiation
   Oncology
Technische Universität Dresden
Dresden, Germany

Leonard Wee
Department of Radiation Oncology (MAASTRO Clinic)
GROW—School for Oncology and Development Biology
Maastricht University Medical Centre
Maastricht, the Netherlands

Catharine M.L. West
Division of Cancer Sciences
The University of Manchester
Manchester Academic Health Science Centre
Christie Hospital
Manchester, United Kingdom

Lei Xing
Department of Radiation Oncology
and
Department of Electrical Engineering
Stanford University
Stanford, California

Yong Yang
Department of Radiation Oncology
Stanford University
Stanford, California

Yinyi Ye
Department of Electrical Engineering
and
Department of Management Science and Engineering
Stanford University
Stanford, California

Big data in radiation oncology: Opportunities and challenges

1

Jean-Emmanuel Bibault

Contents

1.1 What Is Big Data? 2
  1.1.1 The Four V's of Big Data 2
  1.1.2 The Specificities of Medical Data 2
    1.1.2.1 Data Relevance 2
    1.1.2.2 Data Granularity (Surveillance, Epidemiology and End Results Database versus EHRs) 2
    1.1.2.3 Structured Data 3
    1.1.2.4 Unstructured Data: The Challenge of EHRs and the Role of Natural Language Processing 3
  1.1.3 From Big Data and Dark Data to Smart Data 4
1.2 Opportunities of Big Data in Radiation Oncology: Data-Driven Decision Making 4
  1.2.1 Accelerating Treatment Planning 4
    1.2.1.1 Contouring 4
    1.2.1.2 Dosimetry Optimization 4
  1.2.2 Evaluating New Treatment Techniques 4
  1.2.3 Personalized Radiation Oncology 4
    1.2.3.1 Predicting Disease Progression and Treatment Response 4
    1.2.3.2 The Learning Health System 5
1.3 Challenges of Big Data in Radiation Oncology 5
  1.3.1 The Need for a Common Language and Collaborations 5
    1.3.1.1 The Role of Ontologies 5
    1.3.1.2 Existing National and International Collaborative Initiatives 6
  1.3.2 Curation and Storage of Data through Warehousing 6
    1.3.2.1 Data Volume 6
    1.3.2.2 Data Access 6
  1.3.3 Data Mining, Modeling, and Analysis through Machine Learning 7
    1.3.3.1 Support Vector Machine 7
    1.3.3.2 Artificial Neural Network 7
    1.3.3.3 Deep Learning 8
  1.3.4 Ethics and Big Data 8
References 9

The increasing number of clinical and biological parameters that need to be explored to achieve precision medicine makes it almost impossible to design dedicated trials.[1] New approaches are needed for all populations of patients. By 2020, a medical decision will rely on up to 10,000 parameters for a single patient,[2] but it is traditionally thought that our cognitive capacity can integrate only up to five factors in order to make a choice. Clinicians will need to combine clinical data, medical imaging, biology, and genomics to achieve state-of-the-art radiotherapy. Although sequencing costs have significantly decreased,[3,4] we have seen the generalization of electronic health records (EHRs) and record-and-verify systems that generate a large amount of data.[5] Data science has an obvious role in the generation of models that could be created from large databases to predict outcome and guide treatments. A new paradigm of data-driven decision making: The reuse of routine health care data to provide decision support is emerging. To quote I. Kohane, "Clinical decision support algorithms will be derived entirely from data … The huge amount of data available will make it possible to draw inferences from observations that will not be encumbered by unknown confounding."[6]

Integrating such a large and heterogeneous amount of data is challenging. In this first chapter, we will introduce the concept of big data and the specificities of this approach in the medical field. We will show the opportunities of data science applied to radiation oncology as a tool for treatment planning and predictive modeling. We will also explain the main requirements for the implementation of a precision medicine program relying on big data.

1.1 WHAT IS BIG DATA?

This section defines big data and introduces a few key concepts the readers need to be familiarized with before they can proceed.

1.1.1 THE FOUR V'S OF BIG DATA

The fours Vs of big data are volume, variety, velocity, and veracity.[7] A comprehensive EHR for any cancer patient is around 8 GB, with genomic data being much larger than all other data combined (volume). Creating a predictive model in radiation oncology requires a significant heterogeneity in the data types that need to be included (variety). The use of big data for medical decision making requires fast data processing (velocity). As sequencing costs have significantly decreased[3,4,8] and computing power has steadily increased, the only factor preventing us from discovering factors influencing disease outcome is the lack of large phenotyped cohorts. The generalization of the use of EHRs gives us a unique opportunity to create adequate phenotypes (veracity).

1.1.2 THE SPECIFICITIES OF MEDICAL DATA

1.1.2.1 Data relevance

Lambin et al. have described in details the features that should be considered and integrated into a predictive model.[9] They include

- Clinical features: patient performance status, grade and stage of the tumor, blood tests results, and patient questionnaires.
- Treatment features: planned spatial and temporal dose distribution, associated chemotherapy. For this, data could be extracted directly from the record-and-verify software for analysis.
- Imaging features: tumor size and volume, metabolic uptake (more globally included into the study field of "radiomics").
- Molecular features: intrinsic radiosensitivity,[10] hypoxia,[11] proliferation, and normal tissue reaction.[12] Genomic studies play a key role in determining these characteristics.

1.1.2.2 Data granularity (Surveillance, Epidemiology and End Results database versus EHRs)

Big data in radiation oncology means studying large cohorts of patients and integrating heterogeneous types of data. Using these types of data through machine learning holds great promises for identifying patterns beyond human comprehension. Oncology is already moving away from therapies based on anatomical and histological features and focusing on molecular abnormalities that define new groups of patients and diseases. This evolution induces an increasingly complex and changing base of knowledge that ultimately will be not usable by physicians. The other consequence of this is that, as we individualize molecular traits, designing clinical trials will become more and more difficult to the point where it will become statistically impossible to achieve sufficient power. The financial and methodological burdens of designing these clinical trials will eventually become unsustainable. EHR use in most institutions is an elegant and easy way to digitally capture large amounts of data on patient characteristics, treatment features, adverse events, and follow-up. This wealth of information should be used to generate new knowledge. The quality and nature of the data captured is important because poor data will generate poor results ("garbage in, garbage out") and big data should not be seen as a magical box able to answer any question with ease and trust. Clinical trials are designed to avoid confounding factors and gather detailed data that are not always available in EHRs.[13] Several Surveillance, Epidemiology and End Results (SEER) studies have generated fast results on important questions.[14–18] However, when studying radiation treatments, a major limitation of big data is the lack of detailed information on treatment characteristics. Integrating these features straight out of the record-and-verify systems will provide faithful dosimetric and temporal data. Several teams have already published studies using prediction to better adapt radiation treatments.[19–24] None of these approaches have reached clinical daily use. A simple, easy-to-use system would need to be directly implemented into the treatment planning system to provide decision support. The best achievable treatment plan based on a patient's profile would be given to the dosimetrist or physicist. The same system would be used to monitor patients during treatment and notify physicians whenever an adverse event outside of the predicted norm would happen. The data generated by each patient and treatment would be integrated into the model. We are, however, very far from this vision and in order to achieve it several methodological challenges will need to be addressed (e.g., how to capture core radiation oncology data into EHRs, integrate clinical, dosimetric, and biologic data into a single model and validate this model in a prospective cohort of patients).

1.1.2.3 Structured data

In the field of radiation oncology, medical data is already highly structured through the use of oncology information and record-and-verify systems. Data can be easily extracted with the precise features of treatment planning (dosimetry) and delivery; however, this data can have very heterogeneous labels that require time-consuming curation. This is particularly true for anatomical and target volumes labeling. Using routine radiation oncology data requires respecting a set of principles to make it more accessible. These principles, known as the Findable, Accessible, Interoperable, Re-Usable (FAIR) Data Principles,[25] initially developed for research data, are now being extended to clinical trials and routine care data. Data must be Findable, Accessible, Interoperable, and Reusable for research purposes. Behind Findable, Accessible, Interoperable, Re-Usable (FAIR) principles is the notion that algorithms may be used to search for relevant data, to analyze the data sets, and to mine the data for knowledge discovery. EHR data cannot be fully shared, but efforts can be made to make vocabularies and algorithms reusable and enable multi-site collaborations. To achieve that goal, the radiation oncology community must pave the road for semantic frameworks that the sources and the users could agree upon in the future. Besides usual quantitative data (e.g., dose), standard representation of anatomical regions and target volumes is required to study, for example, radiation complications. There are currently several domain-specific software packages for radiation oncology planning: Elekta (MOSAIQ©), Varian (ARIA©), Accuray (Multiplan© and Tomotherapy Data Management System©), and BrainLab (iPlan©). Each of these treatment planning and record-and-verify systems has its own anatomical structure labeling system, and these systems are not consistent across platforms, making it difficult to extract and analyze dosimetric data on a multicenter large scale. Using knowledge management with concept recognition, classification, and mapping, an accurate ontology that is dedicated to radiation oncology structures can be used to unify data in clinical data warehouses, thus facilitating data reuse and study replication in cancer centers.[26]

1.1.2.4 Unstructured data: The challenge of EHRs and the role of Natural Language Processing

Each physician has a specific way of reporting and writing medical notes. To leverage this kind of data, natural language processing (NLP) is required in order to make sense of stored files and extract meaningful data. NLP is a part of machine learning that can help in understanding, segmenting, parsing, or even translating text written in a natural language.[27] It can be used to repurpose electronic medical records (EMR) to automatically identify postoperative complications,[28] create a database from chest radiographic reports,[29] or even rapidly create a clinical summary from data collected for a patient's disease.[30] This kind of technology will be essential for big data analytics in radiation oncology, mostly for clinical information.

1.1.3 FROM BIG DATA AND DARK DATA TO SMART DATA

Radiation oncology is one of the most interesting fields of medicine for big data analytics because treatment planning and delivery data are very structured. However, this type of data is rarely used for analytics (dark data). Data integration approaches are necessary in order to effectively curate clinical unstructured data and this highly structured data. Collecting and repurposing these data into an automatic smart data system will be necessary before any medical use can be made.

1.2 OPPORTUNITIES OF BIG DATA IN RADIATION ONCOLOGY: DATA-DRIVEN DECISION MAKING

This part will highlight a few examples of the potential of big data applications in radiation oncology and cite the main studies that have already used data mining methodologies for technical or clinical questions.

1.2.1 ACCELERATING TREATMENT PLANNING

1.2.1.1 Contouring

The contouring of a large number of organs at risk before treatment planning is very time-consuming. Although manual segmentation is currently viewed as the gold standard, it is subject to interobserver variation and allows fatigability to come into play at the risk of lowering accuracy. A potential way to spare time would be automatic segmentation, with numerous industrial and homemade solutions being developed. Very few of them have been evaluated in clinical practice. Most of the existing solutions use atlases as a basis for automatic contouring. In 2016, DeepMind, a Google-owned startup, announced a project to use deep learning for automatic structures segmentation in head and neck cancer through a partnership with the National Health Service (NHS) in the United Kingdom.[31]

1.2.1.2 Dosimetry optimization

Machine learning has been used to predict radiation pneumonitis after conformal radiotherapy,[32] local control after lung stereotactic body radiation therapy (SBRT),[33] and chemoradiosensitivity in esophageal cancer.[34] In these studies, dose–volume histograms were used as predictive factors. They were also used to predict toxicity after radiotherapy for prostate cancer[35–37] and lung cancer.[38,39] Future treatment planning systems will need to directly integrate machine learning algorithms in order to automatically predict efficacy or toxicity to help the physician choose the optimal dosimetry.[19–24]

1.2.2 EVALUATING NEW TREATMENT TECHNIQUES

Big data studies can help in evaluating new treatment techniques. It is highly unlikely that we will see studies comparing three-dimensional (3D) conformal radiotherapy and intensity-modulated radiotherapy (IMRT). However, IMRT is now used in almost all contexts, even if it was only proven superior to 3D for head and neck cancer.[40] For future treatment technology improvements, big data studies could be used to generate hypothesis that will need to be ideally validated in a prospective trial.

1.2.3 PERSONALIZED RADIATION ONCOLOGY

1.2.3.1 Predicting disease progression and treatment response

Predictive modeling is a two-step process involving qualification followed by validation. Qualification consists of demonstrating that the data are indicative of an outcome. Once predictive or prognostic factors have been identified, they should be validated on a different data set. Once a model has been qualified and validated, further studies must be conducted in order to assess whether treatment decisions relying on the model actually improve the outcome of patients.

Kang et al. have proposed seven principles of modeling[41] in radiation oncology:
1. Consider both dosimetric and non-dosimetric predictors.
2. Manually curate predictors before automated analysis.
3. Select a method for automated predictor selection.
4. Consider how predictor multicollinearity is affecting the model.

5. Correctly use cross-validation to improve prediction performance and generalization to external data provide model generalizability with external data sets when possible.
6. Assess multiple models and compare results with established models.

1.2.3.2 The learning health system

The task of creating and validating a truly integrative model in radiation oncology to guide treatment will require multicenter sharing of data and scientists. However, these models and the methodology used to create them can be used regardless of tumor localization. They will underpin decision support system that will use big data in every radiation oncology department in 10–15 years. These systems will need to be updated almost in real time with dynamic programming and reinforcement learning techniques. They will guide decisions at the time of initial consultation for the best treatment options according to the patient's feature and the state of knowledge. Optimal dose distribution, treatment time, associated chemotherapy, targeted therapy, or immunotherapy will be chosen not by the physician, but by an algorithm. Private initiatives, such as IBM's Watson, are already used in some institutions such as the Memorial Sloan Kettering Cancer Center in New York.[42,43] The same system could also guide treatment decisions for adverse event management as well as after the treatments for follow-up and early detection of any relapse. This "learning health system" will certainly be a game changer in oncology if it can actually be achieved. Follow-up will have to integrate all the data collected by wearable devices and connected objects that are being adopted by a large proportion of the population.[44,45] Continuous, real-time monitoring of abnormal events will lead to earlier detection of relapse and optimization of a salvage treatment's efficiency and cost. Eventually, overall survival will be impacted by such approaches.[46]

1.3 CHALLENGES OF BIG DATA IN RADIATION ONCOLOGY

This section will explore the challenges data scientists face: Means of heterogeneity management and data curation, storage, and access are all mandatory before analysis can actually start. We will also briefly explore the methods available to create predictive models, as an introduction to the following chapters. We will conclude by reminding the regulations and ethics that must be followed when using big data.

1.3.1 THE NEED FOR A COMMON LANGUAGE AND COLLABORATIONS

1.3.1.1 The role of ontologies

Ontologies organize, in a formal logical format, the standardized terms that are both human-readable and machine-processable. Most of them are based on description logics to ensure consistency. They are distributed as open-source components of information systems that can be maintained separately from software and, therefore, shared among many different users and applications. The use of ontology is already widespread in biomedical domains outside of radiation oncology. It has also been recognized as a necessary tool in the basic sciences, for example, the Gene Ontology provides the foundation for annotating genes. The Foundational Model of Anatomy (FMA) was developed by the University of Washington to serve as an ontology of anatomical structures that could be used for multiple purposes.[47] It has been used as a basic ontology in several projects developed by the World Wide Web consortium (W3C) including the NeuroImaging Model. However, it was created as a reference ontology for anatomical entities and is not adapted to represent the anatomical volumes and delineation features specific to radiation oncology. Drawing similar conclusions for medical imaging, the radiology community has developed RadLex, an application ontology that incorporates and accommodates all salient anatomical knowledge necessary to manage anatomical information related to radiology.[48] RadLex has been used, for example, to annotate positron emission tomography-computed tomography (PET-CT) images and support studies on these semantically enriched data.[49] These terminologies can be organized in repositories, such as the Bioportal or the Unified Medical Language System (UMLS) Metathesaurus©.[50] However, the fact that no ontology included radiation oncology–specific terms led to the creation of the Radiation Oncology Ontology (ROO),[51] which reused other ontologies and added specific radiation oncology terms such as *region of interest* (ROI), *target volumes* (Gross Tumor Volume [GTV],

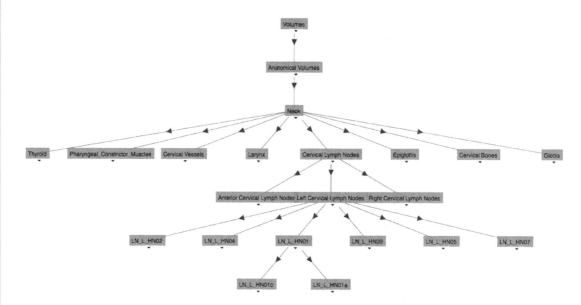

Figure 1.1 Map of the Radiation Oncology Structures (ROS) ontology from the first superclass to the cervical lymph nodes Area I class.

Clinical Target Volume [CTV], Planning Treatment Volume [PTV]), and *dose–volume histogram* (DVH). Still, the ROO does not provide enough anatomical or target volume concepts for an easy use of routine practice data. For example, lymph nodes levels are essential for planning nodal CTV in radiotherapy but are not included.[52,53] A new ontology dedicated to radiation oncology structures was created in 2017. That ontology is available online on Bioportal[26] and GitHub[54] in .owl, .csv, and .rdf formats. Figure 1.1 shows the map from the first superclass to the cervical lymph nodes area I class.

1.3.1.2 Existing national and international collaborative initiatives

Several national and international data curation initiatives are underway. In the United States, the Radiation Therapy Oncology Group (RTOG), National Surgical Adjuvant Breast and Bowel Project (NSABP), and Gynecologic Oncology Group (GOG) have already created a cloud to gather radiation oncology data,[55] along the existing platforms: Radiation Oncology Incident Learning System (RO-ILS),[56] the National Radiation Oncology Registry,[57] and John Hopkins' Oncospace (https://oncospace.radonc.jhmi.edu/). The National Institutes of Health Personalized Medicine Initiative will gather data from a million patients.[58] Finally, American Society of Clinical Oncology (ASCO) has created its own initiative, CancerLinQ (Cancer Learning Intelligence Network for Quality; http://www.cancerlinq.org) that American Society for Radiation Oncology (ASTRO) joined in 2017. In Europe, the German Cancer Consortium (DKTK), the Eurocan Platform,[59] and the EuroCAT[60] have also been created with the same goals.

1.3.2 CURATION AND STORAGE OF DATA THROUGH WAREHOUSING

1.3.2.1 Data volume

The volume of data that needs to be collected and managed is rapidly growing. Today, we can estimate that data for a single patient would amount to about 8 GB, including the raw genomic data that would account for roughly 70% of it (Table 1.1).

1.3.2.2 Data access

Health data security and accessibility is a major challenge for any institution. Health data should be accessible with ease and velocity from anywhere, without compromising their safety. Access to the data requires that the architecture take into account high-security constraints, including a strong user authentication and methods that guarantee traceability of all data processing steps. Login procedures for relevant health care professionals require a scalable process with a significant cost, but they should certainly not be overlooked.[61]

Table 1.1 **Data types and approximate sizes for a single patient**

DATA TYPE	FORMAT	APPROX SIZE
Clinical features	Text	10 MB
Blood tests	Numbers	1 MB
Administrative	ICD-10 codes	1 MB
Imaging data	DICOM	450 MB
Radiation Oncology data (planning & on-board imaging)	DICOM, RT-DICOM	500 MB
Raw genomic data	BAM : Position, base, quality	6 GB
Total		**7.9 GB**

Medical record linkage and data anonymization are very often necessary steps in providing data for research, and they often require a trustworthy third party to take care of these procedures. In general, to provide health care data for research, the data must be moved from the care zone, where data is under the control of the trusted relationship between physician and patient, to the none-care zone, where data is under the control of special data governance bodies, to be anonymized and made available for analysis.

Several translational research platforms are available to integrate large data sets of clinical information with omics data.[62] Despite technological advances, some authors believe the increases in data volume could be outstripping the hospitals' ability to cope with the demand for data storage.[58] One solution would consist of managing this data as most hospitals manage old medical files (i.e., moving the oldest and biggest files to external storage). For digital data, in order to maintain fast and easy access, we would need to move the most voluminous data to a secondary storage-optimized platform, separate from the query platform.

1.3.3 DATA MINING, MODELING, AND ANALYSIS THROUGH MACHINE LEARNING

Several machine learning algorithms have been used in oncology:
- Decision trees (DTs),[63] where a simple algorithm creates mutually exclusive classes by answering questions in a predefined order.
- Naïve Bayes (NB) classifiers,[64,65] which output probabilistic dependencies among variables.
- k-Nearest neighbors (k-NN),[66] where a feature is classified according to its closest neighbor in the data set, is used for classification and regression.
- Support Vector Machine (SVM),[67] where a trained model will classify new data into categories.
- Artificial neural networks (ANNs),[68] where models inspired by biological neural networks are used to approximate functions.
- Deep learning (DL),[69] a variant of ANNs, where multiple layers of neurons are used.

Each of these methods has advantages and limitations, with different computation power requirements.

1.3.3.1 Support vector machine

Logistic regression defines a linear threshold for a limited number of features. If the model needs to integrate a higher number of variables that cannot be separated linearly, SVM can be used to find complex patterns. Similarity functions (or kernels) are chosen to perform a transformation of the data and choose data points or "support vectors." Patients with a combination of vectors are used to compare new patients and predict their outcome. SVMs have been used in several studies to predict radiation pneumonitis after conformal radiotherapy,[32] local control after lung SBRT,[33] and chemoradiosensitivity in esophageal cancer.[34] In these studies, the authors classified the input parameters as dose (dose-volume histogram [DHV], equivalent uniform dose [EUD], biological equivalent dose [BED]) or non-dose features (clinical or biological features). It should be noted that the exact number and nature of features used is not always provided, which might limit the impact and applicability of the results.

1.3.3.2 Artificial neural network

In ANNs, several layers of "neurons" are set up. Each neuron has a weight that determines its importance. Each layer receives data from the previous layer, calculates a score, and passes the output to the

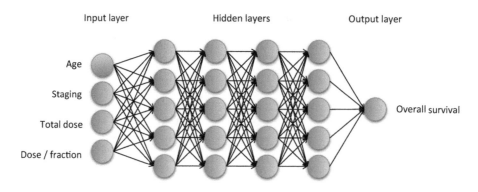

Figure 1.2 An example of a neural network for overall survival prediction.

next layer (Figure 1.2). Using an ANN requires weighting neurons and connections correctly. One method to achieve this is to assign random weights to neurons and iteratively calculate and adjust these weights to progressively improve the correlation. ANN has been used to predict survival in advanced carcinoma of the head and neck treated with irradiation with and without chemotherapy.[70] A three-layer feed-forward neural network integrating 14 clinical parameters was trained through a thousand iterations. Bryce et al. showed that an ANN was more reliable than a Logistic Regression (LR) and used more predictive variables. Six years later, Gulliford et al. used an ANN to predict biological outcome and toxicity after radiotherapy for prostate cancer.[35] They used dosimetric parameters (e.g., DVH) and three separate ANNs on nocturia, rectal bleeding, and PSA measurement. They showed that ANNs were able to predict biochemical control and specific bladder and rectum complications with sensitivity and specificity above 55%. Other studies performed on larger data sets further improved sensitivity and specificity.[36,37] In lung radiotherapy, ANNs have been used to predict pneumonitis.[38,39] In the study by Chen et al., six input features were selected: lung volume receiving >16 Gy (V16), generalized equivalent uniform dose (gEUD) for the exponent a = 1 (mean lung dose), gEUD for the exponent a = 3.5, free expiratory volume in 1 s (FEV1), diffusion capacity of carbon monoxide (DLCO%), and whether or not the patient underwent chemotherapy prior to radiotherapy. All features were then removed one by one from the model to assess their relevance. Except for FEV1 and whether or not the patient underwent chemotherapy prior to radiotherapy, all of them were required for optimal prediction. In another study, ANNs have been used to predict survival in uterine cervical cancer treated with irradiation.[71] In that study, the predictive model used only seven parameters (age, performance status, hemoglobin, total protein, International Federation of Gynecology and Obstetrics [FIGO] stage, and histological aspects and grading of the radiation effect that were determined by periodic biopsy examination).

1.3.3.3 Deep learning

Deep learning is a variant of the ANN. Although ANNs commonly feature one or two hidden layers and is considered as supervised machine learning, DL differentiates itself with a higher number of hidden layers and is able to perform supervised or unsupervised learning. Although DL is gaining interest in medical imaging[72,73] for classification or segmentation, it has not yet been used to predict the outcome after radiotherapy.

1.3.4 ETHICS AND BIG DATA

Big data can be used for a variety of goals, from skin lesions classification[74] to cancer diagnosis prediction, even 1 year before it appears.[75] This powerful tool can be used with both positive and negative intents. As health care providers and researchers, we must prevent its inappropriate use. Internal review boards should be involved whenever a study using stored clinical data is discussed. Patient consent should be obtained before storing health data into a clinical data warehouse. In 2014, a group stole 4.5 million health records from Community Health Systems, which operates 206 hospitals across the United States.[76] Using this kind of nominative data for prediction could have destructive consequences for many patients. Doctor–patient confidentiality, data access security, and theft protection should also be key issues in any institution managing these data warehouses.

REFERENCES

1. Chen, C., He, M., Zhu, Y., Shi, L. & Wang, X. Five critical elements to ensure the precision medicine. *Cancer Metastasis Rev.* **34**, 313–318 (2015).
2. Abernethy, A. P. *et al.* Rapid-learning system for cancer care. *J. Clin. Oncol. Off. J. Am. Soc. Clin. Oncol.* **28**, 4268–4274 (2010).
3. Mardis, E. R. A decade's perspective on DNA sequencing technology. *Nature.* **470**, 198–203 (2011).
4. Metzker, M. L. Sequencing technologies—The next generation. *Nat. Rev. Genet.* **11**, 31–46 (2010).
5. Bibault, J.-E., Giraud, P. & Burgun, A. Big data and machine learning in radiation oncology: State of the art and future prospects. *Cancer Lett.* (2016). doi:10.1016/j.canlet.2016.05.033.
6. Kohane, I. S., Drazen, J. M. & Campion, E. W. A glimpse of the next 100 years in medicine. *N. Engl. J. Med.* **367**, 2538–2539 (2012).
7. Lehmann, C. U., Séroussi, B. & Jaulent, M.-C. Big3. Editorial. *Yearb. Med. Inform.* **9**, 6–7 (2014).
8. DNA Sequencing Costs. Available at: http://www.genome.gov/sequencingcosts/. (Accessed: March 12, 2016.)
9. Lambin, P. *et al.* Predicting outcomes in radiation oncology—Multifactorial decision support systems. *Nat. Rev. Clin. Oncol.* **10**, 27–40 (2013).
10. Bibault, J.-E. *et al.* Personalized radiation therapy and biomarker-driven treatment strategies: A systematic review. *Cancer Metastasis Rev.* **32**, 479–492 (2013).
11. Le, Q.-T. & Courter, D. Clinical biomarkers for hypoxia targeting. *Cancer Metastasis Rev.* **27**, 351–362 (2008).
12. Okunieff, P., Chen, Y., Maguire, D. J. & Huser, A. K. Molecular markers of radiation-related normal tissue toxicity. *Cancer Metastasis Rev.* **27**, 363–374 (2008).
13. Chen, R. C., Gabriel, P. E., Kavanagh, B. D. & McNutt, T. R. How will big data impact clinical decision making and precision medicine in radiation therapy. *Int. J. Radiat. Oncol.* (2015). doi:10.1016/j.ijrobp.2015.10.052.
14. Berrington de Gonzalez, A. *et al.* Proportion of second cancers attributable to radiotherapy treatment in adults: A cohort study in the US SEER cancer registries. *Lancet Oncol.* **12**, 353–360 (2011).
15. Virnig, B. A. *et al.* Studying radiation therapy using SEER-Medicare-linked data. *Med. Care.* **40**, IV–49 (2002).
16. Darby, S. C., McGale, P., Taylor, C. W. & Peto, R. Long-term mortality from heart disease and lung cancer after radiotherapy for early breast cancer: Prospective cohort study of about 300,000 women in US SEER cancer registries. *Lancet Oncol.* **6**, 557–565 (2005).
17. Du, X., Freeman, J. L. & Goodwin, J. S. Information on radiation treatment in patients with breast cancer: The advantages of the linked medicare and SEER data. Surveillance, Epidemiology and End Results. *J. Clin. Epidemiol.* **52**, 463–470 (1999).
18. Song, Y. *et al.* Survival benefit of radiotherapy to patients with small cell esophagus carcinoma—An analysis of Surveillance Epidemiology and End Results (SEER) data. *Oncotarget.* (2015). doi:10.18632/oncotarget.6764.
19. Wu, B. *et al.* Fully automated simultaneous integrated boosted-intensity modulated radiation therapy treatment planning is feasible for head-and-neck cancer: A prospective clinical study. *Int. J. Radiat. Oncol. Biol. Phys.* **84**, e647–e653 (2012).
20. Wu, B. *et al.* Data-driven approach to generating achievable dose-volume histogram objectives in intensity-modulated radiotherapy planning. *Int. J. Radiat. Oncol. Biol. Phys.* **79**, 1241–1247 (2011).
21. Petit, S. F. *et al.* Increased organ sparing using shape-based treatment plan optimization for intensity modulated radiation therapy of pancreatic adenocarcinoma. *Radiother. Oncol. J. Eur. Soc. Ther. Radiol. Oncol.* **102**, 38–44 (2012).
22. Appenzoller, L. M., Michalski, J. M., Thorstad, W. L., Mutic, S. & Moore, K. L. Predicting dose-volume histograms for organs-at-risk in IMRT planning. *Med. Phys.* **39**, 7446–7461 (2012).
23. Zhu, X. *et al.* A planning quality evaluation tool for prostate adaptive IMRT based on machine learning. *Med. Phys.* **38**, 719–726 (2011).
24. Robertson, S. P. *et al.* A data-mining framework for large scale analysis of dose-outcome relationships in a database of irradiated head and neck cancer patients. *Med. Phys.* **42**, 4329–4337 (2015).
25. Wilkinson, M. D. *et al.* The FAIR Guiding Principles for scientific data management and stewardship. *Sci. Data.* **3**, 160018 (2016).
26. Bibault, J.-E. Radiation Oncology Structures Ontology—Summary|NCBO BioPortal. Available at: http://bioportal.bioontology.org/ontologies/ROS/?p=summary. (Accessed: April 15, 2017.)
27. Chowdhury, G. G. Natural language processing. *Annu. Rev. Inf. Sci. Technol.* **37**, 51–89 (2003).
28. Murff, H. J. *et al.* Automated identification of postoperative complications within an electronic medical record using natural language processing. *Jama.* **306**, 848–855 (2011).

Big data in radiation oncology

29. Hripcsak, G., Austin, J. H., Alderson, P. O. & Friedman, C. Use of natural language processing to translate clinical information from a database of 889,921 chest radiographic reports 1. *Radiology.* **224**, 157–163 (2002).

30. Warner, J. L., Anick, P., Hong, P. & Xue, N. Natural language processing and the oncologic history: Is there a match? *J. Oncol. Pract.* **7**, e15–e19 (2011).

31. Applying machine learning to radiotherapy planning for head & neck cancer. *DeepMind.* Available at: https://deepmind.com/blog/applying-machine-learning-radiotherapy-planning-head-neck-cancer/. (Accessed: May 17, 2017.)

32. Chen, S., Zhou, S., Yin, F.-F., Marks, L. B. & Das, S. K. Investigation of the support vector machine algorithm to predict lung radiation-induced pneumonitis. *Med. Phys.* **34**, 3808–3814 (2007).

33. Klement, R. J. *et al.* Support vector machine-based prediction of local tumor control after stereotactic body radiation therapy for early-stage non-small cell lung cancer. *Int. J. Radiat. Oncol. Biol. Phys.* **88**, 732–738 (2014).

34. Hayashida, Y. *et al.* Possible prediction of chemoradiosensitivity of esophageal cancer by serum protein profiling. *Clin. Cancer Res. Off. J. Am. Assoc. Cancer Res.* **11**, 8042–8047 (2005).

35. Gulliford, S. L., Webb, S., Rowbottom, C. G., Corne, D. W. & Dearnaley, D. P. Use of artificial neural networks to predict biological outcomes for patients receiving radical radiotherapy of the prostate. *Radiother. Oncol. J. Eur. Soc. Ther. Radiol. Oncol.* **71**, 3–12 (2004).

36. Pella, A. *et al.* Use of machine learning methods for prediction of acute toxicity in organs at risk following prostate radiotherapy. *Med. Phys.* **38**, 2859–2867 (2011).

37. Tomatis, S. *et al.* Late rectal bleeding after 3D-CRT for prostate cancer: Development of a neural-network-based predictive model. *Phys. Med. Biol.* **57**, 1399–1412 (2012).

38. Chen, S. *et al.* A neural network model to predict lung radiation-induced pneumonitis. *Med. Phys.* **34**, 3420–3427 (2007).

39. Su, M. *et al.* An artificial neural network for predicting the incidence of radiation pneumonitis. *Med. Phys.* **32**, 318–325 (2005).

40. Nutting, C. M. *et al.* Parotid-sparing intensity modulated versus conventional radiotherapy in head and neck cancer (PARSPORT): A phase 3 multicentre randomised controlled trial. *Lancet Oncol.* **12**, 127–136 (2011).

41. Kang, J., Schwartz, R., Flickinger, J. & Beriwal, S. Machine learning approaches for predicting radiation therapy outcomes: A clinician's perspective. *Int. J. Radiat. Oncol. Biol. Phys.* **93**, 1127–1135 (2015).

42. Parodi, S. *et al.* Systems medicine in oncology: Signaling network modeling and new-generation decision-support systems. *Methods Mol. Biol. Clifton NJ.* **1386**, 181–219 (2016).

43. Watson Oncology. *Memorial Sloan Kettering Cancer Center.* Available at: https://www.mskcc.org/about/innovative-collaborations/watson-oncology. (Accessed: March 10, 2016.)

44. The coming era of human phenotyping. *Nat. Biotechnol.* **33**, 567 (2015).

45. Savage, N. Mobile data: Made to measure. *Nature* **527**, S12–S13 (2015).

46. Denis, F. *et al.* Improving survival in patients treated for a lung cancer using self-evaluated symptoms reported through a web application. *Am. J. Clin. Oncol.* (2015). doi:10.1097/COC.0000000000000189.

47. Noy, N. F., Musen, M. A., Mejino, J. L. V. & Rosse, C. Pushing the envelope: Challenges in a frame-based representation of human anatomy. Available at: http://www.sciencedirect.com/science/article/pii/S0169023x03001253. (Accessed: April 25, 2017.)

48. Rubin, D. L. Creating and curating a terminology for radiology: Ontology modeling and analysis. *J. Digit. Imaging.* **21**, 355–362 (2008).

49. Hwang, K. H., Lee, H., Koh, G., Willrett, D. & Rubin, D. L. Building and querying RDF/OWL database of semantically annotated nuclear medicine images. *J. Digit. Imaging.* **30**, 4–10 (2017).

50. Fact SheetUMLS® Metathesaurus®. Available at: https://www.nlm.nih.gov/pubs/factsheets/umlsmeta.html. (Accessed: March 7, 2016.)

51. Radiation Oncology Ontology—Summary|NCBO BioPortal. Available at: http://bioportal.bioontology.org/ontologies/ROO. (Accessed: March 7, 2016.)

52. Grégoire, V. *et al.* Delineation of the neck node levels for head and neck tumors: A 2013 update. DAHANCA, EORTC, HKNPCSG, NCIC CTG, NCRI, RTOG, TROG consensus guidelines. *Radiother. Oncol. J. Eur. Soc. Ther. Radiol. Oncol.* **110**, 172–181 (2014).

53. Rusch, V. W. *et al.* The IASLC lung cancer staging project: A proposal for a new international lymph node map in the forthcoming seventh edition of the TNM classification for lung cancer. *J. Thorac. Oncol. Off. Publ. Int. Assoc. Study Lung Cancer.* **4**, 568–577 (2009).

54. Bibault, J.-E. jebibault/Radiation-Oncology-Structures-Ontology. *GitHub.* Available at: https://github.com/jebibault/Radiation-Oncology-Structures-Ontology. (Accessed: April 15, 2017.)

55. Rosenstein, B. S. *et al.* How will big data improve clinical and basic research in radiation therapy? *Int. J. Radiat. Oncol. Biol. Phys.* **95**, 895–904 (2016).

56. Benedict, S. H. *et al.* Overview of the American Society for Radiation Oncology-National Institutes of Health-American Association of Physicists in Medicine Workshop 2015: Exploring opportunities for radiation oncology in the era of big data. *Int. J. Radiat. Oncol. Biol. Phys.* **95**, 873–879 (2016).

57. Benedict, S. H., El Naqa, I. & Klein, E. E. Introduction to big data in radiation oncology: Exploring opportunities for research, quality assessment, and clinical care. *Int. J. Radiat. Oncol. Biol. Phys.* **95**, 871–872 (2016).

58. Huser, V. & Cimino, J. J. Impending challenges for the use of big data. *Int. J. Radiat. Oncol.* (2015). doi:10.1016/j.ijrobp.2015.10.060.

59. Skripcak, T. *et al.* Creating a data exchange strategy for radiotherapy research: Towards federated databases and anonymised public datasets. *Radiother. Oncol. J. Eur. Soc. Ther. Radiol. Oncol.* **113**, 303–309 (2014).

60. Roelofs, E. *et al.* Benefits of a clinical data warehouse with data mining tools to collect data for a radiotherapy trial. *Radiother. Oncol. J. Eur. Soc. Ther. Radiol. Oncol.* **108**, 174–179 (2013).

61. Li, M., Yu, S., Ren, K. & Lou, W. Securing personal health records in cloud computing: Patient-centric and fine-grained data access control in multi-owner settings. in *Security and Privacy in Communication Networks* (eds. Jajodia, S. & Zhou, J.) 89–106 (Springer, Berlin, Germany, 2010).

62. Canuel, V., Rance, B., Avillach, P., Degoulet, P. & Burgun, A. Translational research platforms integrating clinical and omics data: A review of publicly available solutions. *Brief. Bioinform.* **16**, 280–290 (2015).

63. Quinlan, J. R. Induction of decision trees. *Mach. Learn.* **1**, 81–106 (1986).

64. Langley, P., Iba, W. & Thompson, K. An analysis of Bayesian classifiers. *InAaai.* **90**, 223–228 (1992).

65. Langley, P. & Sage, S. Induction of selective Bayesian classifiers. in *Proceedings of the tenth international conference on Uncertainty in artificial intelligence.* (eds. De Mantaras, R. L. & Poole, D.) 399–406 (Morgan Kaufmann Publishers, San Francisco, CA, 1994).

66. Patrick, E. A. & Fischer III, F. P. A generalized k-nearest neighbor rule. *Inf. Control.* **16**, 128–152 (1970).

67. Vapnik, V. *Estimation of Dependences Based on Empirical Data.* (Springer-Verlag, New York, 1982).

68. Rumelhart, D. E. & McClelland, J. *Parallel Distributed Processing: Explorations in the Microstructure of Cognition.* (Cambridge, MA, MIT Press, 1986).

69. LeCun, Y., Bengio, Y. & Hinton, G. Deep learning. *Nature.* **521**, 436–444 (2015).

70. Bryce, T. J., Dewhirst, M. W., Floyd, C. E., Hars, V. & Brizel, D. M. Artificial neural network model of survival in patients treated with irradiation with and without concurrent chemotherapy for advanced carcinoma of the head and neck. *Int. J. Radiat. Oncol. Biol. Phys.* **41**, 339–345 (1998).

71. Ochi, T., Murase, K., Fujii, T., Kawamura, M. & Ikezoe, J. Survival prediction using artificial neural networks in patients with uterine cervical cancer treated by radiation therapy alone. *Int. J. Clin. Oncol.* **7**, 294–300 (2002).

72. Hua, K.-L., Hsu, C.-H., Hidayati, S. C., Cheng, W.-H. & Chen, Y.-J. Computer-aided classification of lung nodules on computed tomography images via deep learning technique. *OncoTargets Ther.* **8**, 2015–2022 (2015).

73. Guo, Y., Gao, Y. & Shen, D. Deformable MR prostate segmentation via deep feature learning and sparse patch matching. *IEEE Trans. Med. Imaging.* (2015). doi:10.1109/TMI.2015.2508280.

74. Esteva, A. *et al.* Dermatologist-level classification of skin cancer with deep neural networks. *Nature.* **542**, 115–118 (2017).

75. Miotto, R., Li, L., Kidd, B. A. & Dudley, J. T. Deep patient: An unsupervised representation to predict the future of patients from the electronic health records. *Sci. Rep.* **6**, 26094 (2016).

76. Pagliery, J. Hospital network hacked, 4.5 million records stolen. *CNNMoney* (2014). Available at: http://money.cnn.com/2014/08/18/technology/security/hospital-chs-hack/index.html. (Accessed: May 17, 2017.)

2 Data standardization and informatics in radiation oncology

Charles S. Mayo

Contents

2.1 Potential for Big Data in Health Care Informatics and Analytics   13
2.2 Importance of Exchange Unit Standardization in Transactional Systems   14
2.3 Practice Process Standardizations and Cultural Shifts   15
    2.3.1 Diagnosis and Staging   15
    2.3.2 As Treated Plan Sums   16
    2.3.3 Patient-Reported Outcomes   17
2.4 Principles for Development of New Standardizations   18
    2.4.1 Skilled Community Approach   18
    2.4.2 Practical Constraints in Electronic Systems   19
    2.4.3 Templates and Automation in Workflow   20
    2.4.4 Extensibility   20
2.5 Summary   21
References   22

2.1 POTENTIAL FOR BIG DATA IN HEALTH CARE INFORMATICS AND ANALYTICS

A wealth of information is entered on a daily basis into electronic health records (EHRs), radiation oncology information systems (ROISs), treatment-planning systems (TPSs), picture archiving and communications systems (PACSs) and various other electronic systems in the course of treating patients. That should mean that information is readily available as a source to learn what has and has not worked for treating patients today so that we are better informed for treating patients tomorrow.

However, there are common barriers to reaching this vision of data use. In practice, there is substantial variability into which systems data is entered, how it is presented and quantified, and how complete it is. A lack of standardization in clinical processes, key data elements routinely gathered, or methods to demark the data can render attempts at accurate electronic aggregation of this information ineffective.

As a result, clinics commonly resort to accessing this information by manually combing through electronic records, interpreting narrative and explicit indications of key data elements (e.g., toxicity, recurrence, survival, diagnosis, staging, dosimetric metrics) and recording values in spreadsheets. The work is often carried out by research associates hired for the purpose, residents, other staff members, or the principle researchers themselves. This approach is expensive, invites transcription errors, and is not easily extensible from one study to another. It limits the number of patients that can be used in studies to 10 and occasionally to hundreds. Having to rely on manual methods for extraction renders examination of thousands and tens of thousands of patients to more completely represent actual clinical practice as impractical.

By contrast, relatively small amounts of time by highly skilled individuals are required to navigate access to source systems and to build the extraction, transformation, and loading systems (ETLs) that take advantage of standardizations used in entry of data into electronic system to extract this information.

Once the standards are incorporated into routine practice by all providers, ETLs can be created to use the standardization to automate aggregation and analysis as part of routine practice. In addition, the use of standards enables creation of automated electronic curation algorithms to highlight inconsistencies for cross checking. Developing and utilizing standardizations open the door to less effort, lower cost, higher accuracy, and greater speed by enabling accurate automated aggregation of data for all patients treated. Despite the benefits, standardizations for key data elements are currently the exception rather than the rule.

Recently, there has been a rapid increase in the number of researchers interested in constructing and using database systems to incorporate big data efforts into radiation oncology [1–8]. With those efforts, there has been growing awareness of benefits of standardizations, including enabling development of software systems that improve clinical processes and automatable, retrospective analysis of practice norms [9–18]. Implications for technical, cultural, and professional collaboration factors in design and implementation of standardizations are a topic of growing interest [19–21].

In this chapter, we will explore some of the hurdles encountered in attempting to use data stored in electronic systems and the practical approaches that may be employed to overcome those barriers. We will examine the role that standardizations can take in improving the utilization of data for radiation oncology.

2.2 IMPORTANCE OF EXCHANGE UNIT STANDARDIZATION IN TRANSACTIONAL SYSTEMS

The ability to promote order in the exchange units of transactional systems is vital to realizing value from the aggregation of these units and through the expansion of the systems by application of the units to add new capabilities. Disordered transactional units hobble progress by limiting focus to issues of managing entropic units, instead of directing efforts to larger goals for utilizing the underlying value.

Consider a few examples from finance. The early move by the United States in 1792 following the ratification of the Constitution to establish a single stable and regulatable currency was essential for the growth of the united political and economic systems, rather than their dissolution into an impotent array of factious systems. In the commodities markets, the standard of a bushel as a transactional unit with the implied consistencies of quality, form, and accessibility of the constituent grain, enables practical connection and growth of both agrarian and manufacturing systems. For web-based merchants, standardizations in payment information data elements (e.g., credit card number, CSV number expiration date) and in the manifestation schemas for these elements are fundamental to the viability of automatable high-volume, low-cost transactions.

In health care informatics systems, data generated during treatment and follow-up of our current patients is a commodity with the potential to add value for future patients by increasing knowledge of the means to detect and cure disease. Enabling the use of large-scale statistical methods, such as machine learning methods, to explore and elucidate interactions among data elements for large numbers of patients rather than for small subsets requires the development and application of categorization systems as part of routine practice. Standardizations exist for a few of the many data elements required to develop a comprehensive view of covariates that are needed to be built. These have seen widest adoption when required as the basis for financial transactions. Three commonly encountered standardizations are the International Classifications of Diseases (ICD), Current Procedural Terminology (CPT), and Logical Observation Identifiers Names and Codes (LOINC) coding systems.

The Centers for Disease Control and Prevention (CDC) maintains the International Classifications of Diseases Clinical Modification system (ICD-CM) used to classify diseases treated in U.S. health care systems. Versioning is based on the World Health Organization (WHO) ICD system. Thus, ICD-10 is the basis of ICD-10-CM. For example, the ICD-9 and ICD-10 codes used in diagnosis of prostate cancer are 185 and C61, respectively.

The American Medical Association (AMA) maintains the CPT codes used by Centers for Medicare & Medicaid Services (CMS) to categorize procedures used during treatment. For example, treatment planning using intensity modulation radiation therapy (IMRT) or volumetric modulated arc therapy (VMAT) is designated with a CPT code of 77301. CPT codes are reasonably stable but do change as new technologies are introduced or procedures are bundled.

The terminology of LOINC (https://loinc.org/), created by Dr. Clem McDonald working with the Regenstrief Institute, is an open-access system that can be used in EHRs to identify tests and observations. LOINC entries define 18 elements, including distinct numeric and short letter codes for laboratory values and observations and specific conditions of their measurement. For example, albumin measured in urine in units of mass per unit volume has a numerical code of 1754-1 and a letter code of Albumin Ur-mCnc. Albumin in serum or plasma has numeric and letter codes of 1751-7 and Albumin SerPl-mCnc. With the various distinctions, more than 56 entries list albumin as the sole component and 216 reference albumin as part of the component or measurement.

Where these systems have been applied, they bring a valuable capability for ordering information. However, lack of standardization for how they are applied and used in translational research and practice quality improvement efforts can create other difficulties. For example, using the code for prior originating site (e.g., prostate C61) when currently treating a subsequent metastasis (e.g., bone C79.5) creates substantial problems. It means that when using electronic searches of ICD codes linked to treatment courses to identify patient cohorts for a specific originating site (e.g., prostate, lung, breast), the results will incorrectly include metastatic treatments (e.g., bone, brain, lung).

For many key data elements, practical standardizations have not been developed, promulgated, and widely adopted. Without standardizations, clinics operating individually or as part of trials routinely fall back to manual methods to extract data from the EHR, ROIS, and TPS. Consider an example. The general concept of recurrence is key to measuring patient outcomes. Unfortunately, specific values have not been categorized in a standardization widely adopted for use in the clinic and research and then propagated into electronic records systems.

We are both the consumers (translational research, practice quality improvement) and creators (clinical notes, treatment plans) of information in the electronic records. The reality for today's clinical practitioners is that to reach the ability to use the wealth of information to support their own goals, and those of health systems, becoming proactive in development and use of standardizations is needed. Our objective in this chapter is to explore a few of the principles for making this process practical.

2.3 PRACTICE PROCESS STANDARDIZATIONS AND CULTURAL SHIFTS

The emergence of exchange unit standardization is driven in active transactional systems when the systems are sufficiently mature that their participants perceive that the balance between the loss of flexibility and autonomy is offset by significant gains in increased transactional volume and added utilization value. In health care, this perception is fostered by demonstrations of standardizations leading to enhanced data access capabilities and providing tangible benefits to translational research (TR), practice quality improvement (PQI), and resource utilization metrics (RUMs). These demonstrations come with an investment in staff to combine domain-specific knowledge of the needed key data elements, participation in clinical processes that generate the data that is needed, and the informatics skills needed to extract and aggregate data from multiple systems. In other words, to free the data by putting well-designed chains on the processes for entry, it is necessary to demonstrate what gains offset the losses imposed by standardization.

Shifting clinical culture to think of data not just as a by-product of doing what needs to be done to get the day's patients treated but also as information that needs to be aggregated to improve our knowledge of how to better treat tomorrow's patients is challenging. Payments and staff productivity metrics (e.g., RVUs) rarely promote steps to improve ability to automate accurate extraction of the key data elements needed to support TR, PQI, and RUMs. For the moment, mandates to make the needed standardizations in clinical practice process come from us. Let us examine three areas where primary challenges to aggregating data are related to practice process standardization.

2.3.1 DIAGNOSIS AND STAGING

Diagnosis and staging information are central elements of many practice quality improvement and translational research efforts. With these, the ability to accurately identify patient cohorts needed to address specific clinical and research questions is significantly improved. However, when diagnosis or staging is entered, it is often as an unstructured free text form in the EHR that is not effective for later, automatable electronic extraction. Because the EHR may be the electronic system in use by the physician at the point of care, it also becomes the point of data entry. The choice may be expedient but is not optimal for later extraction when unstructured free text is used.

The ROIS in use may have the ability to categorize and quantify the elements of staging in the context of ICD codes selected and the staging system (e.g., American Joint Committee on Cancer [AJCC] or International Federation of Gynecology and Obstetrics [FIGO]), providing a much better system for aggregation and curation of the data. In addition, because it is based on a relational database, the ROIS may be able to quantify relationships that are not easily mirrored in EHR text fields. For example, the ICD code–specific staging, including multiple systems (AJCC7, AJCC6, FIGO)—definies which ICD codes among the several associated with the patient are specifically applicable to a specific course of a treatment, linkages between originating disease and subsequent metastatic disease ICD codes, pathology information to related diagnosis codes, and reflecting history diagnosis that change. In spite of these advantages, because the ROIS may not be open at the point of care, the EHR, which is open, becomes the point of data entry using free text fields.

With process changes, it is possible to assure that information on four key factors is available for accurate automated electronic extraction.

1. Correct diagnosis code for originating disease and for metastatic disease
2. Correct staging
3. Linkage of diagnosis code to course of treatment and treatment plans used to address it
4. Linkage of metastatic disease to originating disease

A motivational factor for changing that cultural norm in the practice pattern is demonstrating value.

By piloting efforts to enter quantified diagnosis and staging information into the ROIS and then extracting it in bulk for all patients for which it was entered, the technical ability to close the loop from entry to use is proven and the potential value to TR, PQI, and RUM efforts is demonstrated.

In a clinical setting, there are many consumers of this type of key data element: physicians, physicists, administrative staff, and therapists. Their applications may range widely. For example, administrative staff may need this information (e.g., RUMs) to respond to requests from state agencies, insurers, or hospital administrations. Physicians and physicists may use the information in constructing computational models for response and survival (e.g., TR and PQI). Therapists may use the information for projecting utilization of technologies based on disease site categories (PQI, RUM).

Carrying out a pilot test as a group effort with these stakeholders gains impetus, once value is demonstrated, to change the practice pattern norms so that the data can be aggregated. The pilot test serves to identify a viable clinical process for data entry, curation, aggregation, and reporting. With stakeholder groups' engagement and data from the pilot test, skepticism barriers can be reduced for transitioning the newly piloted process to a routine process for all patients.

2.3.2 AS TREATED PLAN SUMS

Patient outcomes models incorporate dosimetric measures of the dose distributions delivered to the target and normal structures in a treatment course. Considering first course, boost, and revisions, many plans may be generated in a course of treatment. For example, a breast patient may have tangents, supraclavicular, and posterior axillary boost plans plus an electron boost plan. A head and neck patient may be started with VMAT intended for 25 fractions, then transitioned to a different revised plan after fraction 12 to treat the remaining 13 fractions, and finally treated with a different VMAT plan boosting dose to the primary target. To assess the dosimetric measures, an as-treated-plan sum (ATPS) must be created in the treatment planning system, providing the best practically possible representation of the cumulative contributions of the plans treated.

The cumulative plan sum cannot be perfect. Variations in actual dose distribution due to variations in daily patient and organ treatment position or differences when patients are rescanned ideally could be accurately and automatically represented with deformable registrations. Many researchers and vendors are working toward making this a practical reality for routine care, but it is not currently a viable option. Instead, projecting all plans onto the most representative CT scan is currently the most practical approach. Datum with error bars is preferable to no data.

Changing practice norms to include the construction of ATPS for all patients is a cultural shift. Valuing the ability to automate the extraction of dose–volume histogram (DVH) curves for all treated patients is distinct from regarding treatment plans solely as objects needed to enable patient treatments to proceed. For example, in some clinics, electron boost plans may not be planned on the CT scan at all. This means that when questions arise around cumulative dose distributions (e.g., heart and lung doses for breast patients), an extensive effort is required to return to charts and manually create the ATPS. Generally, this manual effort is only carried out for certain subsets of patients. Shifting practice norms to include creation of ATPS as part of routine practice means that the information is available for all patients.

As with diagnosis and staging, pilot efforts demonstrating value for wider efforts in the clinic are important for motivating changes in practice pattern standardizations. Once the ATPS is constructed, most TPSs enable scripting methods that can be used to extract DVH curves that can be aggregated into databases for later use in supporting TR and PQI efforts. If scripting is not an option, then structure and dose Digital Imaging and Communications in Medicine (DICOM) objects can be exported from the TPS and used to programmatically extract the DVH curves. The technical challenges of constructing software to extract DVH curves are much easier to overcome than the cultural challenges required to incorporate the routine creation of the ATPS used by the extraction software. Demonstrating the ability to automate aggregation and analysis of DVH curves for large sets of patients is very valuable to the multi-stakeholder discussions needed to standardize the creation of ATPS as part of routine care.

2.3.3 PATIENT-REPORTED OUTCOMES

The value of patient-reported outcomes (PROs) as a prognostic measure is an active area of exploration. Clinics interested in embarking on use of PROs typically begin with paper forms. Often the database of stacks of paper never makes it into an accessible electronic database. Electronic alternatives may include the existing EHR, which simplifies subsequent security, hardware, and maintenance discussions. Other possibilities include off-the-shelf survey system applications, custom applications, or specialized vended systems.

In all cases, substantial effort will be required to standardize the clinical processes required to ensure that the data is collected and reviewed when PROs are extended to become a part of routine for patients in a practice. Determining who will help the patients set up the electronic accounts needed to log onto a portal to take the survey and interact with them in the clinic during treatment and follow-up visits to take the surveys on tablets or computers may take much more effort than the technical efforts to implement a survey system.

In addition, substantial dedicated time is required from clinicians for selecting a standardized set of PROs to administer to all patients in the clinic, patients by disease site, or both. Paper forms of many standard instruments may have logic flows or formatting characteristics that do not translate well to electronic survey systems. Judicious consideration of the balance between the number of survey questions asked and the number of answers received without overburdening patients must be considered.

As in the preceding examples, pilot efforts that demonstrate the ability to close the loop are valuable motivational touch points for the culture and clinical process changes needed. In our own clinic, we recently ran a pilot effort that introduced electronic PROs gathered in the EHR as part of routine practice for all head and neck patients. The effort was multidisciplinary, involving physicians, medical assistants, therapist data curators, research administrators, physicists, and information technology database abstractors. By demonstrating the ability to review the PRO results for individuals in the EHR and the ability to batch extract over 12,000 longitudinal question/answer pairs for three PRO instruments administered to 460 patients during the 8-month pilot from the EHR (EPIC) that could support TR and PQI efforts, the value of the investment in clinical processes to gather the data was demonstrated. Instead of being locked in a pile of paper, the standardized process enabled having the data be electronically available.

Currently, there is a lack of standardization of PRO question/response designations. As a result, the ability to accurately and succinctly exchange data on these key elements for analysis is compromised, undercutting their value as more widely utilized instruments in routine practice.

Big data in radiation oncology

2.4 PRINCIPLES FOR DEVELOPMENT OF NEW STANDARDIZATIONS

For many key data elements in radiation oncology, standardizations have not been defined. For example, disease recurrence is a well-known concept central to the literature of outcomes modeling; however, no standardized taxonomy of categories has been widely adopted for use in clinical electronic systems. As a result, when this information is gleaned, it is typically by manual inspection of the EHR.

Increasingly, clinics are interested in developing big data analytics resource systems to enable the combination and use of a wide range of data element categories for TR and PQI efforts. The development of standardizations, where they do not exist, becomes a fundamental part of this effort. In this section, we will discuss four principles:

1. Skilled community approach: Development of standardizations by teams that complete the cycle of data entry, automated electronic data extraction, and data use in TR, PQI, and RUM improves the likelihood of creating viable, practical approaches.
2. Practical constraints on electronic system: Standardizations should be implementable within the constraints of the current electronic systems selected for use for data entry.
3. Templates and automation in work flow: Consider clinical data entry processes when designing standardization to attempt to minimize added work.
4. Extensibility: Consider the potential to enable standardizations developed to address a specific issue to aid a broader scope of issues by adding minor modifications to the schema.

Among the examples considered in examining these principles will be the work of the American Association of Physicists in Medicine (AAPM) Task Group 263 (TG-263): Standardization of Nomenclature for Radiation Oncology. The work of AAPM TG-263 addressed all four principles.

2.4.1 SKILLED COMMUNITY APPROACH

Initial forms for standardizations are generally developed by a few individual researchers while solving data accessibility issues to address their particular clinical and research issues. These researchers have felt the pain of not having a standard and developed practical perspectives of requirements needed to relieve the pain. Subsequent collaboration among these groups under the auspices of professional organizations to revise and extend their work for wider applicability in the community leads to viable standardizations that can be applied across many clinics.

For example, the AAPM TG-263 defined standards for naming of target and normal structures as well as for defining a schema to represent DVH metrics [9]. Foundational work by several authors has demonstrated the plausibility of developing a broadly applicable standardization [10–13]. The task group had 57 members, including foundational authors and representatives from a broad range of stakeholder groups including

1. Clinical role—physician, physicist, vendor
2. Professional societies—AAPM, American Society of Therapeutic Radiation Oncology (ASTRO), European Society of Therapeutic Radiation Oncology (ESTRO)
3. Clinic types—academic and community practices, large and small practices
4. Codependent specialty groups—NRG Oncology; Radiation Therapy Oncology Group (RTOG); Integrating Healthcare Enterprise-Radiation Oncology (IHE-RO), DICOM

With the multi-stakeholder groups, the task group could demonstrate the viability of the proposed nomenclature standardization by piloting an application across many clinics and with vended systems before finalizing recommendations. By including individuals who combine clinical domain knowledge of the data entry processes with information technology skills to extract data from the electronic ROIS and TPS systems, the ability to prove the viability of the standard in all phases of the data entry, extraction, and use cycle was demonstrated. The report and recommendations of the task group are extensive. An illustration of one recommendation, nomenclature for DVH metrics is illustrated in Figure 2.1.

Recently ASTRO published a white paper with recommendations for standardizations for how prescription information is presented [14]. The prescribed dose is a vital key data element for many TR and PQI questions, but it is difficult to accurately extract from many electronic systems owing to incomplete

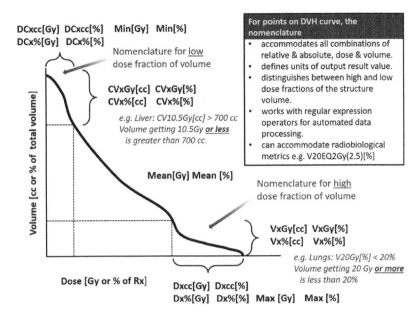

Figure 2.1 TG-263 recommendation for nomenclature standardization for DVH metrics.

information and variability in how data are entered. For example, with IMRT and VMAT technologies multiple areas may be treated to differing dose levels over the several courses of a treatment. Harmonizing the means of defining the fractionation groups (e.g., 1st course, revision, boost), number of fractions, dose to each of the prescription target structures, and treatment modality (e.g., VMAT, particles, brachytherapy) is needed to enable accurate, automated extraction, and exchange of prescription information. Developing these standardizations through a multi-institutional effort promoted by a professional society increases the likelihood of subsequent adoption by a large number of clinics.

2.4.2 PRACTICAL CONSTRAINTS IN ELECTRONIC SYSTEMS

The value of standardizations can be severely compromised if they cannot be practically implemented in existing systems. When designing a standard, testing data entry and extraction across a wide range of systems increases the likelihood of detecting and correcting for unanticipated constraints.

For example, in designing nomenclature for TG-263, the limitations in number and type of characters that can be stored and displayed was an overriding factor in the design of the nomenclature. It was a reason that some foundational work could not be generalized to all vended systems. Preexisting ontologies such as the Foundational Model of Anatomy (FMA) provide important frameworks for categorizing anatomic structures and interrelationships. However, in addition to not being extensible so that radiation oncology structure concepts could be incorporated (e.g., primary and nodal clinical target volumes [CTVp, CTVn], fiducials, bowel bag), many structure names did not meet the character length and type constraints needed to be functional with the current vended systems. In this case, the numerical FMA ID codes were incorporated, where applicable, to the structure names developed for the TG-263 nomenclature to promote interoperability.

Characters used in standardizations (e.g., spaces, forward and backward slashes, octothorps, periods, dashes) require careful consideration by team members skilled with electronic storage, extraction, and analysis. Technical requirements for database, document, and streaming formats likely to be used in transactions should be considered for incompatibilities. For example, if transactions are likely to be carried out with XML documents, then several characters—including the forward slash (/), greater than (>), and less than (<) characters—should not be allowed in the standardization. Spaces are problematic for HTML representations and some versions of UNIX and R.

For some key elements, current vended systems do not provide a practical means of recording key data. In those cases, users may be compelled to create custom software applications until vended systems catch up. In that case, development with intent to interoperate with exiting standardizations and vended systems as much as possible will facilitate the eventual transition to more widely used standardized systems. Prescribed dose, discussed in the previous section, is one example where current vended systems are generally inadequate for the task of motivating the creation of custom applications.

2.4.3 TEMPLATES AND AUTOMATION IN WORKFLOW

The ability to use templates in the natural workflow of point of care/point of data entry increases the ability to promote the adoption of standards. Making the right answer the easy answer is more likely to be successful than relying upon policy to make it the required answer. Often vended systems have some functionality for incorporation of templates that eliminate the need to manually enter standard values and constrain choices to standardized values. Increasingly TPS vendors are incorporating scripting tools that enable adding logic and streamlined interfaces for generating standardized values. Creating these templates and scripts that can be shared among clinics as part of developing standardizations is valuable to increasing the likelihood of adoption.

For example, as part of the pilot phase, users developed scripts and templates to enable automated creation of standardized structures. By reducing the work required to create the structures, the likelihood that they would be created increased.

Approaches may differ depending upon the electronic system that fits most naturally into point of care/point of data entry. For example, many ROISs include well-developed tools for entry of provider-assessed toxicities using common terminology criteria for adverse events (CTCAE). However, because physicians, nurses, and other providers who may enter this information may have the EHR rather than the ROIS open at the point of care, the EHR becomes the system selected for point of data entry.

For example, CTCAE is published jointly by the Department of Health and Human Services (DHHS), the National Institutes of Health (NIH), and the National Cancer Institute (NCI), providing a standard numeric grading scale for toxicities, including xerostomia. However, not all medical staff members (physicians, nurses, physician assistants, mid-level providers) always enter pairs of toxicity items and numerical grade into the EHR in a consistent fashion that will support accurate, automatable electronic extraction. Symptoms may be discussed in a narrative form (Figure 2.1). Toxicities may be mentioned while other text is used to describe the symptoms (Figure 2.1). When data-savvy providers have pioneered standardizing the format used to indicate CTCAE toxicity and grade (Figure 2.1), it becomes possible to leverage that effort make a demonstration of value for the standardization. For this particular case, the standardization consistently applied by the physician's team enabled accurate extraction of thousands of toxicity records within a few seconds once the encounter notes had been extracted from the EHR.

EHRs may contain tools for constructing standardized lists of values that can be incorporated into encounter notes. Using these tools to construct standardized structured text embedded in the note along with narrative free text makes it possible to key data elements very efficiently and accurately using pattern-matching regular expressions once standardized schemas are in consistent use. Recently Mayo et al. described a method using Smart Lists and Smart Phrases in the EPIC EHR to implement a standardized schema for demarcation of key data elements, values, and attributions, enabling an efficient workflow for clinicians in the EHR and subsequent accurate, automatable extraction.

2.4.4 EXTENSIBILITY

When developing standards to address a specific issue, investing time with stakeholders interested in TR, PQI, and RUM to understand how they would use the standardization and if small modifications could improve functionality of the standardization is valuable.

For example, a significant problem with the current implementation of ICD codes is the inability to convey information about interrelationships. For example, when trying to locate a patient treated for bone metastasis subsequent to treatment of concurrent lung and prostate cancer as the originating diagnosis, the use of the ICD code for the current treatment is ineffective. Linkage to prior treatments is not captured in the current ICD coding schema.

With a few small modifications to enable extensibility, the situation could be improved by, for example, (1) separating related time-ordered diagnoses with underscore characters, (2) separating concurrent diagnoses with a plus sign, and (3) ordering diagnosis codes from left to right as from current to preceding diagnosis. With these extensions, the example diagnosis combination could be coded as C79.51_C61+C34.90. The single modified code could more accurately reflect the cohort group than the current approach of entry of diagnosis separately and then relying on checking relative diagnosis dates to intuit a likely connection. Other schemas for the capture of the linkages can be envisioned, but the development of the extension begins with the recognition of the need to explore use cases.

Often the one constant in a process is that what we do will change. In defining standards, providing for means to allow for the incorporation of additional information enables evolution. In DICOM, users are able to create custom tags. Another example is the specification of the caret symbol (^) in TG-263 to separate standardized nomenclature on the left from custom values on the right. With this approach, PTV_High^60 indicates a standard expression for the PTV target receiving the highest prescribed dose (PTV_High) that is further qualified with the nonstandard value of 60 used to indicate the dose in units of Gy. The standard dose units recommended in both the ASTRO prescription white paper and in TG-263 are cGy. We see from this example that provision of a schema for expansion-facilitated adoption of the TG-263 standard in a clinic currently using units of Gy as a standard while logistics to transition to cGy are developed. Defining standardization schemas to specify the means to extend the standardization is a practical approach to incorporating the reality that iterations will be required in response to experience with implementation. It also facilitates the ability of clinics to adopt standardizations, increases ability to incorporate additional information during implementation.

2.5 SUMMARY

The development and implementation of standardizations for key data elements reflecting concepts core to efforts in practice quality improvement, translational research, and resource utilization metrics are important enablers for bringing the potential of big data and machine learning methods into routine clinical practice. Standardizations enable interoperability of custom and vended electronic systems as well as the development of software to improve clinical process flow, curation, and analysis.

Development, promotion, and adoption may be impeded by cultural or procedural norms when the value to participants in the system is perceived to be asymmetric but requirement for utilization of the standard is not. Engagement of multidisciplinary stakeholder groups to pilot viability tests and prove value is a useful means to incorporate new standardizations into routine clinical practice.

When key data elements are entered as notes in the EHR, transitioning away from the cultural norm of a free text narrative description toward a standardized schema representing quantified values enables subsequent extraction with using regular expression pattern matching.

Changes in clinical practice patterns to better utilize existing stems to assure availability of key data elements that are common in a wide range of TR, PQI, and RUM efforts is needed. These include diagnosis and staging, as treated plan sums, AAPM TG-263 nomenclature, and patient-reported outcomes. Where standards are lacking multi-institutional collaborations, including vendors of electronic systems used in radiation oncology and engagement with professional societies is needed to define exchange unit standards that will support the harmonization of transactions and analysis. These elements include

- Recurrence
- Genomics and biospecimens data for disease assessment
- Electronically implementable patient reported outcomes
- Imaging data including, for example, findings, technique/sequences, and contrast agents
- Outcome- and process-relevant treatment details

With the rise of machine learning–based approaches positioning health care professionals to improve modeling of interactions affecting patient outcomes, the need for standardizations leading to improved ability to feed these data-hungry algorithms from electronic records generated as part of routine practice will increase. Proactive engagement by professional societies and clinical stakeholders is needed to build the data culture supporting this growth.

REFERENCES

1. Mayo CS, Kessler ML, Eisbruch A et al. The big data effort in radiation oncology: Data mining or data farming? *Adv Radiat Oncol.* 2016;1(4):260–271.
2. Palta JR, Efstathiou JA, Rose CM et al. Developing a national radiation oncology registry: From acorns to oaks. *Pract Radiat Oncol.* 2012;2(1):10–17.
3. Robertson SP, Quon H, McNutt TR et al. A data-mining framework for large scale analysis of dose-outcome relationships in a database of irradiated head and neck cancer patients. *Med Phys.* 2015;42(7):4329–4337.
4. Chen RC, Gabiel PE, Kavanagh BD, McNutt TR. How will big data impact clinical decision making and precision medicine in radiation therapy? *Int J Radiat Oncol Biol Phys.* 2016;95(3):880–884.
5. Skripcak T, Belka C, Bosch W, Baumann M et al. Creating a data exchange strategy for radiotherapy research: Towards federated databases and anonymised public datasets. *Radiother Oncol.* 2014;113(3):303–309.
6. Nyholm T, Olsson C, Montelius A et al. A national approach for automated collection of standardized and population-based radiation therapy data in Sweden. *Radiother Oncol.* 2016;119(2):344–350.
7. Roelofs E, Dekker A, Lambin P et al. International data-sharing for radiotherapy research: An open-source based infrastructure for multicentric clinical data mining. *Radiother Oncol.* 2014;110(2):370–374.
8. Mayo CS, Matuszak MM, Jolly S, Schipper M, Hayman J, Ten Haken RK. Big data in designing clinical trials: Opportunities and challenges. *Front Radiat Oncol.* 2017;7:187.
9. Mayo C, Moran JM, Xiao Y, Matuszak M et al. AAPM Taskgroup 263 tackling standardization of nomenclature for radiation therapy. *IJORBP.* 2015;93(3):e383–e384.
10. Santanam L, Hurkmans C, Mutic S et al. Standardizing naming conventions in radiation oncology. *Int J Radiation Oncol Biol Phys.* 2012;83(4):1344–1349.
11. Yu J, Straube W, Mayo C et al. Radiation therapy digital data submission process for national clinical trials network. *Int J Radiat Oncol Biol Phys.* 2014; 90(2):466–467.
12. Yu J, Straube W, Mayo C et al. Radiation therapy digital data submission process for national clinical trials network. *Int J Radiat Oncol Biol Phys.* 2014;90(2):466–467.
13. Mayo CS, Pisansky TM, Petersen IA et al. Establishment of practice standards in nomenclature and prescription to enable construction of software and databases for knowledge-based practice review. *Pract Radiat Oncol.* 2016;6(4):e117–e126.
14. Evans SB, Fraass BA, Berner P et al. Standardizing dose prescriptions: An ASTRO white paper. *Pract Radiat Oncol.* 2016;6(6):e369–e381.
15. Sandström H, Chung C, Jokura H, Torrens M, Jaffray D, Toma-Dasu I. Assessment of organs-at-risk contouring practices in radiosurgery institutions around the world—The first initiative of the OAR Standardization Working Group. *Radiother Oncol.* 2016;121(2):180–186.
16. Fedorov A, Clunie D, Ulrich E et al. DICOM for quantitative imaging biomarker development: A standards based approach to sharing clinical data and structured PET/CT analysis results in head and neck cancer research. *Peer J.* 2016;4:e2057.
17. Meldolesi E, van Soest J, Damiani A et. al. Standardized data collection to build prediction models in oncology: A prototype for rectal cancer. *Future Oncol.* 2016;12(1):119–136.
18. Mayo C, Conners S, Miller R et al. Demonstration of a software design and statistical analysis methodology with application to patient outcomes data sets. *Med Phys.* 2013;40(11):111718.
19. Kessel KA, Combs SE. Review of developments in electronic, clinical data collection, and documentation systems over the last decade—Are we ready for big data in routine health care? *Front Oncol.* 2016;6:75.
20. Mayo CS, Deasy JO, Chera BS et al. How can we effect culture change toward data-driven medicine? *IJORBP* 2016;95(3):916–921.
21. Sloan JA, Halyard M, El Naqa I, Mayo C. Lessons from large-scale collection of patient-reported outcomes: Implications for big data aggregation and analytics. *IJORBP* 2016;95(3):922–929.

3

Storage and databases for big data

Tomas Skripcak, Uwe Just, Ida Schönfeld, Esther G.C. Troost, and Mechthild Krause

Contents

3.1	The Role of Big Data Management in Translational Radiation Oncology	24
3.2	Hypothesis and Data-Driven Research Workflows	26
3.3	Toward Sustainable Big Data Research Platforms	28
	3.3.1 Storage System Requirements	29
	3.3.1.1 Distributed File System	30
	3.3.1.2 In-memory Databases	30
	3.3.2 Data Modeling Requirements	30
	3.3.2.1 Key–Value Stores	30
	3.3.2.2 Column Family Databases	31
	3.3.2.3 Document Databases	31
	3.3.2.4 Graph Databases	32
	3.3.2.5 NewSQL Databases	32
	3.3.3 Processing Requirements	32
	3.3.3.1 Low-Level API	32
	3.3.3.2 MapReduce	32
	3.3.3.3 Query Languages	33
	3.3.4 Portal Requirements	33
	3.3.4.1 Data Management	34
	3.3.4.2 Dashboards	34
3.4	Recent Big Data Technologies	34
	3.4.1 Volume versus Real-Time Response	34
	3.4.1.1 Hadoop and Hadoop Distributed File System	34
	3.4.1.2 In-memory Databases	35
	3.4.2 NoSQL Databases	35
	3.4.2.1 HBase	35
	3.4.2.2 MongoDB	35
	3.4.2.3 AllegroGraph	36
	3.4.3 Data Processing Platforms	36
	3.4.3.1 Pig	36
	3.4.3.2 Hive	36
	3.4.3.3 Spark	36
	3.4.4 Portal Implementations	37
	3.4.4.1 TranSMART	37

3.5 Big Data Systems Security 38
    3.5.1 NoSQL Default Security 38
    3.5.2 Mitigation via Network Level Security 38
    3.5.3 NoSQL Attack Vectors 38
3.6 Integration of Big Data Technologies into Translational Research Platforms 39
References 39

3.1 THE ROLE OF BIG DATA MANAGEMENT IN TRANSLATIONAL RADIATION ONCOLOGY

Today, one of the main objectives for translational research in radiation oncology throughout the world is to accelerate the application of health care innovations into day-to-day care of the cancer patient. Instead of a one-size-fits-all treatment, "personalized medicine" is presented as a novel strategy that takes into account the individual patient's characteristics and, based thereupon, strives for better treatment outcomes. Personalization of oncological treatment approaches will define smaller and smaller patient cohorts as candidates for the same treatment, thus requiring large data sets and new approaches of accumulation of clinical knowledge beyond randomized trials. A new computerized electronic health record (EHR) database infrastructure needs to emerge to enable real-time learning from source medical data collected in large volumes across a diverse spectrum of information systems that one can find in modern radiotherapy clinics or research institutes.

Overall this type of data-driven research in radiation oncology is still in the early-adoption stages while the hypothesis-driven approach that is prevailing in randomized clinical trials had enough time to mature. Both can benefit from data management guidelines for efficient medical evidence discovery. The field of clinical research informatics—which in the past has been able to deliver methods and tools for proficient conduction of clinical trials—is now facing new challenges that come with the implementation of the advanced data-driven analytical workflow of translational researchers. The early practical application of machine learning techniques in radiation oncology presented by EL Naqa et al. (2015) serves as an example of the analytical methods used in data-driven research. In this light, the ability to rapidly prepare data sets (training data) to feed real-time analytical tools, and to prospectively improve models, represents a basic prerequisite for learning health systems.

It becomes obvious that without a proper data management strategy in place, one is not able to sustain the increasing demands for creation of special-purpose data repositories that contain patient information aggregated across multiple clinical data domains. However, the technical feasibility of cross-domain data linkage is not the only obstacle. Once such expressive data sets are created, the conventional database management systems based on the relational model that were introduced decades ago by Codd (1970) stop scaling well enough for the increase in volume (size), variety (diversity of domains), velocity (rate of data flow), and variability (change of meaning), also known as the four Vs of data. Harrison (2015) described this era as the third database revolution, which started when technological leaders represented by Google, Facebook, and Amazon encountered a similar data explosion problem that was mostly triggered by the uprise of cloud computing, social networking, and the Internet of Things (IoT).

Despite the fact that big data in radiation oncology is driven by very different sources—the utilization of EHRs, medical imaging (Digital Imaging and Communications in Medicine [DICOM]), and radiation treatment (DICOM-RT) archival; the generation of large comprehensive omics data sets (e.g., radiomics, genomics, proteomics, metabolomics), patient-reported outcomes (ePRO), and clinician outcome assessment (eCOA); the utilization of wearable sensors in telemedicine; and the broader acceptance of mobile health (mHealth)—it is very likely that the industrially established big data technologies can present key solutions to fulfil the needs of translational research as well.

A problem that prevails is related to the deployment of next-generation database systems to support the specific translational research processes. The radiation oncology community unfortunately does not possess a complete manual that would help to light the way through the big data marketing buzzwords and pragmatically introduce the possibilities of novel underlying database storage technologies. Software providers are trying to sell big data systems as a coherent product. However, in practice, one has to carefully count the pros and cons of specific systems to pick the appropriate toolset for long-lasting information technology (IT) infrastructure. There is no one-size-fits-all solution and it is not possible to simply buy a big data database system and hope that all the data integration and performance scaling problems will be solved on its own. Now it is more important than ever to delineate the guidelines for big data in translational research and set up the data management strategies accordingly. The dynamics of this field also suggest that changes will occur and change management should be considered from the very beginning. It is exciting to watch the early adopters' progress in bringing the radiation oncology big data management components, depicted in Figure 3.1, to the wider audience for critical discussion.

An attempt to conceptually review the research databases of leading radiotherapy institutions for the feasibility of large-scale multicenter data exchange was presented by Skripcak et al. (2014). Here, optimization of institutional data acquisition workflows, which requires the full engagement of research institution personnel, is presented as the foundation for high-quality data pooling regardless of the central, decentral, or hybrid storage model. Highlighted is also the need for linking medical data with rich metadata to provide a solution for the high variability of medical information and make the semantic interoperability for cross-institution data exchange feasible, although this work does not cover database deployment details that are left open for further investigation.

A different example examining practices of curating the local data collection to prospectively create bigger data (data farming) was presented by Mayo et al. (2016). This work also stresses the importance of local data management. In addition, it promotes the idea of standardizing data acquisition elements in order to reduce variability up front to guarantee the later straightforward availability for advanced analytics. Nonetheless, the characteristics of the resulting database solution are based on the conventional relational model and next-generation databases are not discussed in further detail.

There is a great potential within next-generation database systems in all their diversity to solve some of the data management challenges that translational research field is facing, but one should not forget

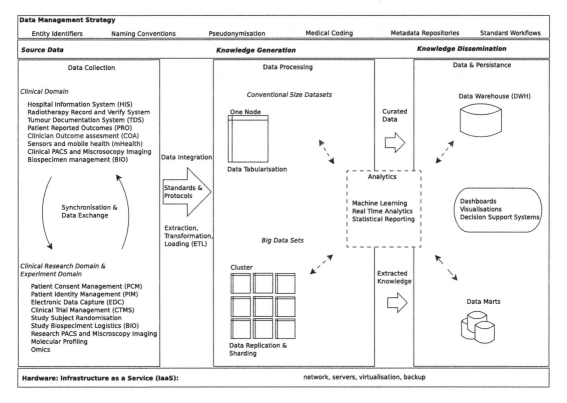

Figure 3.1 Components of the radiotherapy clinical research IT platform as an environment to support translational data-driven research.

that they are merely software tools designed to meet specific requirements of the technological giants that created them. If these should be applied for a specific field of radiation oncology, the scientific community needs to be presented with realistic scenarios of what can be accomplished and the leaders of translational radiation oncology need to share their stories and best practices. It is in the hands of translational research informatics practitioners to build vivid communities to bring a successful and interoperable IT infrastructure that enables collaboration and large-scale multicenter research.

3.2 HYPOTHESIS AND DATA-DRIVEN RESEARCH WORKFLOWS

The traditional, hypothesis-driven approach for generation and validation of medical evidence for clinical practice is the conduction of randomized clinical studies. Such medical knowledge has been deemed necessary to, for example, verify novel treatments aimed at improving outcome.

In hypothesis-driven clinical research, a hypothesis is formulated at the very beginning of the data workflow. According to Richesson and Andrews (2012), the clinical research environment consists of execution-oriented processes classifiable into discrete set of phases that sequentially occur one after another and effectively define the traditional clinical research data life cycle (Figure 3.2). A specific selection of database information systems supports each phase within this waterfall model. These databases have clearly defined objectives, specialized domain data models, and a dedicated class of users who interact with them via predefined user interfaces. If there is a need for some of the data that was captured in one phase to be propagated to a subsequent phase, it is usually copied (data duplication) or referenced via some sort of linkage mechanism. With a data management strategy in place, important domain entities (e.g., patients, biospecimens, imaging scans) use a consistent mechanism with an identity (ID) as a shared key for entity identification and referencing. Sometimes the more advanced information systems are integrated closely together, allowing automatic propagation of information. Prime examples are applications of the Health Level-7 (HL7) standard and the protocol that supports, for instance, patient data synchronization within the integrated health care environment, the DICOM standard for medical imaging data exchange, or the Clinical Data Interchange Standards Consortium Operational Data Model (CDISC ODM) standard for representation of electronic case report forms (eCRF) and their data. In this workflow, it is very uncommon to require data access across all phases of clinical research at once, making scaling problems very unlikely to occur while using the well-established relational database systems. Furthermore, data sets for hypothesis

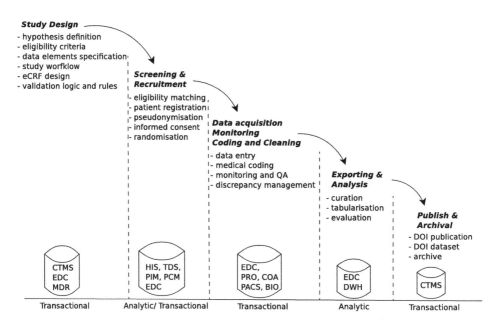

Figure 3.2 Waterfall model for hypothesis-driven clinical research data workflow and the classification of workloads typical for each phase.

testing are created prospectively as the study conduction progresses. Even in the case of retrospective studies, where the source data is available in paper or electronic form, information still has to be extracted and organized, free text codified and cleaned (discrepancies resolved), and tabulated for further analytical processing.

In translational science, the data-driven research methodology is often additionally used to perform a novel analysis of available frozen data sets. This is why the term secondary use of data is often associated with data-driven research scenarios. It simply refers to the fact that the primary data was collected during clinical practice or clinical studies and the consent exists that allows further scientific exploration to examine the potential causal relationships between variables. The comprehensive data sets that are used for data-driven research aggregate data elements from multiple structured, semi-structured, or unstructured primary data domains. As mentioned at the beginning of this chapter, what counts is the ability to rapidly generate cross-domain data sets that often meet the four Vs characteristic of big data. Specialized users participating in data-driven research are data engineers primarily responsible for the data extraction, transformation, and loading (ETL) into the big data environment and the data scientists who develop analytical models and generate structured knowledge. This extracted information represents data-driven research outcome and again often persists in traditional databases. Dedicated user interfaces of visualization or decision support systems afterward provide access to the extracted knowledge for radiation oncologists or medical physicists (Figure 3.3). The choice of the specific data model to be used for the information representation in the big data environment greatly depends on the structure of the source data and the type of analysis to be performed. Big data data sets are normally not meant for long-term archiving; they have the life span of a research project and are designed to scale the storage and parallel processing demands of dedicated use cases.

With Figures 3.2 and 3.3 in mind, one can see why the incentive for the utilization of big data technologies is lesser in the scope of conventional clinical research. There, the data is separated across a diverse set of databases with a well-defined relational data model that allows taking advantage of structured query language (SQL) for data definition and manipulation. A wide variety of available literature sources have

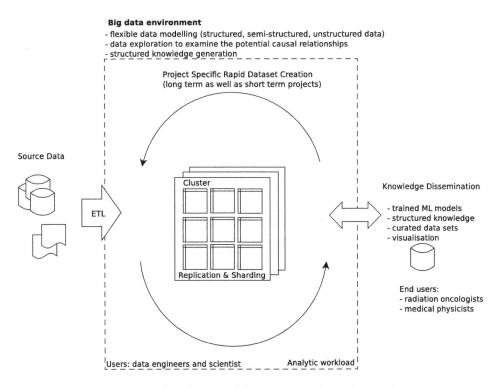

Figure 3.3 Rapid data set creation in data-driven workflow within translational research.

Table 3.1 **Attributes of transactional and analytic workloads**

	TRANSACTIONAL WORKLOAD	ANALYTIC WORKLOAD
Data access	Read/write	Read
Data scope	Small record set	Large record set
Concurrency	Large number	Small number
Latency	Small (real time)	Higher (long-running jobs)
Persistence	Long-term archive	Project time archive
Data model	Fixed, stable schema	Flexible/schemaless (analysis dependent)

covered the relational model in detail (e.g., Codd 1970, 1991). For the purpose of this chapter, it is useful to recapitulate that the relational database consists of a set of tables that defines classes of stored entities, where each table defines a primary key, identifying an entity record and a set of columns (attributes) of an entity. Each row of a table represents an instance of an entity (record). Relations can be represented by referring to the primary key of the specific entity within another foreign entity or relationship table.

Furthermore, it is useful to differentiate between two typical use cases for data workloads. The first type of workload focuses on scenarios where users are performing actions (creating, reading, updating, and deleting [CRUD] operations) on a relatively small record set at a time. This is known as transactional workload or online (real-time) transaction processing (OLTP). The second type of workload, also known as online analytical processing (OLAP), covers scenarios when users are querying a big record set for analytical reporting (such as a query that requires reading from every record in the database). Both of these scenarios are supported by conventional relational databases, the difference lies in how the design of the logical data model is handled. OLTP systems are usually recommended for work with normalized databases (third normal form introduced by Codd [1971]) and OLAP systems use various denormalization techniques to improve the query performance. Table 3.1 summarizes important differences in the attributes of the transactional and analytic workloads.

Relational databases are known for strictly following the atomicity, consistency, isolation, durability (ACID) properties for database transactions and they act as ACID-compliant transaction managers. This prevents concurrent users from working with data in a conflicting way; however, this also comes with a performance penalty and limits the system-scaling ability. The approaches for scaling the database resources fall into two categories: vertical scaling (scale up) and horizontal scaling (scale out). This analogy originates from the typical positioning of hardware in a rack within a data center. Scaling up refers to the addition of new resources (e.g., storage space, memory) to the existing server within one rack, and scaling out refers to the addition of new server nodes to neighboring racks in the server room. Conventional relational databases normally scale up well, but their ability to scale out is limited. Moreover, scaling out relational databases often means losing ACID compliance, which is their biggest advantage in transactional workload scenarios.

A database management system that supports data-driven research will be primarily concerned with analytic data workloads. The main motivation for introducing big data technologies is to prevent performance scaling problems that result from research data growth and usage. In other words, they can help to handle more efficiently the situations when it is no longer technically or financially feasible to scale up the database system, and thus the scale out approach needs to be followed. In the scope of database technologies, scale out is a synonym for distributed databases, where the data is replicated (to answer the need for a large number of concurrent requests and improve the availability) and divided (data is partitioned to multiple nodes according to a control key in order to distribute its volume or to load balance the operations).

3.3 TOWARD SUSTAINABLE BIG DATA RESEARCH PLATFORMS

From the previous sections, it becomes clear that in order to support the institutional efforts for a sustainable big data analysis platform in radiation oncology, it is beneficial to define a structured set of technical requirements essential for the data-driven research workflow. These are summarized in Table 3.2 and further described in this section.

Table 3.2 **Requirements for a big data platform in radiation oncology**

CATEGORY	SOLUTION	COMMENT
Storage system	Distributed file system	Does not require any data model (file storage)
		Very large data volumes
		No real-time analysis
	In-memory database	Requires some sort of data model in place
		Memory size is the limitation
		Could be used for real-time analysis
Data modeling	Key–value stores	Very fast
		Type agnostic (usually restriction on key part)
		Only key lookup (complex processing on client side)
	Column family databases	Efficient for sparse data
		Very fast aggregation operations across key range
		Ideal for time dimensional data (measurements)
	Document databases	Document formatted complex data
		Server-side data processing operations
		Analytic and transactional workloads (without ACID)
	Graph databases	Graph-formatted, complex linked data
		Server-side object traversal
		Standards for graph representation and processing
	NewSQL databases	Relational data model
		SQL for data processing
		Transactional scale-out scenarios
Processing	Low-level API	Supported server-side operations only
		Data processing logic on the client side
		Programming language or Web service
	MapReduce	Parallel MapReduce on server side
		Client submits the jobs as MapReduce functions
	Query languages	Procedural = how to retrieve data
		Declarative = what data to retrieve
Portal	Data management	Analysis project management
		User authentication and authorization
		ETL or analytical job triggers
		Knowledge and model persistence
	Dashboards	Interactive visualizations
		Clinician-facing components

3.3.1 STORAGE SYSTEM REQUIREMENTS

The big data environment should be able to satisfy diverse needs for data modeling and storage of different data-driven analytical projects. The ability to set up a new data storage system on demand can be crucial measure for a big data IT infrastructure, because it is always important to choose the right tool for the specific research scenario. This subsection introduces storage types used for next-generation databases.

3.3.1.1 Distributed file system

In big data environments, distributed file systems (DFS) are designed to store data collections reliably in a cluster consisting of a large number of nodes, typically operating on standard commodity servers with directly attached storages. This solution replicates the data to multiple nodes in order to deliver a highly fault-tolerant system and to enable parallel querying and processing of stored data. The nodes in DFS are likely to communicate with each other through a Transmission Control Protocol/Internet Protocol (TCP/IP)–based protocol, which allows wide area network (WAN) or Internet-wide distribution of servers within a cluster. Normally, file system metadata and data are stored separately on dedicated nodes. Data can be stored on DFS in files the same way files can be stored on a file system of local disk drives. This also means that there is no data model that would force a structure of stored data. It gives the flexibility to store any data format; however, querying and processing is limited to low-level application programming interfaces (APIs) that do not understand the structure of stored files. DFS on its own is also not suitable for real-time analytical projects. Processing jobs are defined as long-time running tasks where processing time for input/output (I/O) operations is not predictable or consistent enough to satisfy real-time feedback.

3.3.1.2 In-memory databases

For situations in which high I/O rates and low latencies are of top priority, as it is common for real time analytical tasks, the in-memory database storage solutions provide an interesting solution if the database size does not exceed the available memory space. Currently, these systems cannot cope with biggest data volumes; however, they are very comfortable with accommodating small-to-medium sized databases while still providing the best possible I/O that is a prerequisite for real-time data processing algorithms. This is fitting for use in cases where processing jobs are triggered from end-user facing components and results should be delivered in a reasonable time. In-memory systems usually provide some alternative storage mechanism to prevent data loss (backup) in case of memory cell failure. It can be implemented as replication of data to other cells in the cluster or persisting to the disk (database images or transactional log). In-memory database systems usually require some sort of data model (vary on database implementation) in order to structure the data for organization and processing.

3.3.2 DATA MODELING REQUIREMENTS

The data model is a formalized representation of data structures that are going to be managed within the database. In the realm of big data technologies, the schemaless data model or schema on write is often discussed. This describes the ability of big data systems to work with data models that support all kinds of non-structured, semi-structured, and structured information. The choice of the data model should reflect the ways of interaction that are going to be required by data engineers and scientists and the level of composition (complexity of data structure) that needs to be persisted. A less complex representation of a data structure is more likely to scale better but may require nontrivial processing algorithms in order to retrieve all related records for analysis. Structures that are more complex cause the composition or relationship information within the model to persist and may result in a decrease of the performance of the system.

3.3.2.1 Key–value stores

Key–value stores are based on the idea of distributed hash tables searchable by a key that identifies the corresponding value (binary data), see Figure 3.4a. In computer science, these sorts of data structures are also known as associative arrays. Intrinsic operations that are supported by a key–value store itself are limited to the management of key–value pairs such as insert new pair, reassign value, remove pair, or lookup according to a key. Any other data processing (including reference integrity checking) has to be performed on the client site by the power-user who has the knowledge about the internal structure of values within the key–value store and the key naming conventions. Key–value store technology is often used as a low-level modeling scheme even for more complex data models. Different implementations of key–value stores can provide extended support for specialized value types declared as lists and sets or documents using Extensible Markup Language (XML) or JavaScript Object Notation (JSON) encoding. Often they use an in-memory storage mechanism but can also operate on a distributed file system.

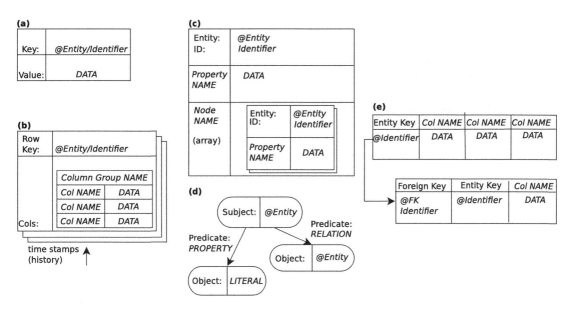

Figure 3.4 Data modeling archetypes of distributed NoSQL systems (from left to right) utilizing the schema-on-write approach are simpler and less structured than the relational (tabular) model: (a) key–value stores, (b) column family databases, (c) document databases, (d) graph databases, and (e) SQL and NewSQL databases.

3.3.2.2 Column family databases

Column family databases extend the idea of key–value stores by a concept in which the value represents one or more column families. Each column family group contains a set of columns, where each column is a triplet consisting of a column name, value, and timestamp capturing the moment of data insertion into the database, as depicted in Figure 3.4b. Similar to the key–value store, column family databases provide an effective way for storing sparse data because columns without values do not need to occupy any storage space. No general purpose query language is supported out of the box. The column family data can be retrieved by row key, and columns are sorted according to column name at the time of their insertion into the column family. Thus, it is critical to consider the key and column naming aspects during data model design to achieve good query performance. This type of database performs very well in aggregation queries (e.g., sum, count, avg) that need to access column values across the whole row key range. The embedded time dimension allows for easy storage of time-series data or repeating measurement instruments data. Column family databases are normally deployed on distributed file system and can leverage parallel processing paradigms (such as MapReduce), which are discussed later in this chapter.

3.3.2.3 Document databases

The core of a document database is also based on the principle of a key–value store. The difference lies in its data representation, where in contrast to pure key–value stores, the document databases will expect the data to be hierarchically formatted and encoded into a specific format (most commonly XML or JSON or its binary variant BSON [Binary JSON]) (Figure 3.4c). A document database knows about the structure of stored data and this knowledge allows the implementation of features that require server side data processing (e.g., queries that need to access specific property values within a document). Stored documents need to represent a valid document instance of the chosen encoding format although they are not enforced to comply with a specific scheme (schema is defined on write). Documents within the database can refer to one another and thus create relations. However, the database does not check the integrity of those references on its own and does not allow server-side document traversal. These data management systems are ideal candidates for a situation where a complex and flexible data model is needed, such as a complete patient EHR or study subject audit trail. They also fit the requirements of transactional workload systems that do not call for the full ACID compliance.

3.3.2.4 Graph databases

Graph databases operate on mathematical principle known as graph theory. Data elements in graph databases are either nodes (entities) or edges (relationships), and both of them can have defined property fields. This data modeling technique gives the possibility of describing complex data structures and provides full support for object linkage. The understanding of structure and relations allows the database engine to support server-side operations that require graph traversal (navigation from one entity to another); however, it is important to say that this comes with a performance penalty. Triplestores represent a specialized type of graph databases often used in life sciences to share biomedical data. The biomedical research community utilizes the semantic web to develop and maintain special-purpose information models (ontologies), which are used to semantically annotate research data sets. Graphs in triplestores are expressed via the Resource Description Framework (RDF) that captures information in the form of a triple (i.e., subject, predicate, and object) as shown in Figure 3.4d. RDF, together with the Simple Protocol and RDF Query Language (SPARQL), make triplestores an excellent choice for storing medical knowledge that should be shared across a wider audience.

3.3.2.5 NewSQL databases

All previously described models are commonly known under the term "not only SQL" (NoSQL). They oppose the relational data stores to resolve the limitation in scale-out scenarios and do not give explicit ACID guarantees. They also excel in analytic workloads; however, their usage in transactional workloads is not always optimal. Each NoSQL solution normally comes with its own data processing APIs, and there is very little standardization within this field (except SPARQL). NoSQL systems are not trivial to use and the missing standard query language leads to a fractioned user base, each often with a small set of specially trained individuals. As a reflection on these issues, NewSQL databases have been developed. They originate on the resurrection of the relational logical data model (Figure 3.4e), ACID compliance, and SQL as a standard interface. Looking at the low-level implementation details, they are, in fact, closer to their NoSQL cousins than to their SQL predecessors in providing competitive horizontal scalability. NewSQL databases can be considered as alternatives to NoSQL systems for transactional as well as analytic workloads, but because they respect the relational logical model of data, they may not fit scenarios where schema on write is required.

3.3.3 PROCESSING REQUIREMENTS

The analytic workload in the big data environment predetermines the importance of querying and processing tasks. Most of the information loaded into a big data database is there for the purpose of data aggregation and its separation from source databases. Analytical operations can be then performed on a single data store without the risk of impairing the productive systems that manage the source data. This subsection explores the data querying and processing paradigms available within big data environments.

3.3.3.1 Low-level API

NoSQL databases have been designed for programmers or data scientists and often provide only low-level API for a specific set of supported programming languages as a method to query and manipulate data. Such APIs cover the set of operations that are supported by the database backend; however, these can be quite limited (e.g., allowing query only by key or value aggregation across key range) for some of the NoSQL database types. This means that data processing logic that cannot be executed on the server side (such as document traversal in document databases or value operations in key–value stores) has to be programmed on the client side by a data scientist. Sometimes APIs based on Web service technologies are implemented in order to provide a generic API layer for any programming language capable of handling hypertext transfer protocol (HTTP).

3.3.3.2 MapReduce

MapReduce is a data processing paradigm that was originally created by Dean and Ghemawat (2008) at Google but is also utilized in other NoSQL database systems and allows parallel execution of tasks across

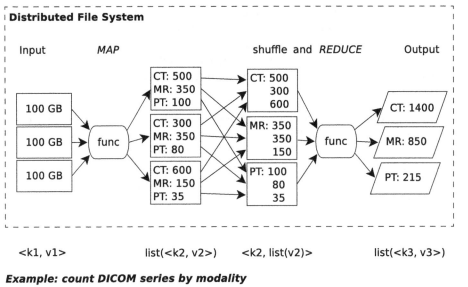

Distributed File System

<k1, v1>   list(<k2, v2>)   <k2, list(v2)>   list(<k3, v3>)

Example: count DICOM series by modality

| Key:
- DicomSeriesUID
Value:
- Modality | Key:
- Modality
Value:
- Count | Aggregate:
sum for each key | Key:
- Modality
Value:
- Sum |

Figure 3.5 MapReduce processing paradigm applied for counting of DICOM series by modality within hypothetical distributed file system based research Picture Archiving and Communication System (PACS).

many nodes of storage clusters. Nevertheless, this is often a nontrivial task because the processing problem needs to be decomposited to a set of MapReduce jobs. Both Map and Reduce functions operate on data formatted as key–value pairs. The job of the Map function is to convert input data stored in a big data environment into a list of key value pairs (Figure 3.5). The Reduce function is executed once for each key generated by the Map and it processes the list of values assigned to that key from all available mappers. The Reduce function performs certain aggregation operations (e.g., sum) on all values for a specific key and generates the list of output key–value pairs. It is worth mentioning that MapReduce processing is not particularly fast and, therefore, should be considered as a long-time running job (not ideal for real-time analytical processing).

3.3.3.3 Query languages

There is ongoing development focused on establishing higher-level query languages that would be able to generate low-level data processing instructions as described above and to provide more people with the possibility of querying big data databases. In general, it is possible to recognize two trends. The first trend follows the procedural query language paradigm where the set of operations describe how the data should be retrieved and processed to produce output. The second trend favors establishing the declarative query language (similar to SQL), which would allow the description of what data should be queried (given some logical model) and let the execution engine underneath prepare the most optimal execution plan. The fact is that both of these developments open the field of NoSQL databases to a wider audience of power-users.

3.3.4 PORTAL REQUIREMENTS

Until now, the core low-level aspects of big database systems have been described. These components do not necessarily provide a graphical user interface (GUI), but are controlled via dedicated APIs. Such an approach, although it can work well for end users with computer science specialization, is not very useful as a direct interface presentable to clinical scientists or clinicians. A clinical facing GUI layer is necessary in order to bring analytical tools and models to broader everyday use. For the purpose of this chapter, the term portal will be used to describe these kinds of components although it is worthwhile to point out that some clinical research data warehouses (CRDW) have the ability to provide portal features.

3.3.4.1 Data management

Data management is crucial for effective conduction of the increasing number of translational research projects. Portals should provide the possibility to persist projects metadata not only to help maintain an overview of running projects but also to allow the linkage with the publication of results. Another mandatory feature is the user administration (authentication and authorization) for restricted access to project data and privileges for executing specified sets of actions. These actions can include, for example, ETL jobs moving and aggregating data from clinical source systems or triggering long-running analytical jobs to be performed within the big data environment. Finally, they should provide a means for maintaining the analytical results or generated knowledge for further use once the big data analysis project is finished.

3.3.4.2 Dashboards

When it comes to dissemination of new knowledge and trained decision support models, even the standard generic GUI of portal can turn out to be too complicated for certain scenarios such as the busy everyday life of clinicians. It is advantageous when the portal has a possibility of dedicated dashboards development and deployment. They can serve as a single-purpose clinician-facing component. Often, they take the form of interactive data visualizations with predefined base criteria or a simple group of input fields necessary for the utilization of train models.

3.4 RECENT BIG DATA TECHNOLOGIES

This section covers some of the currently available implementations of big data systems that follow the technical requirement categories introduced in the previous sections.

3.4.1 VOLUME VERSUS REAL-TIME RESPONSE

When it comes to the decision whether to employ file system or memory as a storage backend, it narrows down to the ability to understand the size of data volume to be persisted in the big data environment and the possibility to perform real-time queries. Only file system–based technologies can provide the necessary storage space for the biggest data sets and, on the other hand, only in-memory optimized solutions will give the best real-time response rate.

3.4.1.1 Hadoop and Hadoop Distributed File System

The Hadoop software framework was developed as an open-source implementation of Google's big data technology stack. Ghemawat et al. (2003) published details about the underlying distributed Google File System (GFS), which triggered work on the Hadoop core storage part, known as the Hadoop Distributed File System (HDFS). HDFS manages the retrieval and storing of data across large clusters of commodity hardware server nodes successfully forming Hadoop distributed data infrastructure. There are already two major versions of Hadoop available. The second iteration extends the processing abilities that are no longer limited to only MapReduce jobs; it now allows the integration of different processing paradigms. It is worth mentioning that nowadays Hadoop can operate also on third-party file systems such as the IBM General Parallel File System (GPFS) that brings the ability to use standard UNIX utilities for accessing and working with data (Portable Operating System Interface [POSIX] compliance). Hadoop by default does not provide data modeling features and is consuming raw data as they are provided by the client. However, as it is described later, there is one column family database (HBase) that can natively run on top of HDFS. Furthermore third party processing platforms exist that can overcome some of Hadoop main limitations such as long time running queries. Hadoop has been proven to handle the storage and processing of world's largest amounts of data. Yahoo! maintains a prime example of one such Hadoop cluster that currently stores over 455 petabytes of data.

3.4.1.2 In-memory databases

Database systems that utilize in-memory storage usually implement one of the NoSQL data modeling paradigms. From a hardware perspective, they require only commodity hardware with sufficient aggregated memory capacity to accommodate small-to-medium sized data collections.

A typical NoSQL representative for in-memory systems is the Remote Dictionary Server (Redis). It is a high-performance key–value store that has the capacity of holding databases larger than available memory by allocating virtual memory on disk, although this safety mechanism will cause significant performance degradation. Redis is an open-source solution that, together with its ability to work on less-expensive hardware, makes it an ideal choice for storing key–value modeled data and is often used as an in-memory cache component for disk-based databases.

SAP HANA is characterized as an implementation of a NewSQL in-memory database system that requires deployment on HANA-certified server hardware equipped with fast solid-state drives (SSDs). HANA's NewSQL nature allows the accommodation of relational data models (row-oriented) as well as columnar storage (column family). As such, it supports both OLTP and OLAP workloads. HANA keeps all relational data in-memory and allows configuring of which columnar store data will be loaded for in-memory access on the database startup. ACID compliance is obtained by each commit synchronously written to disk (transaction logging) that limits the otherwise in-memory speed of I/O operations. HANA represents a flexible enterprise grade next-generation database solution that naturally comes with significant investment costs.

3.4.2 NoSQL DATABASES

The other real-world examples of NoSQL databases presented in this section are not classified as in-memory optimized stores because they normally operate on a special-purpose distributed file system such as HDFS or have the cluster orchestration behavior embedded within their core data management subsystem.

3.4.2.1 HBase

Shortly after Chang et al. (2008) introduced the concept of Bigtable, used in Google's big data technology stack, HBase appeared as an open-source alternative that can run natively on Hadoop HDFS. It is a typical example of a column family database. When deployed to Hadoop, the underlying HDFS takes care of data replication within the cluster while HBase allows structuring of the persisted data using the column family modeling techniques. The Hadoop environment manages the location of HBase tables within the cluster and also provides API-level access to stored data and features MapReduce distributed data processing abilities.

3.4.2.2 MongoDB

MongoDB is one of the most popular JSON-oriented document databases that have been released under the open-source license. Internally, it uses the binary-encoded format called BSON. The BSON format extends the JSON specification about low-level representation of data types such as Dates and allows document parsing operations to be much more efficient. MongoDB utilizes JavaScript as a special-purpose query language that can be used to perform ad hoc queries on document collections and for execution of user-defined functions on the server side. It is known for its developer friendliness that allows quick entry into to world of big data technologies; however, it does not currently provide the best robustness and scalability. Furthermore, the utilization of JavaScript makes MongoDB vulnerable to NoSQL injection attacks discussed in the following section. Even considering these flaws, MongoDB could be a good candidate for big data prototyping scenarios. However, in its present state, it is not a best choice for sustainable productive big data infrastructure in translational radiation oncology research.

3.4.2.3 AllegroGraph

When it comes to the storage and reasoning of a huge number of graph-formatted linked data, AllegroGraph can provide an enterprise-grade closed-source RDF triplestore. AllegroGraph is one of the best performing graph databases currently available on the market. It represents an ACID-compliant system, which makes it suitable for transactional workload scenarios. One of the main advantages that RDF stores are offering compared with other solutions is the existence of SPARQL as a standardized query language for linked data. AllegroGraph is especially interesting for medical research use as an industry-proven, high-performance graph store because its license agreements allows it to operate the free version of the database with up to 5 million triples. Such a structured database is ideal for representing rules that come from treatment guidelines or new knowledge generated from data analysis. It provides embedded support for Prolog reasoning and integrates with other third-party components such indexing applications powering a full text search functionality.

3.4.3 DATA PROCESSING PLATFORMS

An implementation of more advanced interfaces for data processing is important to encourage a broader acceptance of new-generation storage systems. Subsequently listed are some of the higher-level languages that allow decoupling from low-level programming APIs and processing paradigms that are limited for programmers and big data technology experts.

3.4.3.1 Pig

Pig Latin (Pig) was designed by Yahoo as a high-level procedural query language that would allow executing a sequential set of operations on data collections loaded from the Hadoop environment. Effectively, it provides a scripting language as a high-level abstraction from low-level MapReduce jobs that otherwise would have to be programmed using verbose Java Hadoop API. Pig is often used by data scientists who have experience with scripting languages like Python and can embed Pig statements directly into their Python data processing scripts.

3.4.3.2 Hive

Facebook originally developed Hive as another data processing platform that can operate on top of Hadoop clusters. It allows for defining metadata that describe the content (properties and data types) of files stored within the HDFS file system of Hadoop environment (including HBase tables). Using this metadata schema, one is able to compose declarative SQL such as Hive Query Language (HiveQL) queries that are internally translated into Hadoop processing tasks most often represented as a collection of MapReduce jobs. Hive is successfully opening access to the data stored in the Hadoop environment for users with existing SQL experience without dealing with low-level APIs. However, users have to respect the long-running processing nature that is so characteristic of Hadoop. HiveQL still cannot satisfy typical real-time data processing requirements.

3.4.3.3 Spark

Spark was developed as a part of Berkeley Data Analysis Stack (BDAS) at the University of California, Berkeley in order to provide an in-memory data processing framework for Hadoop environments and, therefore, overcome the non–real-time processing limitations of Hadoop alone. It can efficiently create a distributed in-memory file system using the server nodes of a Hadoop cluster. The data loaded into Spark's in-memory storage are represented as immutable resilient distributed data sets (RDDs). All high-level APIs that Spark implements operate on RDDs. The core of Spark processing is based on in-memory implementation of MapReduce. The existence of high-level interfaces in the form of GraphX (Graph Compute Engine), Spark SQL, and MLBase/MLlib (Machine Learning Library) efficiently shield users from the processing core details. Data within RDDs are usually modeled using the key–value paradigm because functions embedded in Spark work specifically on key–value oriented data sets. Although Spark does not support transactional workloads, it provides the possibility to load data from third-party transactional systems such as conventional relational databases on the fly.

Listing 1: Examples of Pig and Hive and Spark processing syntax

```
// Pig Filtering
mr_dicom_series = FILTER dicom_series BY modality == 'MR';
// Pig Sorting
sorted_dicom_series = ORDER dicom_series
BY series_date DESC, patient_id, study_instance_uid;

// Hive Filtering
SELECT * FROM dicom_series WHERE modality = 'MR';
// Hive Sorting
SELECT * FROM dicom_series
ORDER BY series_date DESC, patient_id, study_instance_uid;

// Spark Filtering
val mrDicomSeries = hiveContext.table("dicom_series")
      .filter("modality = 'MR'");
// Spark Sorting
val sortedDicomSeries = hiveContext.table("dicom_series")
      .sort(
          $"series.date".desc,
          $"patient_id",
          $"study_instance_uid"
      );
```

3.4.4 PORTAL IMPLEMENTATIONS

Portal (sometimes referred to as platform) implementations in biomedical research started as life sciences–tuned integrated data warehouses built on top of conventional relational databases. They were designed to bring data-driven research closer to scientists by providing a standard user interface for data management and organization as well as hypothesis testing and data visualization. Nowadays, these software systems more often present a facade (entry point) to big data analysis for nonprogrammer users.

3.4.4.1 TranSMART

At the beginning of 2017, the open-source foundations governing the development of the two most often used portal systems for translational medicine research (i2b2 and tranSMART) joined forces and established the unified i2b2 tranSMART foundation. Scheufele et al. (2014) originally introduced the installation of tranSMART as a fork of i2b2 dedicated for translational science use cases. The recent merge under the wings of one foundation promises closer collaboration between both systems and the ultimate upgrade of the core database model of tranSMART to correspond to the recent version of i2b2. TranSMART supports the storage of various data types, including clinical data and metadata, derived imaging, and biobanking data and high-dimensional data such as gene expressions, copy number variation data, small genomic variants, and peptide or protein and metabolite quantities. All of these are stored within separated database schemes. However, they are linked together via common identifiers. All data that is stored within TranSMART need to be curated and mapped to those schemes. TranSMART implements a user interface for patient cohort exploration and comparison, although its full potential arises from the ability to execute advanced workflows that can be programmed using R language for statistical computing. Restful API is also available for bringing third-party components to the core environment to perform specialized custom analyses. These important features make it possible to integrate TranSMART with specific big data environments such as Hadoop and allow the predefined interaction of nonprogrammer scientists with big data environments.

3.5 BIG DATA SYSTEMS SECURITY

Although NoSQL databases are trying to prove their readiness for industry strength deployment in performance, scaling, and robustness properties, they often lack the enterprise-level security features known from conventional relational solutions. It is true that the shortage of NoSQL API and languages standardization make it difficult for a potential attacker to create technology-wide breaches; however, this also suggests that the world will more often witness specialized attempts to break into systems with the highest adoption rates.

3.5.1 NoSQL DEFAULT SECURITY

Role management, authentication, and authorization are just a couple of security features that are often missing in next-generation database systems. Even when some solutions are slowly introducing security features, they are often considered as advanced and are disabled in default deployment configurations or they are only available via third-party commercial plug-ins (e.g., Elasticsearch JSON-oriented document database). This alone does not need to be critical when the big data infrastructure is only used internally, once the access to the servers is exposed to the Internet, it is necessary to employ countermeasures to mitigate the security risks. In 2017, two massive attacks of different popular NoSQL databases were prominent in the news. The first being a ransomware attack that encrypted publicly available MongoDB databases that had an exposed administration port, which is by default enabled without password protection. Elasticsearch, with its no default security policy, was the second victim of malicious attempts to collect ransom. Often these NoSQL database instances are deployed on public cloud infrastructure (e.g., Amazon Web Services [AWS]) by untrained users that are not aware of proper data protection and security strategies. The rule of thumb should be that one possess full knowledge about security options of the chosen big data system before its productive deployment.

3.5.2 MITIGATION VIA NETWORK LEVEL SECURITY

The most effective approach for dealing with NoSQL systems without any configurable security is via network-level security (Open Systems Interconnection [OSI] network layer). This allows for configuring the allowed and denied network hosts and data flows in detail. Such a configuration focuses on the underlying networks' infrastructure that provides the transportation services and is completely independent from application software. A common way of securing database systems deployed within local networks from unwanted network traffic from the public Internet is by establishing a demilitarized zone (DMZ). A DMZ allows for the separation of the local network that is accessible within the organization from the public Internet by creating another perimeter network that will guard allowed communication to the local network. Multiple variants of DMZ topologies exist, but the dual firewall setup where the front-end firewall protects the Internet-to-DMZ communication and the back-end firewall controls communication between the DMZ and the local network is very common.

3.5.3 NoSQL ATTACK VECTORS

Ron et al. (2016) characterized the five main classes of possible NoSQL attack vectors, which are further described in Table 3.3. It is possible to consider these risks when deploying big data infrastructure with features that allow advanced data querying and processing via dedicated query language or high-level restful Web API. The developers of big data applications first of all need to be aware of these techniques because the most effective way of mitigating them is by early secure software design, prototyping, and testing.

Table 3.3 **Summary of five main NoSQL attack vector categories**

NAME	DESCRIPTION
Tautologies	Bypassing authentication by expressions that are always true
Union queries	Tautologies that lead to non-authenticated data extraction
JavaScript injections	Passing malicious user input into JavaScript queries
Piggybacked queries	Utilizing special characters to insert malicious code
Origin violation	Using trusted clients to perform malicious site actions

3.6 INTEGRATION OF BIG DATA TECHNOLOGIES INTO TRANSLATIONAL RESEARCH PLATFORMS

The adoption rate of the presented big data technologies is growing slowly in the field of translational radiation oncology research. The utilization of novel databases is currently very often limited to special-purpose prototypes dealing with medical imaging data as shown by Kestelyn (2016) or transcriptomic data as presented by Wang et al. (2014). Relatively small standardization of next-generation databases leads to the necessity of onsite experts with sufficient hands-on experience. This also means that the learning curve to master those systems is quite steep for newcomers. Although it was shown that NoSQL systems can outperform the conventional relational databases when it comes to storage and management of big data, their target audience is mostly limited to data engineers or scientists with a sufficient computer science background. Even with newly developed processing platforms that hide the low-level APIs and MapReduce processing jobs, it remains necessary to think about their key–value based structural foundation to utilize the power of embedded data processing functions. Portals and clinical research data warehouses are established in the field of translational research as good data-driven research solutions for clinical scientists. It is only expected that the upcoming years will bring a tighter integration of the big data system into translational research data portals. This will eventually unleash the full potential of big data analytical models as additional decision support technologies in clinical practice, which will contribute to the practical implementation of personalized medicine.

REFERENCES

Chang, F., J. Dean, S. Ghemawat, W. C. Hsieh, D. A. Wallach, M. Burrows, T. Chandra, A. Fikes, and R. E. Gruber (2008). Bigtable: A distributed storage system for structured data. *ACM Transactions on Computer Systems 26*(2), 1–26.

Codd, E. F. (1970). A relational model of data for large shared data banks. *Communications of the ACM 13*(6), 377–387.

Codd, E. F. (1971). Further normalization of the data base relational model. *IBM Research Report, San Jose, California RJ909*.

Codd, E. F. (1991). *The Relational Model for Database Management: Version 2* (Reprinted with corr ed.). Reading, MA: Addison-Wesley. OCLC: 832203055.

Dean, J. and S. Ghemawat (2008). MapReduce: Simplified data processing on large clusters. Simplified data processing on large clusters. *Communications of the ACM 51*(1), 107.

EL Naqa, I., R. Li, and M. J. Murphy (2015). *Machine Learning in Radiation Oncology: Theory and Applications*. Cham, Switzerland: Springer. OCLC: 906656999.

Ghemawat, S., H. Gobioff, and S.-T. Leung (2003). *The Google File System*. pp. 29. New York, ACM Press.

Harrison, G. (2015). *Next Generation Databases: NoSQL, NewSQL, and Big Data. The Expert's Voice in Oracle*. New York: Apress, IOUG. OCLC: ocn920849271.

Kestelyn, J. (2016). Processing and Indexing Medical Images With Apache Hadoop and Apache Solr.

Mayo, C. S., M. L. Kessler, A. Eisbruch, G. Weyburne, M. Feng, J. A. Hayman, S. Jolly, I. et al. (2016). The big data effort in radiation oncology: Data mining or data farming? *Advances in Radiation Oncology 1*(4), 260–271.

Richesson, R. L. and J. E. Andrews (Eds.) (2012). *Clinical Research Informatics. Health Informatics.* London, UK: Springer. doi:10.1007/978-1-84882-448-5.

Ron, A., A. Shulman-Peleg, and A. Puzanov (2016). Analysis and mitigation of NoSQL injections. *IEEE Security & Privacy 14*(2), 30–39.

Scheufele, E., D. Aronzon, R. Coopersmith, M. T. McDuffie, M. Kapoor, C. A. Uhrich, J. E. Avitabile, J. et al. (2014). TranSMART: An open source knowledge management and high content data analytics platform. *AMIA Joint Summits on Translational Science proceedings. AMIA Joint Summits on Translational Science 2014*, 96–101.

Skripcak, T., C. Belka, W. Bosch, C. Brink, T. Brunner, V. Budach, D. Büttner, J. et al. (2014). Creating a data exchange strategy for radiotherapy research: Towards federated databases and anonymised public datasets. *Radiotherapy and Oncology 113*(3), 303–309.

Wang, S., I. Pandis, C. Wu, S. He, D. Johnson, I. Emam, F. Guitton, and Y. Guo (2014). High dimensional biological data retrieval optimization with NoSQL technology. *BMC Genomics 15*(Suppl 8), S3.

4 Machine learning for radiation oncology

Yi Luo and Issam El Naqa

Contents

4.1 Introduction to Machine Learning 41
    4.1.1 Overview 41
    4.1.2 The Art of Learning from Data 42
    4.1.3 Application in Cancer Prognosis and Prediction 42
4.2 Machine Learning Approaches 43
    4.2.1 Supervised versus Unsupervised 43
    4.2.2 Discriminant versus Generative 44
    4.2.3 Static versus Dynamic 44
    4.2.4 Deep Learning 45
4.3 Validation of Machine Learning Algorithms 45
    4.3.1 Fitting versus Regularization 45
    4.3.2 Information Theoretic Approaches 46
    4.3.3 Statistical Resampling Approaches 46
    4.3.4 Significance Testing 47
4.4 Application of Machine Learning in Radiotherapy Big Data 48
    4.4.1 Data Categories 48
    4.4.2 A Novel Bayesian Network Approach 49
    4.4.3 Results of the BN Approach 51
    4.4.4 Comparison to the Random Forest (RF) Approach 53
    4.4.5 Discussion of Results 56
4.5 Conclusions 57
References 57

4.1 INTRODUCTION TO MACHINE LEARNING

4.1.1 OVERVIEW

The term "machine learning" was originally coined by Arthur Samuel while he was working at IBM in 1959. Machine learning evolved from the study of pattern recognition and computational learning theory in artificial intelligence, and it explores the study and construction of algorithms that can learn from data and make data-based predictions [1]. Machine learning can be formally defined as "A computer program is said to learn from experience E with respect to some class of tasks T and performance measure P if its performance at tasks in T, as measured by P, improves with experience E." [2]

Machine learning tasks are typically classified into supervised learning and unsupervised learning as shown in Figure 4.1, where the former is intended to learn a general rule that maps inputs to outputs from known labels and the latter aims to find structure from inputs with no labels provided. Considering the desired output category of a machine learning system, machine learning can be categorized into

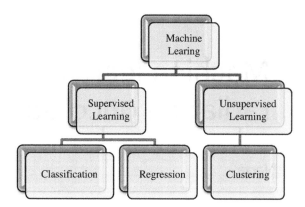

Figure 4.1 The classification of machine learning tasks.

classification, regression, clustering, and so forth. Machine learning is closely related to computational statistics, and mathematical optimization provides methods, theory, and application domains to it. Machine learning is sometimes conflated with data mining, where the former focuses on prediction based on known properties from the training data and the latter focuses on the discovery of unknown properties in the data. [3,4]

Machine learning has witnessed tremendous increased use in radiation oncology. Early applications have focused on treatment planning optimization and predicting normal tissue toxicity; however, currently its application has branched into almost every area in the field, including tumor–response modeling, radiation physics quality assurance, contouring and treatment planning, image-guided radiotherapy, and respiratory motion management, as reported in detail in the literature. [5] In the following, we will focus on the role of machine learning to devise complex models and algorithms that lend themselves to accurate prediction. These analytical models allow us to produce reliable relationships between patients' properties and radiation-induced outcomes from varying treatment plans by uncovering possibly hidden insights in the data.

4.1.2 THE ART OF LEARNING FROM DATA

The entities that we measure in a data analytic study are called the *subjects*. The population is the set of all the subjects of interest. In practice, we usually have data for only *some* of the subjects who belong to that population. These subjects are called a *sample*. "Random" is often thought to refer to a chaotic or haphazard phenomenon, but randomness is an extremely powerful tool for generating good samples and conducting experiments. A sample tends to be a good reflection of a population when each subject in the population has the same chance of being included in that sample. This is referred to as having good *sample coverage*. That is the basis of random sampling, which is designed to ensure that the sample is representative of the population. Basically, random sampling allows us to make powerful inferences about populations, and randomness is also crucial to performing experiments well. [6] The combination of good sample coverage and sample size is essential for successful machine learning (supervised or unsupervised) training.

4.1.3 APPLICATION IN CANCER PROGNOSIS AND PREDICTION

With the advent of new technologies in the field of medicine, large amounts of cancer data are being collected and are becoming available to the medical research community. However, the accurate prediction of a disease outcome is one of the most interesting and challenging tasks for physicians. Consequently, machine learning methods have become a popular tool for medical researchers with which they can discover and identify complex patterns and relationships between factors and outcomes from heterogeneous data sets and effectively predict future outcomes of a cancer type with a certain treatment.

An obvious challenge in these studies includes the integration of mixed data, such as clinical, imaging, and genomics. Recent advances in omics technologies has paved the way to further improve our understanding of a variety of diseases; however, more accurate validation results are needed before gene expression signatures can be used routinely in the clinics. It is clear that the application of machine learning methods could improve the accuracy of cancer susceptibility, recurrence, and survival predictions. The accuracy of cancer prediction outcome has significantly improved by 15%–20% in the last few years with the application of machine learning techniques. [7]

A common problem in several reported works in the literature is the lack of external validation or independent testing regarding the predictive performance of the proposed models. Other potential drawbacks include the experimental design, the limited sample size, the quality of data samples, and the penetration of clinical practice. [8] Feature selection is intended to be used to choose the most informative feature subset for training a model with a certain learning capacity. Due to the lack of static entities when dealing with clinical variables, it is important for a feature selection technique to be adjusted to different feature sets over time. The choice of the most appropriate algorithm depends on many parameters, including the types of data collected, the size of the data samples, the time limitations, and the type of prediction outcomes and their desired accuracy for clinical practice. [9,10]

New methods should be studied for the future of cancer modeling to overcome the aforementioned limitations. Better statistical analysis of the heterogeneous data sets used would provide more accurate results and would improve the understanding of disease outcomes. Further research is required based on the construction of more public databases that would collect highly curated cancer data sets of all patients who have been diagnosed with the disease. The exploitation of these data sets by researchers would facilitate their modeling studies and result in more valid results and better integrated clinical decision making. [8]

4.2 MACHINE LEARNING APPROACHES

There are many machine learning approaches that can learn from and make predictions based on data. In this chapter, we will present them by comparing supervised and unsupervised methods, discriminant and generative models, and static and dynamic methods.

4.2.1 SUPERVISED VERSUS UNSUPERVISED

According to whether there are example inputs and their desired outputs given by a "teacher," machine learning tasks are typically classified into supervised and unsupervised learning categories. Although a learning "signal" or "feedback" is available for the former, no labels are given to the latter.

The general steps of supervised learning include the identification of the type of training examples; the collection of a training data set; feature selection; algorithm development for function structure learning and parameter tuning; and the evaluation of the accuracy of the learned function. The trade-off between bias and variance is the main issue in developing many supervised learning methods. Suppose we have available several different and equally good training data sets. In order to achieve low bias, a learning algorithm must be flexible to fit these data sets well. However, if the learning algorithm is too flexible, it will fit each training data set differently, resulting in high variance. Because the prediction error of a learned classifier is related to the sum of its bias and its variance, a good supervised learning algorithm is designed to be able to adjust this trade-off, as shown in Figure 4.2.

The most widely used supervised learning algorithms are linear regression, logistic regression, linear discriminant analysis (LDA), support vector machines (SVMs), naïve Bayes, decision trees, random forests (RFs), k-nearest neighbor (KNN), and Neural Networks (multilayer perceptron) (NNs). Although these learning algorithms were developed for classification and/or regression, their prediction performance for an application may not be the same due to their different objectives and mathematical formulations. Whether choosing a learning algorithm or not depends on the properties of the data set. For example, whereas decision trees can easily handle heterogeneous data, supervised learning algorithms developed based on distance functions such as KNN and SVM with Gaussian kernels are likely to be more sensitive.

Compared with supervised machine learning, there is no evaluation of the accuracy of the structure that is output by the relevant algorithm in unsupervised learning. Although unsupervised learning does not figure out the right output, it explores the data and can draw inferences from data sets to describe hidden structures from unlabeled data. In fact, a central case of unsupervised learning is the problem of density estimation in statistics. [11] Methods in the unsupervised learning category include, for example, principal component analysis (PCA), clustering, and self-organized neural networks.

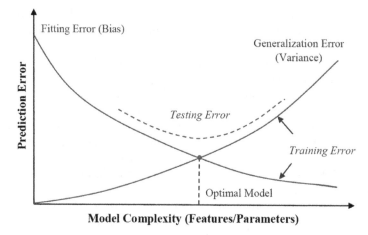

Figure 4.2 The trade-off between bias and variance with the increment of model complexity.

4.2.2 DISCRIMINANT VERSUS GENERATIVE

From a probabilistic perspective, machine learning algorithms can be divided into discriminant or generative models. In supervised classification, it is assumed that inputs x and their labels y arise from an unknown joint probability $p(x, y)$. Although discriminative methods model the posterior $p(y|x)$ *directly*, or learn a direct map from inputs x to the class labels, generative methods learn a model of the joint probability, $p(x, y)$, of the input x and the label y, and make their predictions by using Bayes' rule to calculate $p(y|x)$ and picking the most likely label y. [12] If $p(x|y)$ is Gaussian and $p(y)$ is multinomial, then the corresponding generative–discriminative pair is normal discriminant analysis and logistic regression. For the case of discrete inputs, the naïve Bayes and logistic regression form a generative–discriminative pair. [13,14]

Generative classifiers have a smaller variance than discriminant models. The generative approach converges to the best model for the joint distribution $p(x, y)$, but the resulting conditional density is usually a biased classifier. [15] Although the assumed generative model is rarely exact, an asymptotically and discriminative classifier should typically be preferred. [16] The key argument is that the discriminative estimator converges to the conditional density that minimizes the negative log-likelihood classification loss against the true density $p(x, y)$. [17] It turns out that there is a bias–variance trade-off between generative and discriminative classifiers. Literature shows that for the naïve Bayes model and its discriminative analog, logistic regression, the generative model does have a higher asymptotic error than the discriminative model. However, the former may also approach its asymptotic error much faster than the latter, suggesting that as the number of training examples is increased, the generative model has already approached its asymptotic error and is thus doing better, while the discriminative model approaches its lower asymptotic error and may also do better. [12]

4.2.3 STATIC VERSUS DYNAMIC

Static machine learning can be defined as follows: Given a set of inputs and outputs, find a static map between the two during supervised "training" and use this static map for study purposes during "operation." Static machine learning is adequate for one-off detection (and subsequent offline intervention) if your study has a high tolerance for false positives. However, dynamic machine learning is now entering the "walk" stage of the "crawl-walk-run" evolution of machine learning. The dynamic machine learning solution involves state–space data models, where states are "meta"-level descriptions of the machine.

The main distinction between static and dynamic machine learning algorithms is between their data models and learning methods. Static machine learning trains the static model without time variables for multiple types of work pieces separately and switches the map when the work piece has switched or trained the static model map for all potential work piece types. The data model of dynamic machine learning is a time-series model and uses delayed time to provide "memory" to the model. Then, the output is a variable over time.

Figure 4.3 The relationship among deep learning, machine learning, and big data.

There are three classes of learning methods: block, recursive, and real-time recursive. Block methods such as the pseudo-inverse method is a way to solve for constant coefficients of a multiple linear regression *static* model. The recursive Bayesian method has some additional useful properties [18] that update the model using new pieces of information as they arrive, providing an on-line learning solution. When data is processed only once, the recursive learning methods are called "Real-time Recursive Learning" methods and can be used to model the effect of time on the parameters of the state–space data model for dynamic machine learning. [19]

4.2.4 DEEP LEARNING

Deep learning is a class of machine learning algorithms that can learn in both supervised and unsupervised manners, where the former is considered as a classification method and the latter is treated as pattern analysis tool. [20,21]. The presentation of deep learning is loosely based on interpretation of information processing and communication patterns in biological sensory/nervous systems, including deep neural networks, deep belief networks, and recurrent neural networks. Layers used in deep learning include hidden layers of an artificial neural network and sets of propositional formulas, [22] and these layers of factors are assumed to correspond to different levels of *abstraction* or the composition of the data. Varying numbers of layers and layer sizes can provide different degrees of abstraction, and deep learning exploits the idea that higher level with more abstract concepts are learned from the lower level ones. Deep learning helps disentangle these abstractions and pick up which features are useful for improving performance without explicitly handcrafting such features. [20] For supervised learning, deep learning methods translate the data into compact intermediate representations akin to principle components and derive a layered structure that removes the redundancy in representation. [23] Deep learning algorithms can also be applied to unsupervised learning tasks, and deep structures can be trained in an unsupervised manner including deep belief networks [20] and neural history compressors. [23] The relationship among deep learning, machine learning, and big data is illustrated in Figure 4.3.

4.3 VALIDATION OF MACHINE LEARNING ALGORITHMS

4.3.1 FITTING VERSUS REGULARIZATION

Fitting is the process of applying a statistical model or a machine learning algorithm to adequately capture the underlying structure of the data. If the model or algorithm does not fit the data well enough and shows low variance and high bias, underfitting happens due to an excessively simplistic model. However, if the model or the algorithm is developed too closely or exactly to a particular set of data points, it may have more parameters than can be justified by the data and fail to fit additional data or predict future observations reliably. Thus, overfitting occurs because the model unknowingly extracts some of the residual variation (noise) as if that variation represented underlying model structure. [24]

Regularization can be motivated as a technique to improve the generalizability of a learned model. It is a process of introducing additional information in order to prevent overfitting and can be implemented by applying penalties to objective functions of optimization models in ill-posed problems. The learning algorithm is intended to find a function that fits or predicts the outcome that minimizes the expected error over all possible inputs and labels. However, only a subset of input data and outcomes are available, and they are measured with some noise. Then the expected error is unmeasurable, and the best surrogate available is the empirical error over the available samples. If there are no bounds on the complexity of the function space, a model will be learned that incurs zero loss on the surrogate empirical error. When measurements are made with noise, the model may suffer from overfitting and display poor expected error. In order to improve generalization, regularization introduces a penalty for exploring certain regions of the function space used to build the model.

4.3.2 INFORMATION THEORETIC APPROACHES

The Akaike information criteria (AIC) is an estimator of the relative quality of machine learning models for a given set of data, and it is developed based on information theory. The AIC estimates the relative information lost when a given model is used to represent the process that generated the data and mainly deals with the trade-off between the goodness of fit to the model and the complexity of the model. For a statistical model of some data, the value of AIC can be obtained from the difference between the number of estimated parameters in the model and the log function of the maximal value of the likelihood function for the model. [25,26]

As we know, using a candidate model to represent the "true model" will almost always result in information loss. We would like to choose the model with the minimal information loss, and the preferred model is the one with the minimal AIC value. Then, AIC encourages goodness of fit described by the likelihood function, but it also includes a penalty to avoid overfitting indicated by an increasing function of the number of estimated parameters. However, the AIC reveals nothing about the absolute quality of a model, only its quality relative to other models.

The Bayesian information criterion (BIC) is also a criterion for model selection among a finite set of models, and the model with the lowest BIC is preferred. The BIC is closely related to the AIC and is intended to resolve overfitting by introducing a penalty term for the number of parameters in the model. They have a different penalty for the number of parameters. Although the penalty of AIC is related to the number of estimated parameters, the penalty of BIC is related to its product with the log function of the sample size. It is assumed that a "true model" is in the set of candidates and that BIC will select the "true model" with the probability of 1, as $n \to \infty$, but the probability of the selection via AIC is less than 1. [25,27,28] However, a simulation study demonstrates that the risk of selecting a very bad model is minimized with AIC under such an assumption. [27] If the "true model" is not in the candidate set, AIC is appropriate for finding the best approximating model when the approximation is done with regard to information loss. [25,27,28]

4.3.3 STATISTICAL RESAMPLING APPROACHES

Usually, when the ideal condition of accessing the entire population or a lot of representative data cannot be satisfied, the limited data available has to be re-used in smart ways such as resampling to obtain sufficiently large numbers of samples to evaluate the performance of our classifiers. Resampling can be divided into two categories: simple resampling and multiple resampling, which allow the use of the same data point only once and more than once for testing, respectively.

Simple resampling includes cross-validation (CV) and its variants. If the data is not plentiful and we do not want to "waste" data by sampling a fresh validation set, the k-fold CV technique is intended to provide an accurate estimate of the true error without wasting too much data. In k-fold CV, the original training set is partitioned into k subsets evenly. For each subset, the algorithm is trained on the other subsets and the error of its output is estimated using the subset. Then, the estimate of the true error of the algorithm can be obtained from the errors of these subsets.

Typical choices of k are 5 or 10. When k is equal to the total number of the sample, the validation is known as leave-one-out CV (LOO-CV). If the training set has limited samples, the LOO-CV can be used.

Although LOO-CV estimator is approximately unbiased for the expected prediction error, it can have high variance due to similar training sets. On the other hand, at 5- or 10-fold, CV will overestimate the true prediction error. Whether this bias is a drawback in practice depends on the objective. [29] Overall, 5- or 10-fold CVs are a good compromise of the trade-off between the bias and variance. [30,31]

Multiple resampling usually includes the bootstrap, permutation test, and repeated k-fold CV. The basic idea of bootstrapping is to randomly draw data sets with replacement from the training data, and each sample set has the same size as the original training set. The bootstrap is a general tool for assessing statistical accuracy. Bootstrapping is the practice of estimating properties of an estimator by measuring those properties when sampling from an approximating distribution. It is often used as a robust alternative to inference based on parametric assumptions when those assumptions are in doubt or where parametric inference is impossible or requires very complicated formulas for the calculation of standard errors. Bootstrapping is useful in practice when the sample is too small for CV approaches to yield a good estimate. [32]

The permutation test is a process repeated a very large number of times in an attempt to establish whether the error estimate obtained on the true data is truly different from those obtained on large numbers of "bogus" data sets, which are created by taking the genuine samples and randomly choosing to either leave their label intact or switch them. Repeated k-fold CV is intended to perform multiple runs of simple resampling schemes to obtain more stable estimates of an algorithm's performance and enhance the replicability of the results. [33]

Both the CV and the bootstrap estimate the variability of a statistic from the variability of that statistic between subsamples, and the latter can be seen as a random approximation of the former. As with CV, the bootstrap seeks to estimate the conditional error, but typical estimates will only approximate the expected prediction error. [29] The main practical difference between them is that the bootstrap gives different results when repeated on the same data, whereas the CV gives exactly the same result each time. Usually, the CV is easier to apply to complex sampling schemes such as stratification or multiple stages than the bootstrap. However, the bootstrap estimate of model prediction bias is thought to be more realistic and more precise than CV estimates with linear discrimination function or multiple regression. [34]

4.3.4 SIGNIFICANCE TESTING

Statistical significance testing in intended to answer the question: "Can the observed results be attributed to real characteristics of the classifiers under scrutiny, or are they observed by chance?" and it can help us gather evidence of the extent to which the results returned by an evaluation metric on the resampled data sets represent the general behavior of our classifiers. [35]

Hypothesis testing consists of stating a null hypothesis, which usually is the opposite of what we wish to test (for example, classifiers A and B perform equivalently). Then a suitable statistical test and statistic are chosen that will be used to reject the null hypothesis. The significance level defined for a study, α, is the probability of the study rejecting the null hypothesis given that it were true, and the p-value of a result, p, is the probability of the study rejecting the null hypothesis if it were true. The result is statically significant when $p < \alpha$. [36,37] The significance level for a study is chosen before data collection. If the p-value of an observed effect is less than the significance level, an investigator may conclude that the effect reflects the characteristics of the whole population, resulting in rejecting the null hypothesis. If not, we fail to reject the null hypothesis but do not accept it either. Rejecting the null hypothesis gives us some confidence in the belief that our observations did not occur merely by chance. Although hypothesis testing never constitutes a proof that our observation is valid, it provides added support for our observations, yet we can never be 100% sure about them. [33]

Statistical tests come in two forms: parametric and nonparametric, where the former make strong assumptions about the distribution of the underlying data, and the latter make weaker assumptions about the data but are also typically less powerful than the former. Regarding the type of problem, the t-test (parametric), McNemar's test (nonparametric), and sign test are frequently used for the comparison of two algorithms on a single domain; the sign test (nonparametric), and Wilcoxon's signed-rank test (nonparametric) are designed for the comparison of two algorithms on several domains; the Friedman's test (nonparametric) and Nemenyi test are employed for the comparison of multiple algorithms over multiple domains. Although it is often difficult, if not impossible, to verify that all the assumptions hold in these statistical tests, and the results of statistical tests are often misinterpreted, it is always possible to show that a difference between two alternatives, no matter how small, is significant, provided that enough data are used. However, machine learning and data mining researchers should know the applicability and limitations of statistical methods and decide on their own when a statistical test is warranted and when the search for new ideas may be necessary. [33]

4.4 APPLICATION OF MACHINE LEARNING IN RADIOTHERAPY BIG DATA

In order to illustrate how machine learning can be applied for radiation oncology problems and the issues associated with this process, a detailed example of personalized radiotherapy will be introduced and discussed in this section. The example involves developing a Bayesian network (BN) approach, a class of generative machine learning algorithms, to explore the biophysical relationships among the characteristics of patients with non-small-cell lung cancer (NSCLC) and their radiotherapy outcome such as local tumor control (LC) and to find the best radiation treatment plan based on a large-scale heterogeneous data set.

4.4.1 DATA CATEGORIES

Our NSCLC patient data set is from two institutions: the University of Michigan Hospital and the VA Ann Arbor Healthcare System. It can be separated into a discovery data set for model training and a validation data set for model testing, as shown in Table 4.1, based on a patient group in different time periods. Each patient has 361 features as illustrated in Table 4.2, including dosimetric parameters; single-nucleotide polymorphisms (SNPs); microRNAs (miRNAs); clinical factors; pre-, week 2–, and week 4–treatment cytokines; and pre-, week 2–, and week 4-treatment PET radiomics.

The patients had been treated on four prospective protocols under institutional review board approval. Under the first three protocols, patients were treated with standard doses (60–66 Gy); the fourth protocol was a dose escalation study, intensifying doses to persistent PET-avid target volumes during treatment with 2.1–2.85 Gy per fraction up to a total dose of 85.5 Gy over 30 fractions. [38–40] The Analytical Anisotropic Algorithm (AAA) dose algorithm was used to compute radiation dose distributions in the tumor within the Varian Eclipse treatment planning system. [41] All total tumor dose values were converted into their 2 Gy equivalents (EQD2) using a locally developed software tool by the linear-quadric model with an alpha/beta ratio of 10 Gy. The generalized equivalent uniform dose (gEUD) was calculated with various tissue-specific parameters "a" to describe the volume effect, and the values for D5, D90, and D95 were extracted from EQD2-corrected dose–volume histograms.

Blood samples were obtained at baseline and after approximately one- and two-thirds of the scheduled radiation doses were delivered. Pre-treatment blood samples were analyzed for cytokine levels, [42] miRNAs, [43] and SNPs. [44,45] These variables have been identified from the literature as candidates

Table 4.1 **The number of NSCLC patients in the retrospective data set**

RETROSPECTIVE DATA SET	TOTAL NUMBER OF NSCLC PATIENTS	TOTAL NUMBER OF NSCLC PATIENTS WITH LC
Discovery data set	68	48
Validation data set	50	38

Table 4.2 **The number of features for each patient**

CATEGORIES OF FEATURES	DOSE	SNP	MIRNA	CLINICAL FACTOR	PRE, WEEK 2, AND WEEK 4 CYTOKINE	PRE, WEEK 2, AND WEEK 4 PET
Number of features	6	60	62	14	90	129

for being related to lung cancer or inflammatory disease. The slopes (SLPs) of cytokine changes and the relative differences (RD) of PET radiomics features [46] from pre-, week 2–, and week 4–treatment were also determined as patients' responses to radiation treatment. Although the total number of features for pre-treatment is still 215, the total number of features during treatment becomes 288 due to the use of RD and SLP.

The fluorodeoxyglucose (FDG)-PET images were acquired using clinical protocols and the pre-treatment and intra-treatment PET images were registered to the treatment planning CT using rigid registration. Image analysis was performed using customized routines in MATLAB® (Mathworks; Natick, MA). Radiomics textural features were extracted after images were quantized into 32 levels using Lloyd's algorithm, [47] including the widely used gray-level co-occurrence matrix (GLCM), neighborhood gray-tone difference matrix (NGTDM), run-length matrix (RLM), and gray-level size–zone matrix (GLSZM) features. [46]

Hartemink's pairwise mutual information method was employed to discretize continuous variables into three categories, [48] and interval discretization was used for the categorical variables. The SNPs were described by three kinds of genotypes: wild type homozygote, minor allele homozygote, and heterozygote. Although the heterozygote is assigned number "1" in this study, the homozygotes with and without the ancestral allele are assigned numbers "0" and "2," respectively. Statistical methods were implemented in the R-software environment; candidate predictor's selection and the BN learning were based on the "bnlearn" package. [49]

4.4.2 A NOVEL BAYESIAN NETWORK APPROACH

A novel BN building approach was generated here, mainly including the following two steps. First, an extended Markov blanket (MB) approach was developed to determine important factors related to LC from a high-dimensional data set. Second, although bootstrap sampling and Tabu Search were employed to optimize the BN structure for a given number of predictors, the best scale for the BN was learned from predictor pruning and arc selection in terms of the prediction performance of resulting BNs based on CV.

In this study, an extended MB neighborhood approach was developed to find out not only the inner family of interrelated variables as commonly practiced (the variables directly related to LC based on the MB of LC) but also their extended family, that is, next-of-kin variable relationships (the variables directly related to the inner family based on their MBs). For instance, the MB of LC would be the smallest set containing all variables carrying information about LC that cannot be obtained from any other variable (inner family). Then for each member in the LC blanket, a next-of-kin MB for this member was also derived. In addition to finding good feature subsets for BN structure learning, the MB also helps determine causal relationships among various nodes in the BN. Here, the MBs of LC and next-of-kin were found using the Semi-Interleaved Hiton Parents and Children (SI-HITON-PC) algorithm, a fast-forward selection technique for neighborhood detection designed to exclude nodes early based on marginal association. [49]

A purpose for BN structure learning is to explore the possible interactions among the above-selected variables from the existing data set. Before starting to learn the best BN structure, the known causal and logical influence relationships between every two variables were defined as "biophysical rules." Although there are many biophysical interactions related to LC and the relationships among potential variables in each interaction could be complicated, there existed many known prior biophysical relationships related to LC onset that could be exploited to constrain the search space in the process of building the BN structure. If a connection could be identified from literature or the influence of time order, the relationship was considered as an inclusion connection; otherwise, it was an exclusion connection. Tabu Search, a metaheuristic search method that uses a local search for mathematical optimization, [50] was employed to generate a stable BN structure for pre- or during-treatment. It combined the selected biophysical variables through an iterative statistical resampling based on generating bootstrap samples of the original data to ensure robustness. In the meantime, radiobiologically plausible relationships among a set of selected biophysical variables related to LC were built among them by adding constraints to the Tabu Search and enforcing biophysical rules.

Although the interdependence between two associated biophysical factors is represented by an edge/arc in a BN, the degree of their dependence can be described by the strength of the edge/arc between them, which could be measured from score gain/loss caused by an edge/arc's removal. [49] Because the larger strength value represents stronger dependency and BN structure can be adjusted by selecting different edge strength thresholds, the properties of edge strength provide an opportunity to remove the relatively loose interactions among the predictors in the BN in terms of LC prediction by increasing the threshold of edge strength toward a stable BN on bootstrap resampling. A node without any descendant related to LC is considered as a leaf node, and it has limited contribution to LC prediction compared with other predictors. The BN with appropriate biophysical interactions based on the extended MB neighborhood approach may or may not have a good LC prediction performance because when each predictor and edge is able to contribute to biophysical information for LC prediction, they may also cause noise at a certain level. Therefore, the BN's prediction performance could be improved by removing such leaf nodes and keeping the appropriate edge strength threshold to balance the loss of information and noise reduction through an iterative bootstrap resampling. This is the basic idea of how the BN structure learning was to achieve a more robust model.

The BN structure learning was guided by the performance of the resulting BN for LC prediction. CV was employed to prevent overfitting by assessing how a statistical model will generalize to an independent data set, and receiver operating characteristic (ROC) curves were generated to evaluate the prediction power of a prediction model based on CV. The ROC curves show the trade-offs between true-positive and false-positive rates for a given diagnostic task, and the area under the ROC curve (AUC) provides a single measure of the performance of a classifier by assigning 0.5 to a random signal and 1.0 to a perfectly discriminant signal. In the training data, there were only 20 local failure events in 68 patients. Therefore, a stratified CV approach was used to avoid the problem of "inverse CV effect" that could appear due to the imbalanced event rate. [51]

A nested CV was generated to evaluate the performance of the overall BN approach, as shown in Figure 4.4, where the outer stratified CV is conducted based on the discovery data set and inner stratified CV is employed to guide BN structure learning within each iteration of the outer CV. Let N or I be the set of predictors or the set of threshold levels of arc strength to build a BN during an iteration of outer CV, n or i be the index of the number of these predictors ($n \in N$) or the index of these threshold levels ($i \in I$), and a BN generated from the training data set of the outer CV can be described as BN$n, i$. The prediction performance of BN$n, i$ evaluated by the inner CV can be denoted as AUCn, i. During the process of the current BN approach, the values of AUCn, i vary as the number of the nodes and the thresholds of arc

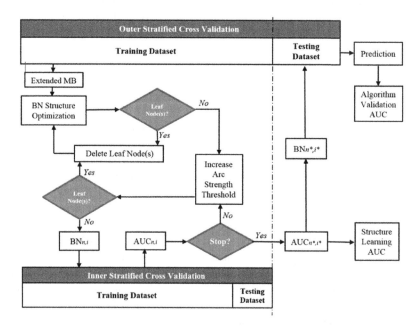

Figure 4.4 Validating the two-step BN approach with nested CV.

strength change. Let AUCn^*, i^* denote the maximal value of AUCn, i, and n^* and i^* indicate the number of predictors and the threshold level associated with it, respectively ($\in N$, $\in I$). Then the best BN guided by the inner CV based on the training data set of the outer CV can be denoted as BNn^*, i^*, and it was used to predict the patients' LC in the reserved testing data set. After stacking the patients' LC prediction from all iterations of the outer CV, the ROC curve for evaluation of the performance of the current BN approach was obtained.

4.4.3 RESULTS OF THE BN APPROACH

The MB of LC based on the discovery data set is formed from "miR-20a-5p," "pre-eotaxin," "GLSZM-LGZE." Moreover, each of these variables may have its own MB neighborhood as shown in Figure 4.5. For example, "miR-17-5p," "miR-93-5p," "miR-124-3p," and "Tumor-gEUD" from the MB of "miR-20a-5p," where "Tumor-gEUD" is used to denote a gEUD with $a = -10$. In this study, the extended MB neighborhoods within two layers of LC were used as potential variables of pre-treatment full BN.

Figure 4.6 shows a stable pre-treatment full BN with an edge strength of ≥ 0.65 where 11 important biophysical predictors are identified. Their relationships in terms of LC prediction are indicated by directed edges, and the thickness of an edge represents the strength of a connection. Whereas the green and red lines represent positive and negative influences between the connected predictors, respectively, the gray lines indicate the mixed positive and negative influences between them.

A balanced seven fold cross-validation (about 10 samples per fold), which maintained the same ratio of event-to-no event in each fold in both training and holdout, was employed to guide predictor pruning and to find a robust BN structure with a high LC prediction power. In the meantime, it was also used to evaluate the performance of pre-treatment full BN in the discovery data set, as shown in Figure 4.7a, where the AUC value is 0.81 (95% confidence interval [CI]: 0.69, 0.90) (DeLong) based on 2,000 stratified bootstrap replicates. In comparison, the main dosimetric predictor (Tumor_gEUD) had an AUC of only 0.61 (95% CI: 0.51, 0.75) from 2,000 stratified bootstrap replicates. For the validation data set, the ROC curve of pre-treatment full BN is illustrated in Figure 4.7b with an AUC = 0.76.

In an extended LC prediction model, the relative changes of PET features before and during-treatment were incorporated as part of the patients' response during the course of radiotherapy. The important biophysical predictors for LC prediction during radiation treatment can be found from the SI-HITON-PC algorithm. The updated extended MB neighborhoods are shown in Figure 4.8.

Local Control (Pre-Treatment)		
	miR-20a-5p	miR-17-5p
		miR-93-5p
		miR-124-3p
		Tumor-gEUD
	pre-eotaxin	pre-MCP-1
		tp53-Rs1042522
		GLCM-Contrast
		pre-IL1R-alpha
		GLSZM-LZLGE
		cxcr1-Rs2234671
		tp53-Rs1625895
	GLSZM-LGZE	GLSZM-GLN
		GLRLM-LGRE
		GLSZM-SZLGE
		GLSZM-SZHGE

Figure 4.5 The extended MB neighborhoods of LC before radiation treatment, where the second and third columns represent the inner family of LC and the next of kin for each of its members, respectively.

Big data in radiation oncology

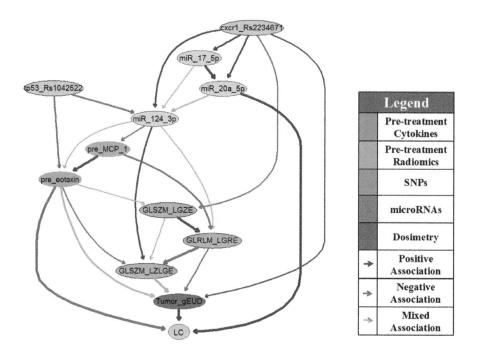

Figure 4.6 Pre-treatment full BN for LC prediction.

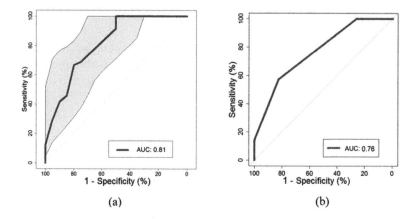

Figure 4.7 The ROC curves of pre-treatment full BN based on (a) discovery and (b) validation data sets, where the shaded area indicates the 95% confidence interval of the AUC.

A during-treatment full BN with edge strength of ≥0.68 was developed via BN structure learning and predictor pruning as illustrated in Figure 4.9. Figure 4.10a shows the ROC curve of the during-treatment BN based on sevenfold CV within the discovery data set, and the AUC of the internal testing CV is 0.85 (95% CI: 0.75, 0.93) (DeLong) based on 2000 stratified bootstrap replicates. Figure 4.10b presents the ROC curve of during-treatment full BN for the validation data set with an AUC = 0.79.

The nested CV, as shown in Figure 4.4, was employed to validate the performance of the two-step BN approach. With pre-treatment discovery data set, the ROC curve of the BN approach can be described in Figure 4.11a, and its performance reaches an AUC = 0.75 (95% CI: 0.64, 0.86) (DeLong) based on 2,000 stratified bootstrap replicates. With additional during-treatment information, the ROC curve of the BN approach is illustrated in Figure 4.11b, and its performance increases to an AUC = 0.80 (95% CI: 0.71, 0.89) (DeLong) from 2,000 stratified bootstrap replicates.

		miR-17-5p
Local Control (During-Treatment)	**miR-20a-5p**	miR-17-5p
		miR-20a-5p
		Stage
		miR-18a-5p
		miR-19b-3p
		miR-224-5p
		RD-GLRLM-GLV
		Tumor-gEUD
	pre-eotaxin	pre-MCP-1
		tp53-Rs1042522
		GLCM-Contrast
		SLP-fractalkine
		pre-IL1R-alpha
		pre-EGF
		cxcr1-Rs2234671
		RD-GLRLM-SRHGE
	GLSZM-LGZE	RD-GLSZM-ZSV
		GLSZM-GLN
		GLRLM-LGRE
		GLSZM-SZLGE
		GLRLM-RLN
		GLSZM-SZHGE

Figure 4.8 The MB neighborhoods of LC during radiation treatment, where the second and third columns show the inner family of LC and the next of kin for each of its members, respectively.

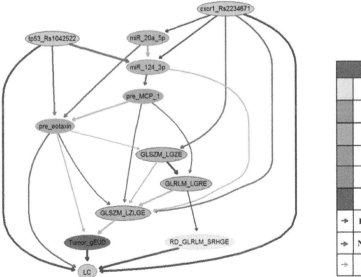

Figure 4.9 During-treatment full BN for LC prediction.

4.4.4 COMPARISON TO THE RANDOM FOREST (RF) APPROACH

As a representative of discriminant machine learning methods, the random forest (RF) is an ensemble decision tree method for classification and regression, [52] and considerable empirical evidence has shown it to be a highly effective approach. [53] Similar to BN, the prediction models developed by the RF are a finite directed graph with no directed cycles, where the nodes and arcs in the graph represent the features

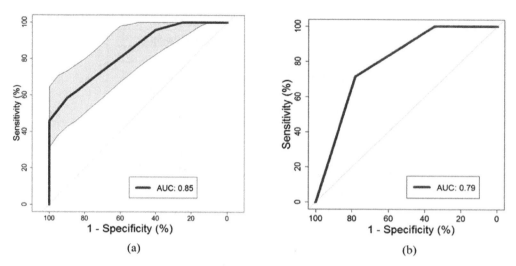

Figure 4.10 The ROC curves of during-treatment full BN based on (a) discovery and (b) validation data sets, where the shaded area indicates the 95% confidence interval of the AUC.

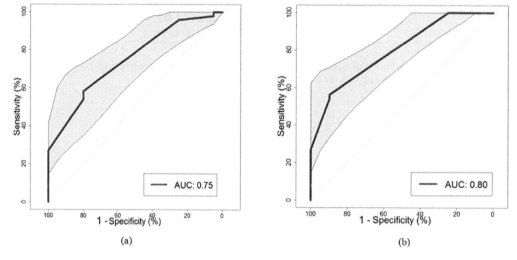

Figure 4.11 The ROC curves of the two-step BN approach with (a) pre-treatment and (b) whole discovery data sets based on nested CV, where the shaded area indicates the 95% confidence interval of the AUC.

and their association for classification, respectively. It is well known that constructing ensembles from base learners, such as decision trees, can significantly improve learning performance, and the ensemble learning can be further improved by injecting randomization into the base learning process. [52] Here the RF approach was employed to compare the performance of our BN analysis for LC prediction, and the R package "randomForestSRC" was used to implement the RF algorithm in our study.

First, an LC prediction model was developed based on all 288 pre- and during-treatment features by using the RF approach with minimal depth variable selection. The ROC curve of its performance based on CV is shown in Figure 4.12, and the value of AUC = 0.63 (95% CI: 0.48, 0.78) based on 2,000 stratified bootstrap replicates. This shows that the RF has difficulties similar to other traditional machine learning methods in directly handling high-dimensional data sets due to the multicollinearity noise of unimportant features in the prediction model. Similarly, it was challenging for our BN analysis to find the BN structure directly from all 288 features because of the exponential increase in computation. The RF and our BN analysis are also compared based on feature selection and prediction performance separately.

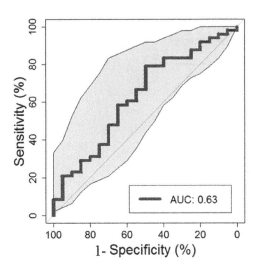

Figure 4.12 The ROC curve of the pure RF model for LC prediction developed from 286 features based on cross-validation, where the shaded area indicates the 95% confidence interval of the AUC.

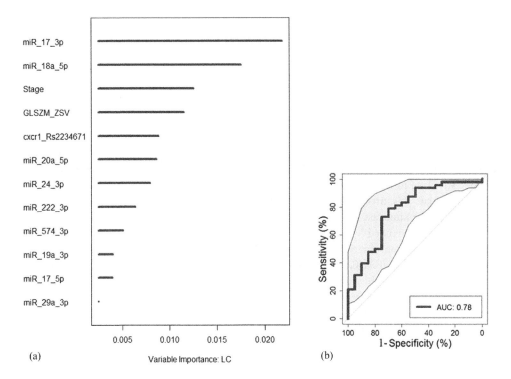

Figure 4.13 (a) The top 12 ranked important features selected from 286 features by the RF approach; (b) the ROC curve of a LC prediction model from these 12 features based on cross-validation, where the shaded area indicates the 95% confidence interval of the AUC.

The RF can be used as a feature selection tool to identify important features for LC prediction based on a permutation-based score. [54] Figure 4.13a shows the top 12 ranked important features selected from the 288 features based on RF.

A new LC predication model was developed from these features via the RF approach. Its performance is illustrated in Figure 4.13b, and the value of AUC increased to 0.78 (95% CI: 0.65, 0.91) based on 2,000 stratified bootstrap replicates. However, although the features in our extended MB include important radiation treatment and during-treatment biological and radiomics information such as "Tumor-gEUD"

Big data in radiation oncology

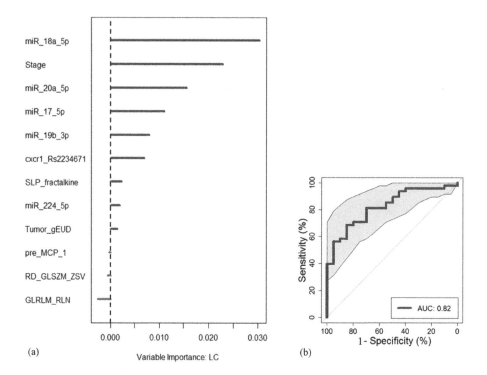

(a)

Variable Importance: LC

(b)

Figure 4.14 (a) The top 12 ranked important features selected from the extended MB by the RF approach; (b) the ROC curve of a LC prediction model from these 12 features based on cross-validation, where the shaded area indicates the 95% confidence interval of the AUC.

and "RD-GLRLM-GLV" as shown in Figure 4.8, the features selected by the RF do not seem to have these features as top contributors. In this regard, the concept of MB seems to make more clinical sense than the ensemble tree structure in terms of providing adaptive decision support for LC prediction.

In order to evaluate this further, the prediction performances of the RF and our BN analysis were compared based on the same features in the extended MB as shown in Figure 4.8. The top 12 ranked features selected from the extended MB via the RF are illustrated in Figure 4.14a, and another LC prediction model can be developed from these features via the RF approach. Figure 4.14b shows the ROC curve of its performance, and the value of AUC becomes 0.82 (95% CI: 0.72, 0.92) based on 2,000 stratified bootstrap replicates. It turns out that in this case the RF approach can achieve a similar prediction performance as our BN analysis approach. However, it is noted that the ensemble tree structure based on the average depth in the forests is still unable to illustrate the biophysical interaction among the patients' clinical, physical, and biological features behind the radiation outcome or to provide a clinically interpretable decision support system for adaptive radiation treatment planning. In addition, another useful advantage of the BN is its ability to handle missing information in clinical practice based on its inherit property of marginalization.

4.4.5 DISCUSSION OF RESULTS

In this example, a large-scale BN analysis was developed to unravel biophysical interactions behind LC through the use of a new extended Markov blanket approach to demonstrate its utility as a feature selection technique when the ratio of the variables to the samples is high. It is noted that mid-treatment information had added value, this could be explained by the fact the initial dose received acts as a probe that would further improve one's understanding of the response that the patient is likely to have. In addition, it was shown that the full BN model has excellent robustness characteristics to missing information in case of the validation data set and had a consistent performance across the discovery and validation data sets.

The developed BN is not only able to predict LC but can also be used to explore an understanding of underlying LC radiobiology, which is essential for understanding molecular mediators of response and the development of targeted interventions as discussed below. Clinically, this would allow one to conduct personalized treatment planning based on an individual's characteristics. Following the arrows entering node "Tumor_gEUD" in Figures 4.6 and 4.9, the appropriate radiation dose to increase the probability of LC could be predicted by the patient's characteristics including SNPs, miRNAs, cytokines, and PET information.

The BN analysis presented here is a data-driven, response-based approach to explore the relationship of biophysical factors leading to radiation-induced LC. The resulting pan-omics Bayesian networks can be used clinically for personalized adaptive radiation treatment planning, However, future studies are necessary to overcome its current limitations such as that the estimated influences are calculated from the adjacent parent and children nodes and may not necessarily reflect the accumulative effect or direct causal interaction of all variables along the network in reality. Other intermediate factors that are not included in the study may need to be considered as well. In the next step, we plan to enhance the current Bayesian network analysis, include known biological prior knowledge, validate the LC prediction model with larger external data set, and develop dynamic BNs to guide our personalized radiation treatment practice.

4.5 CONCLUSIONS

In this chapter, machine learning and its application in radiation oncology were presented and discussed. First, we introduced the background and definition of machine learning, the different classification categories of machine learning algorithms. Then, we presented the different validation approaches to test and evaluate machine learning algorithms performance. Finally, we employed a detailed example to show how to develop machine learning algorithms based on generative learning with BNs from a high dimensional pan-omics data set to predict a relevant radiotherapy outcome (local tumor control) and find the best treatment plans before and during the courses of radiation treatment in a robust and stable way and compared its performance to discriminant learning with RF. Basically, machine learning includes a large family of algorithms to learn patterns from a retrospective data set. However, not all of the algorithms are suitable for adaptive radiation treatment planning. In this chapter, a comparison was made between the generative BN approach and the discriminant RF approach, another popular analytics approaches in machine learning. Although the latter can achieve a prediction performance that is similar to the former based on the selected features from the concept of MB, the ensemble tree structure cannot yet illustrate the nature of the biophysical pathways behind the observed radiation outcome, which is a limitation to providing a clinically interpretable decision support system for adaptive radiation treatment planning.

To recapitulate, the optimal selection of a certain machine learning algorithm in radiation oncology for a particular application would vary from one application to another and will depend on the task at hand.

REFERENCES

1. Samuel, A.L., Some studies in machine learning using the game of checkers. *IBM Journal of Research and Development*, 1959. **3**(3): 210–229.
2. Mitchell, T.M., Does machine learning really work? *AI Magazine*, 1997. **18**(3): 11–20.
3. Mannila, H., Data mining: Machine learning, statistics, and databases. *Eighth International Conference on Scientific and Statistical Database Systems, Proceedings*, 1996: 2–9.
4. Friedman, J.H., Data mining and statistics: What's the connection? *Mining and Modeling Massive Data Sets in Science, Engineering, and Business with a Subtheme in Environmental Statistics*, 1997. **29**(1): 3–9.
5. El Naqa, I., R. Li, and M.J. Murphy, Eds., *Machine Learning in Radiation Oncology: Theory and Application*. 2015, Springer International Publishing: Cham, Switzerland.
6. McClave, A.K., Statistics: The art and science of learning from data. *Journal of Official Statistics*, 2008. **24**(1): 157–159.
7. Cruz, J.A. and D.S. Wishart, Applications of machine learning in cancer prediction and prognosis. *Cancer Informatics*, 2007. **2**: 59–77.
8. Kourou, K. et al., Machine learning applications in cancer prognosis and prediction. *Computational Structural Biotechnology Journal*, 2015. **13**: 8–17.
9. Kim, W. et al., Development of novel breast cancer recurrence prediction model using support vector machine. *Journal of Breast Cancer*, 2012. **15**(2): 230–238.

10. Eshlaghy, A.T. et al., Using three machine learning techniques for predicting breast cancer recurrence. *Journal of Health & Medical Informatics*, 2013. **4**(2): 124–130.

11. Wilson, R.A., The cognitive sciences: A comment on 6 reviews of the MIT Encyclopedia of the Cognitive Sciences. *Artificial Intelligence*, 2001. **130**(2): 223–229.

12. Ng, A.Y. and M.I. Jordan, On discriminative vs. generative classifiers: A comparison of logistic regression and naive Bayes. *Advances in Neural Information Processing Systems*, 2002. **14**: 841–848.

13. Vapnik, V.N., *Statistical Learning Theory*. 1998, John Wiley & Sons: New York.

14. Rubinstein, Y.D. and T. Hastie. Discriminative vs. informative learning. in *Proceedings of the Third International Conference on Knowledge Discovery and Data Mining*. 1997, AAAI Press.

15. Bouchard, G. and B. Triggs, The trade-off between generative and discriminative classifiers. in *COMPSTAT'2004 Symposium*. 2004, Physica-Verlag/Springer.

16. Andel, J., M.G. Perez, and A.I. Negrao, *Estimating the dimension of a linear-model*. Kybernetika, 1981. **17**(6): 514–525.

17. Efron, B., Efficiency of logistic regression compared to normal discriminant-analysis. *Journal of the American Statistical Association*, 1975. **70**(352): 892–898.

18. Sarkka, S., A. Solin, and J. Hartikainen, Spatiotemporal learning via infinite-dimensional Bayesian filtering and smoothing. *IEEE Signal Processing Magazine*, 2013. **30**(4): 51–61.

19. Madhavan, P., *Systems Analytics: Adaptive Machine Learning workbook*. 2016, CreateSpace Independent Publishing; Scotts Valley, CA.

20. Bengio, Y., A. Courville, and P. Vincent, Representation learning: A review and new perspectives. *IEEE Transactions on Pattern Analysis and Machine Intelligence*, 2013. **35**(8): 1798–1828.

21. Schmidhuber, J., Deep learning in neural networks: An overview. *Neural Networks*, 2015. **61**: 85–117.

22. Bengio, Y., Learning deep architectures for AI. *Foundations and Trends in Machine Learning*, 2009. **2**(1): 1–127.

23. Schmidhuber, J., Learning complex, extended sequences using the principle of history compression. *Neural Computation*, 1992. **4**(2): 234–242.

24. Burnham, K.P. and D.R. Anderson, Multimodel inference—understanding AIC and BIC in model selection. *Sociological Methods & Research*, 2004. **33**(2): 261–304.

25. Wagenmakers, E.J., Model selection and multimodel inference: A practical information-theoretic approach. *Journal of Mathematical Psychology*, 2003. **47**(5–6): 580–586.

26. Akaike, H., New look at statistical-model identification. *IEEE Transactions on Automatic Control*, 1974. **19**(6): 716–723.

27. Vrieze, S.I., Model selection and psychological theory: A discussion of the differences between the Akaike Information Criterion (AIC) and the Bayesian Information Criterion (BIC). *Psychological Methods*, 2012. **17**(2): 228–243.

28. Aho, K., D. Derryberry, and T. Peterson, Model selection for ecologists: The worldviews of AIC and BIC. *Ecology*, 2014. **95**(3): 631–636.

29. Hastie, T., R. Tibshirani, and J. Friedman, *The Elements of Statistical Learning Data Mining, Inference, and Prediction*. 2008, Springer: New York.

30. Breiman, L. and P. Spector, Submodel selection and evaluation in regression—The X-random case. *International Statistical Review*, 1992. **60**(3): 291–319.

31. Kohavi, R., A study of cross-validation and bootstrap for accuracy estimation and model selection, in *International Joint Conference on Artificial Intelligence (IJCAI)*. 1995, Morgan Kaufmann, Los Altos, CA. 1137–1143.

32. Japkowicz, N., *Performance evaluation for learning algorithms*. Advances in Artificial Intelligence, 2016. **9673**: xviii–xviii.

33. Japkowicz, N. and M. Shah, *Evaluating Learning Algorithms: A Classification Perspective*. 2011, Cambridge University Press: Cambridge/New York.

34. Verbyla, D.L. and J.A. Litvaitis, Resampling methods for evaluating classification accuracy of wildlife habitat models. *Environmental Management*, 1989. **13**(6): 783–787.

35. Japkowicz, N. and M. M. Shah, Performance evaluation in machine learning. In: *Machine Learning in Radiation Oncology Theory and Applications*, ed. I. El Naqa, R. Li, and M.J. Murphy. 2015, Springer, Cham, Switzerland.

36. Johnson, V.E., Revised standards for statistical evidence. *Proceedings of the National Academy of Sciences of the United States of America*, 2013. **110**(48): 19313–19317.

37. Sham, P.C. and S.M. Purcell, Statistical power and significance testing in large-scale genetic studies. *Nature Reviews Genetics*, 2014. **15**(5): 335–346.

38. Kong, F.M. et al., Radiation dose effect in locally advanced non-small cell lung cancer. *Journal of Thoracic Disease*, 2014. **6**(4): 336–347.

39. Stenmark, M.H. et al., Combining physical and biologic parameters to predict radiation-induced lung toxicity in patients with non-small-cell lung cancer treated with definitive radiation therapy. *International Journal of Radiation Oncology Biology Physics*, 2012. **84**(2): E217–E222.

40. Kong, F.M. et al., High-dose radiation improved local tumor control and overall survival in patients with inoperable/unresectable non-small-cell lung cancer: Long-term results of a radiation dose escalation study. *International Journal of Radiation Oncology Biology Physics*, 2005. **63**(2): 324–333.

41. Sievinen, J., W. Ulmer, and W. Kaissl, AAA photon dose calculation model in eclipse. *Varian Medical Systems*, 2005. **118**: 2894.

42. Fukuyama, T. et al., Cytokine production of lung cancer cell lines: Correlation between their production and the inflammatory/immunological responses both in vivo and in vitro. *Cancer Science*, 2007. **98**(7): 1048–1054.

43. Guo, L.L. et al., MicroRNAs, TGF-beta signaling, and the inflammatory microenvironment in cancer. *Tumor Biology*, 2016. **37**(1): 115–125.

44. Slattery, M.L. et al., Genetic variation in the TGF-beta signaling pathway and colon and rectal cancer risk. *Cancer Epidemiology Biomarkers & Prevention*, 2011. **20**(1): 57–69.

45. Damaraju, S. et al., Association of DNA repair and steroid metabolism gene polymorphisms with clinical late toxicity in patients treated with conformal radiotherapy for prostate cancer. *Clinical Cancer Research*, 2006. **12**(8): 2545–2554.

46. El Naqa, I., The role of quantitative PET in predicting cancer treatment outcomes. *Clinical and Translational Imaging*, 2014. **2**(4): 305–320.

47. Lloyd, S.P., Least squares quantization in PCM. *IEEE Transactions on Information Theory*, 1982. **28**(2): 129–137.

48. Hartemink, A.J., Principled computational methods for the validation and discovery of genetic regulatory networks, in *Department of Electrical Engineering and Computer Science*. 2001, Massachusetts Institute Of Technology, Cambridge, MA.

49. Scutari, M., Learning Bayesian networks with the bnlearn R Package. *Journal of Statistical Software*, 2010. **35**(3): 1–22.

50. Lokketangen, A., Tabu search—Using the search experience to guide the search process—An introduction with examples. *AI Communications*, 1995. **8**(2): 78–85.

51. Perlich, C. and G. Swirszcz, On cross-validation and stacking: Building seemingly predictive models on random data. *ACM SIGKDD Explorations Newsletter*, 2010. **12**–**15**(2): 11–15.

52. Breiman, L., Random forests. *Machine Learning*, 2001. **45**(1): 5–32.

53. Ishwaran, H. et al., Random survival forests for competing risks. *Biostatistics*, 2014. **15**(4): 757–773.

54. Genuer, R., J.-M. Poggi, and C. Tuleau-Malot, Variable selection using random forests. *Pattern Recognition Letters*, 2010. **31**(14): 2225–2236.

5 Cloud computing for big data

Sepideh Almasi and Guillem Pratx

Contents

5.1	Cloud Computing	62
	5.1.1 Introduction	62
	5.1.2 Historical Background	63
	5.1.3 Service Provision Model of Cloud Computing Systems	64
	5.1.4 Physical Implementation of Cloud Computing	66
	5.1.5 Economics of Cloud Computing	66
	5.1.6 Commercial Providers	67
5.2	Software Platforms for Distributed Processing of Big Data	68
	5.2.1 MapReduce	68
	5.2.2 Hadoop	69
	5.2.3 Spark	70
	5.2.4 Other Platforms	70
5.3	Security and Privacy of Cloud Computing	70
	5.3.1 Data Security Risks	70
	5.3.2 Data Loss	71
	5.3.3 Regulatory Issues in Health Care Clouds	72
5.4	Opportunities for Cloud Computing in Radiation Oncology	72
	5.4.1 Information Systems in Radiation Oncology	72
	5.4.2 Cloud Storage of Medical Information	73
	5.4.3 Medical Computation in the Cloud	74
	5.4.4 Clinical Trials	75
	5.4.5 Challenges of Cloud Computing in Radiation Oncology	76
5.5	Final Thoughts	76
	Acknowledgments	77
	References	77

Health care institutions are now collecting patient data in such large quantities that storing and processing these data is becoming a significant challenge. Fortunately, a solution has emerged in recent years with the development of Cloud computing technologies (Figure 5.1).

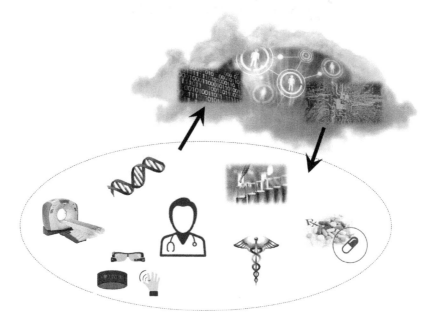

Figure 5.1 Illustration of Cloud computing for handling health care data.

5.1 CLOUD COMPUTING

5.1.1 INTRODUCTION

The emergence and evolution of Cloud computing over the past few decades have been astounding from computer science and commercial perspectives (Armbrust et al. 2010). The basic idea behind Cloud computing is that computation power and data storage can be sold on demand as services via the Internet. This concept treats these computational resources as commodities that users can purchase in precisely metered amounts for their usage, without having to worry about maintaining expensive computing facilities on their own.

Many definitions of Cloud computing have been proposed, but the one most often cited, by the National Institute of Standards and Technology (NIST), defines Cloud computing as "a model for enabling ubiquitous, convenient, on-demand network access to a shared pool of configurable computing resources (e.g., networks, servers, storage, applications, and services) that can be rapidly provisioned and released with minimal management effort or service provider interaction" (Mell and Grance 2011). In general, the term *Cloud* refers to the collection of hardware infrastructure and software that enables these resources.

One of the key technologies enabling Cloud computing is *virtualization*. Virtualization allows Cloud providers to abstract away the hardware infrastructure and present the user with an array of virtual computers (also known as a virtual machine, or VM) with customizable characteristics (e.g., compute power, memory size, and storage size). A VM is essentially a software program that emulates a computer, in other words, a computer simulation running on a physical computer. This paradigm provides capabilities that are not available when running applications directly on the native hardware. Virtual machines can be created, destroyed, replicated, or migrated to any physical data center connected to the Web on demand. Furthermore, snapshots recapitulating the state of a VM can be acquired and stored to easily recall or duplicate a given VM.

The use of virtualization enables another key property of Cloud computing known as *elasticity*. Because computer servers can be allocated in real time and added to a user's Cloud with little overhead, a Cloud system can expand or contract based on the instantaneous level of demand for a service. On the user side, this makes Cloud computing a highly cost-effective solution because a peak in demand can be dynamically met by allocating additional VMs without having to provision a constant level of computing resources. On the provider side, elasticity is made possible by sharing the physical infrastructure between a large number of tenants. From a statistical standpoint, it is unlikely that the tenants of a given datacenter will all experience a surge in demand at the same time. Therefore, thanks to the averaging effect, the relative fluctuations in user demand from the average pool of users are much smaller than that of any single customers, in relative terms. These fluctuations can be handled by provisioning a relatively small number of extra physical servers. Another advantage of virtualization is that it enables a user to share the hardware resources with other users while maintaining a level of privacy and security comparable to using a dedicated server.

The ability of a Cloud computing system to seamlessly expand their physical footprint (i.e., the size of the datacenter that powers it) is another property of Cloud computing known as *scalability*. When a datacenter reaches maximal capacity, additional physical servers can be brought online without any interruption in service. If no additional servers are available on site, resources can be allocated off-site from datacenters that are experiencing low demand. These characteristics of Cloud computing promote efficient use of resource and high service availability while reducing cost.

Last, another important advantage of Cloud computing is its accessibility from any location through a Web connection. This enables new opportunities for long-distance collaborations and telemedicine. Importantly, the user experience will be the same from any location as long as the Web connection to the Cloud is fast enough.

5.1.2 HISTORICAL BACKGROUND

In the 1960s, John McCarthy envisioned the core idea behind Cloud computing as another type of utility provided to the public in the form of computing facilities (Zhang et al. 2010). The term Cloud started to appear in the 1990s in various contexts, but the concept really gained momentum in 2006, following a remark by Eric Schmidt (Google's CEO at the time) who used "Cloud" to describe the business model of providing services across the Internet.

The technological backbone of Cloud computing has evolved from the Grid computing model that is a distributed computing paradigm in which interconnected resources perform computational tasks in parallel (Foster et al. 2008). Being in use since the 1980s, Grid computing itself was primarily developed to enable large-scale scientific computations on static clusters of networked computers. Cloud computing inherited the distributed model of computation and resource utilization of Grid computing, but it has expanded the technology by virtualizing compute resources. Furthermore, Cloud computing platforms are built according to a layered architecture in which different layers of services are implemented on top of virtual servers. These features allow Cloud computing systems to meet the instantaneous demands of a computing task, whereas Grid computing systems have to allocate fixed computing resources independently of the actual need. Because Grid computing systems are engineered to handle the most intensive task possible, they often end up being underused. In contrast, the ability of Cloud computing to scale the utilized infrastructure in real time makes it a more economical approach than Grid computing.

Cloud computing has also been compared with IBM's autonomic computing model that was introduced in 2001 (Kephart and Chess 2003). Autonomic computing systems implement self-management procedures, meaning that they can react to internal and external observations without human intervention. They were originally envisioned to overcome the management complexity of computer systems. The Cloud computing model shares some of the autonomic features such as automatic resource provisioning but lacks the complexity of autonomic computing systems.

In summary, Cloud computing leverages virtualization technology to provide computing resources as a utility. It shares certain aspects with Grid computing and autonomic computing but differs from them in other aspects, especially with respect to the ease of use for the end-user and the economies of scale that can be achieved by the elastic multi-tenancy model.

Big data in radiation oncology

5.1.3 SERVICE PROVISION MODEL OF CLOUD COMPUTING SYSTEMS

Cloud computing is built following a layered, service-driven model. Each layer is implemented as a service to the layer above it or to the public. Conversely, each layer can be perceived as a customer of the layer below, meaning that the functionalities and security of an upper layer derive from the characteristics of the layers below. Advantageously, Cloud layers are loosely coupled with each other, making it possible to upgrade or modify one layer independently of the others. Cloud computing layers are generally grouped into three main categories:

1. **Infrastructure as a service (IaaS)**: Under this model, Cloud providers deliver virtualized computational resources, storage, and networks as services. Amazon Elastic Compute Cloud (EC2) and Simple Storage Services (S3), Joyent, Terremark, GoGrid, and Flexiscale are some examples of this service model. Users are able to request the allocation of virtual computing resources on demand and customize their specifications to match their needs. They are then provided with networked storage space accessible through various interfaces, and IP addresses and login information to an array of virtual servers. The user does not share these virtual resources with anyone else. However, the physical hardware on which the services run is typically shared among multiple users, but this fact is hidden to the users who can only see the virtualized resources that are allocated to them. Any number of operating systems and applications can be run on these servers, but it is the responsibility of the user to set them up. The ability to customize the software platform to meet specific technical requirements is the main advantage of the IaaS model. In practical terms, an IaaS system is implemented by partitioning physical resources using virtualization software such as Xen, KVM, and VMware to implement distributed data storage systems and VMs.

2. **Platform as a service (PaaS)**: Under this model, Cloud providers deliver a platform on which customers can develop, deploy, and manage their own applications and software. The PaaS provider handles the low-level infrastructure and operation of the Cloud, saving developers from the complexities of managing such systems. The PaaS model may be developed on top of an existing IaaS Cloud infrastructure (Figure 5.2). Examples of tools provided on a PaaS system include the operating system (Windows, UNIX, or LINUX), middleware (e.g., JAVA and .NET run time environments), core services such as web services (databases and HTTP server), and security tools. The PaaS layer depends on the virtualization of resources delivered by IaaS for scalability and software migration. For example, the Google App Engine operates at the PaaS layer to provide application program interfaces (APIs) for implementing data storage, databases, and business logic for web applications. Other PaaS providers include Microsoft Azure and Amazon Elastic Beanstalk.

3. **Software as a service (SaaS)**: Under this model, Cloud providers deliver on-demand applications, hosted on the Cloud, as Internet-based services that relieve end-users from the complexity of developing, installing, and managing applications on their own computer. This model may be hosted on top of PaaS or IaaS or directly on the Cloud infrastructure. Google Apps, SalesForce CRM, and business applications from Oracle and SAP, along with consumer-oriented applications such as Gmail, Google Docs, and TurboTax Online are examples of SaaS. In the health care arena, a number of companies are providing SaaS for applications such as electronic health records (EHR), medical coding, payment/reimbursement, human resource management, and picture archiving and communications systems (PACS).

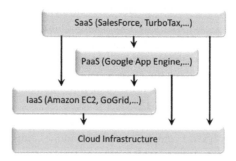

Figure 5.2 Cloud service layers and their interdependencies.

In addition to the three aforementioned layers, Cloud computing services can be provided via public, private, or hybrid deployment models. When a Cloud is made available to the general public, we call it a *public Cloud* and the services being sold are termed utility computing. Public Clouds are available to customers such as hospitals, corporations, governments, or academic organizations, but their operation is exclusively managed by a third-party Cloud provider.

A *private Cloud* refers to virtualized computing resources that are available only to a single organization. This model gives the owner greater control over the configuration and security of the Cloud infrastructure. The private Cloud may be managed by the organization itself, by a third-party provider, or by a combination of the two. Physically, a private Cloud may be located on-site or at a remote location managed by a Cloud provider.

A *hybrid Cloud* is a combination of these two models that tries to address the limitations of each approach. In a hybrid Cloud, part of the service infrastructure runs on a private Cloud only accessible to the organization while the remaining parts run on one or more public Clouds, shared with other organizations or with the public (Zou et al. 2016). Hybrid Clouds offer more flexibility than either public or private Clouds. Critical resources can be hosted onto a private Cloud for tighter control and better security. At the same time, they allow for the information systems of an organization to interface with other systems outside this organization via a public Cloud. In the health care industry, a hybrid Cloud alleviates some of the data security concerns while allowing patients and medical providers to access health care systems through public Cloud interfaces. The main challenge of hybrid Clouds is in their complex design that calls for determining the optimal point at which the balance between public and private Cloud features is fulfilled. It is projected that hybrid Clouds will dominate the market for most consumers for their broader range of features.

Selecting the right Cloud model is highly dependent on the application. Scientific tasks that are performed as short bursts of intense computation are best deployed on public Clouds because these platforms allow users to allocate large amounts of resources on demand, for a reasonable cost. In contrast, applications that handle protected health information should be hosted on a more secure private Cloud.

Finally, Cloud platforms may differ with respect to multi-tenancy. A tenant is defined as one or more users that share the same data, configuration, and features on the system they are accessing. Clouds commonly use one of the four configurations of multi-tenancy shown in Figure 5.3. In the first case, a, each tenant receives an exclusive instance of the Cloud service that is customized based on that tenant's goals

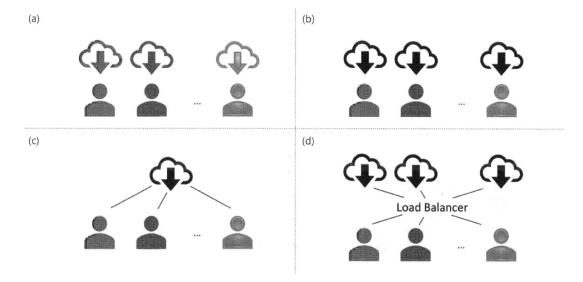

Figure 5.3 Four multi-tenancy configurations. (a) Exclusive and customizable allocation, (b) Exclusive allocation with limited customizability, (c) Shared resource allocation with additional on-demand service pro vision, and (d) Shared and dynamic allocation using load balancer.

and applications. In the next configuration, b, the tenants still have a dedicated instance of the service but with minimal customization. In case c, all tenants share the same instance of the service. The application in this case is divided into a core application component and extra components that are loaded on-demand to the requesting tenant. The last case, d, assigns tenants to a load balancer that redirects requests to a suitable instance based on system load. The cases c and d have to be considered rigorously because they expose tenants to security risks due to shared memory and hardware.

5.1.4 PHYSICAL IMPLEMENTATION OF CLOUD COMPUTING

The lowest layer of a Cloud is the hardware layer, which includes physical resources such as servers, routers, switches, power, and cooling systems. This layer, typically implemented in large datacenters, consists of a collection of stacked server blades laid out in racks and hierarchically interconnected through switches and routers (Figure 5.4). The servers may incorporate various types of hardware including specialty ones such as graphics processing units (GPUs) and large amounts of memory for parallel computation (Pratx and Xing 2011) or memory-intensive applications. An access switch, located on top of each rack, connects the servers of that rack to one another and to the aggregation layer above. For redundancy, access switches are preferably connected to more than one aggregation switch. The role of the aggregation layer is to provide domain services, location services, and server load balancing (Dave et al. 2016). Finally, the core layer connects the aggregation switches together and manages traffic into and out of the datacenter.

5.1.5 ECONOMICS OF CLOUD COMPUTING

Cloud computing, with its unique technical features, has given rise to a new crop of robust and elastic web applications. However, economic incentives have also played a significant role in the rapid adoption and popularity of this technology. Cloud computing technologies make it cheaper and faster to deploy a web-based application for a large user base.

A key enabler of Cloud computing has been the construction and operation of large datacenters across different geographical locations to provide cost-efficient and reliable computation (Armbrust et al. 2010). Perhaps surprisingly, a major cost of Cloud computing is electricity usage. Cloud computing systems use several strategies for minimizing the cost of computation.

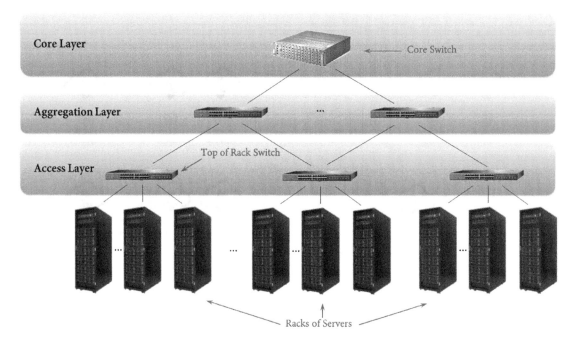

Figure 5.4 Typical layered architecture of a datacenter used for Cloud computing.

First, Cloud computing providers locate their datacenters in regions endowed with abundant and inexpensive natural resources, such as hydropower or wind. In addition, they use geodiversity to exploit fluctuations in electricity prices over multiple geographical time zones for maximal cost efficiency. These two strategies allow them to access electricity at the lowest cost possible.

Second, Cloud providers use virtualization technology extensively to ensure that hardware resources remain fully utilized. Thanks to statistical multiplexing among many users, the demand for computing resources is more predictable and less subject to fluctuations, resulting in greater average utilization of the hardware. Cloud computing systems can also use a dynamic pricing mechanism to shift demand from peak hours and higher-demand locations to off-peak hours and less-popular locations. Such a mechanism ensures a tight balance between supply and demand of computing resources, leading to efficient resource allocation. These economic benefits of Cloud computing are particularly striking for data storage services, the cost of which has dramatically decreased in recent years.

On a side note, it is important to mention that Cloud computing promotes green computing by reducing waste of electricity and hardware and by enabling datacenters to be located in areas that have access to low-carbon sources of electricity, such large hydroelectric power stations.

According to CFO Research (2012), consumers have reported a cost reduction as a result of using Cloud computing by saving on hardware supplies (up to 71%), system backup and recovery (up to 66%), software (66%), and information technology (IT) labor (59%). Organizations (hospitals in particular) are becoming more aware that their substantial capital investments in IT are often underutilized. Recently, it has been noted that in six corporate datacenters, most of the servers were using just 10%–30% of their capacity. In general, 64% of finance executives claim that Cloud computing could reduce their operational costs by up to 20%, and an additional 15% say that they would anticipate operational-cost reductions in excess of 20%. Only one in five finance executives would expect no reduction in operational costs as a result of a complete implementation of Cloud computing. A nearly immediate access to hardware resources without the need for purchasing and setting up new equipment allows companies to respond more quickly to emerging needs. This is especially beneficial for smaller institutions that have become able to employ compute-intensive business analytics thanks to the advent of Cloud computing.

5.1.6 COMMERCIAL PROVIDERS

The landscape of Cloud computing includes many large providers such as Amazon, Microsoft, IBM, and Google as well as smaller companies. The following describes the main providers of Cloud services.

Amazon EC2 provides a virtual computing environment (IaaS) that enables users to run VMs on Amazon's Cloud. Users can either create new Amazon machine images (AMI) containing an operating system (Windows or Linux), applications, and libraries or select from a collection of globally available AMIs. AMIs are stored using the S3, Amazon's distributed Cloud storage, as objects grouped in buckets. Each bucket can be stored in one of several geographical regions. The regions can be specified by users considering latency, cost, and regulatory requirements. Tasks are billed by Amazon EC2 according to the run time (CPU hour used) and the level of performance (CPU speed, memory, disk storage) of the machine selected by the user. Jobs handled by S3, on the other hand, are charged according to the amount of data stored and transferred (GB per month). CloudWatch is another service provided by Amazon that delivers performance metrics such as CPU utilization pattern, network transfer volume, or disk operations out of the data generated by other services like Amazon EC2. Finally, Amazon Virtual Private Cloud (VPC) securely connects a user's private infrastructure to Amazon's Cloud through a virtual private network (VPN). This enables owners of an already setup infrastructure to enhance their network options by connecting to Amazon Web Services (AWS) while being protected by Amazon's security services, firewalls, and intrusion detection systems.

Google App Engine is another platform for developing web applications using Google's infrastructure using programming languages such as Java and Python and web frameworks such as Django, CherryPy, Pylons, and web2py. Specialized APIs, Google Accounts, URL Fetch, and email services are all supported by Google App Engine. Google App Engine also accommodates the need for managing the execution of applications by providing users with a web-based administration console. As of 2017, there was no cost associated with Google App Engine for storage usage of less than 500 MB and about 5 million page views per month.

Microsoft is another popular provider of Cloud computing, thanks to its PaaS platform called Azure. This platform provides different types of Cloud services. The first one is a Windows-based environment for applications and distributed data storage (e.g., SQL servers). Azure applications can be built using languages such as .NET, C#, Visual Basic, and C++. Next is the distributed infrastructure for Cloud-based applications that is supported by .NET services. One advantage of the Azure platform is its flexibility: It can be used for both Cloud-based and locally run applications. The platform also provides developers with a framework for web application development known as ASP.NET.

Finally, IBM provides Cloud computing services called Blue Cloud that provides organizations with management tools for large-scale applications and database. This provider also offers consulting services to help customers integrate their infrastructure into the Cloud.

5.2 SOFTWARE PLATFORMS FOR DISTRIBUTED PROCESSING OF BIG DATA

The advent of Cloud computing has opened new avenues for performing massively distributed computation using large clusters of temporary VMs allocated on demand on the Cloud. This new paradigm has immense appeal for scientific computation due to its cost effectiveness and ease of development, especially compared with other specialty solutions such as GPU clusters (Pratx and Xing 2011) and computer grids. Scientific computation can now leverage new computational frameworks that have been developed by Internet companies to handle big data tasks such as analytics, web indexing, data mining, and machine learning. These new programming paradigms developed for Cloud computing include MapReduce, Hadoop, Spark, NoSQL, Cassandra, and MongoDB. These highly accessible tools are easy to program in addition to being open source in the majority of cases. This is in contrast to previous tools such as the message passing interface (MPI), which were tremendously difficult to scale for big data. In this section, three of the most popular big data–distributed Cloud computing frameworks are described in detail.

5.2.1 MapReduce

MapReduce is a programming framework for processing large data sets that was initially introduced by Google in 2004 (Dean and Ghemawat 2008). The framework was developed based on the observation that many large-scale computations require applying the same sequence of operations to large data sets before aggregating the output into a final result. In this programming paradigm, the user specifies Map and Reduce functions. To run the program, the input data, which are stored on a distributed file system, are first partitioned into independent subsets that are distributed to parallel Map tasks for processing (Figure 5.5).

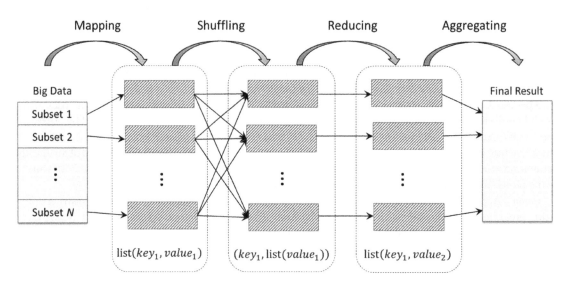

Figure 5.5 MapReduce distributed computation model.

Each Map task processes its assigned subset sequentially. Importantly, no communication is possible between Map tasks. Coordination between Map tasks is achieved by assigning each Map output a key that will enable different subsets of output data to be combined together by Reduce tasks, resulting in multiple final outputs (one per key).

In technical terms, MapReduce follows a master-slave format, with one master node managing an arbitrary number of slave nodes. MapReduce jobs are assigned to the master node by the user's client application. The master node breaks each job into parallel Map tasks that are assigned to slave nodes. Importantly, because data is distributed all over the slave nodes, the master node matches Map tasks to slave nodes that contain the relevant input data. This way, Map tasks rarely have to fetch data from remote nodes; the data they need is stored locally on the node they reside on. This method of matching tasks and nodes increases efficiency by minimizing data transfers, which is important for data-intensive tasks.

Map tasks produce their outputs in the form of *<key, value>* pairs. After shuffling, these key–value pairs are sorted by keys such that pairs with identical keys are grouped and dynamically assigned to a Reduce task for aggregation. The Reduce task aggregates these values into a final output and saves the result back to the distributed file system.

One important benefit of MapReduce is that the system is resilient to most hardware failures. The master node frequently checks the status of active nodes and keeps a record of all the tasks in the distributed filesystem. When a slave node fails, the master node can reassign the corresponding failed tasks to other nodes. Similarly, tasks that fail to complete in the allowed time frame are rescheduled onto a different node. This situation can arise when a node is working but crippled in some way.

Because the outputs of Map tasks are stored on the local disks of the failed machine, they will not be accessible. Therefore, these tasks are reset back to their initial state and then reassigned by the master node to other worker machines to rerun. Completed Reduce tasks do not need to be re-executed because their output is stored in the global file system. This process guarantees fault tolerance of the method.

Google has successfully used the MapReduce framework for a variety of tasks such as index building for Google Search, article clustering for Google News, and statistical machine translation. Facebook has also employed a similar technology for ad placement optimization and spam detection. The success of MapReduce stems specifically from the power of MapReduce for processing very large data sets with no data transfer bottlenecks. MapReduce however is not well suited for compute-intensive tasks or for iterative operations, mainly due to the launching overhead that comes with any round of computations even when repeating the same task.

5.2.2 HADOOP

Hadoop (Reyes-Ortiz et al. 2015) is an open-source software implementation of Google's proprietary MapReduce platform. This software package was developed with significant support from Yahoo for distributed big data processing. Yahoo used this software package for index building for Yahoo Search and spam detection for Yahoo Mail. In 2009, Yahoo set a record by sorting 1 petabyte of data in 16.3 h on 3658 nodes (1 petabyte = 1,000,000 GB).

Hadoop is made up of three main components. The first part is the MapReduce framework that handles work distribution and fault tolerance. A Java-based API allows Hadoop to run the parallel processing across a cluster of nodes. The second component is the Hadoop distributed file system (HDFS), an open-source storage system for large files. HDFS handles the redundant storage of big data on large computer clusters. The final component is a cluster manager that handles the computing resources and schedules jobs. The three components are tightly integrated with one another to optimize performance.

HDFS comprises a namenode (master node) and multiple datanodes (slave/storage nodes). It first divides data into fragments of a given size (e.g., 128 MB) and then distributes them over the datanodes across many networked disks. As required, HDFS delivers a single representation of the data via the HDFS API. Datanodes can communicate with each other to rebalance data distribution or replicate data for redundancy. Each of the data fragments are labeled with a unique identification number kept by the namenode that respectively comprises the current metadata information (e.g., filename and location). When the data is moved from client to datanodes or across datanodes, the new metadata is updated in the namenode. This information is used to access the desired data determined by any application. Thereby, the node that hosts the namenode is the Achilles' heel of the cluster because, when it fails, data becomes inaccessible and the system crashes.

Given that the probability of hardware failure approaches 100% as the number of nodes increases, the HDFS framework was designed to achieve fault tolerance through data replication across multiple datanodes. Typically, two copies of the data are saved in datanodes located on the same rack, and a third copy will be on a different rack. The namenode is responsible for keeping track of the number and location of the replications and for initiating necessary duplications, as well as removal of unnecessary copies left following the repair of damaged nodes. This makes the HDFS specifically resilient against isolated datanode crashing and failure in networking equipment.

The JobTracker task, which runs on the namenode, is a key part of the system that makes parallel applications viable in Hadoop. It determines the files to be processed, assigns tasks among nodes, and monitors the execution of tasks. Independent TaskTrackers also run on every datanode and accept tasks from the JobTracker. These TaskTrackers have to regularly communicate with the JobTracker because the Jobtracker will assume that a silent datanode has crashed and its associated jobs will be reassigned to other nodes. Similarly, tasks running on *straggling* nodes (i.e., nodes that are crippled in some way but still respond to pings from the namenode) will be duplicated elsewhere. In this case, the first task to finish will preempt other tasks from completing.

5.2.3 SPARK

Spark (Zaharia et al. 2010) was developed to efficiently deal with intense computational procedures that recursively perform operations over large data sets, such as supervised machine learning and deep learning algorithms. Spark achieves a high level of performance by running processes in-memory rather than in-disk, thus avoiding data loading and writing. Spark has been shown in practice to outperform MapReduce by up to two orders of magnitude (Xin et al. 2013).

The building blocks of Spark are resilient distributed data sets (RDDs). RDDs are distributed, immutable, and fault-tolerant memory abstractions. RDDs can reside in memory, disk, or their combination. They incorporate elements that are responsible for either producing other RDDs or for performing computations. However, they are only computed on actions following a Lazy Evaluation strategy in order to perform minimal computation and prevent unnecessary memory usage.

5.2.4 OTHER PLATFORMS

Among other platforms, Sawzall is an interpreted language developed by Google that is inspired by MapReduce and improves on its parallelization. Pig is another programming language introduced by Yahoo built on Hadoop. Microsoft has also developed the Dryad processing framework. Haloop is a programming paradigm built on the concept of reusing Mappers and Reducers to support iterative algorithms. This framework inherits the fault tolerance of MapReduce while being compatible with Hadoop.

5.3 SECURITY AND PRIVACY OF CLOUD COMPUTING

In addition to risks common to all Web-based technologies, Cloud computing introduces new security challenges that need to be taken into account.

5.3.1 DATA SECURITY RISKS

Cybersecurity has become one of the defining issues of our time. Malicious computer attacks are nothing new, but the scale of these attacks has reached epic proportions in recent years. These data breaches are now occurring so frequently that anyone's information may be stolen multiple times in any given year. The annual cost of these attacks has been estimated to exceed $100 billion to the U.S. economy.

In theory, Cloud computing systems are designed to prevent such attacks, but the characteristics of these systems make their management vulnerable to faults and their security cannot be completely guaranteed (Subashini and Kavitha 2011). Some of the issues stem from vulnerabilities in data access protocols, virtualization, and web applications (e.g., SQL injection and cross-site scripting); physical access issues, privacy and control issues arising from third parties having physical control of data; issues related to identity and credential management; issues related to data verification, tampering, integrity, confidentiality, and data loss and theft; and issues related to authentication of the respondent device; and IP spoofing.

Cloud systems rely heavily on Internet technologies such as web protocols (HTTP) for SaaS web-based applications, SOAP, REST, and RPC Protocols for PaaS Web-based services and remote connections (VPN, SSH, and FTP) for managing VMs and storage services in IaaS. Importantly, any vulnerability in these protocols affects the security of the whole Cloud system. To minimize risks associated with network connectivity, it is important to maintain a robust firewall, to use end-to-end encryption in any connection to the Cloud, and to patch system vulnerabilities before they can be exploited by cybercriminals.

In addition, essential characteristics of Cloud systems such as multi-tenancy and elasticity make them vulnerable to new types of cyber attacks. This is intensified in public Clouds: The open architecture of these systems makes them a prime target for data thieves and other computer hackers. In a multi-tenant system, different users run code and store data on the same hardware. In theory, each user is cordoned off by the virtualization software and cannot access memory locations or hard-drive blocks beyond that user's assigned range. However, it is conceivable that a malicious user could exploit a vulnerability in the virtualization platform to inject code into off-limit memory locations and, in such a way, gain control of the underlying native hardware. Once a user controls the native hardware, he or she can access any of the VMs running on the same server. This type of exploit compromises the confidentiality and integrity of other tenants' assets, and cannot be detected by tenants who, by design, have no access to the hardware layer.

In fact, during the writing of this book, Google announced the discovery of a major security flaw in Intel, AMD, and ARM chips that could be exploited to bypass the virtualization system and read memory locations beyond those assigned to a given VM. Hence, a malicious program running on one VM could attempt to read critical data, such as passwords and cryptographic keys, from another user's VM running on the same hardware.

Another potential issue in Cloud computing arises from elasticity (scaling up or down) that allows a tenant to reuse the resources freed by other tenants. This method of resource allocation potentially threatens privacy because new resources assigned to a user may contain residues of data from a previous user (in unallocated memory for example). The virtualization platform is supposed to entirely wipe up all data storage devices before allocating them to new users, but it is conceivable that this process could fail.

The architecture of the Cloud as a stack of interdependent layers also complicates the effort to secure the Cloud. Any vulnerability in a given layer impacts all the layers above it. In other words, a Cloud layer is only as secure as the weakest layer below it. These multiple layers offer as many targets for hackers and cybercriminals, and many computer security experts are required to ensure the end-to-end security of a Cloud system and its different layers.

The best approach to protect data stored on the Cloud is to encrypt it. However, it must be noted that computers cannot perform operations on encrypted data; therefore, if the data is to be used, it eventually needs to be decoded and stored unencrypted in volatile computer memory (i.e., RAM). Therefore, no approach can fully protect data against a "red-pill" hostile takeover of the Cloud's hardware layer.

Given the aforementioned risks, one may conclude that Cloud systems are not yet safe enough for storing confidential information. However, this assumption would be incorrect considering that physical theft of employee devices and data remains the main cause of health care data breaches. In recent years, whole-drive encryption has been deployed by many organizations to limit risks of data theft, but compliance with these rules is still far from perfect and many employees store protected data on unsecured laptops or portable drives. Implementation of Cloud computing allows employees to access their data from anywhere, in a safe and protected manner, removing the impetus for carrying data on physical devices. Cloud storage settings can also be set up to monitor data access patterns and to send alerts in case of abnormal access or downloads. Thus, the Cloud system can actually enhance data security for institutions dealing with protected information.

5.3.2 DATA LOSS

In today's digital economy, the loss of critical data can be costly and even life threatening for health care institutions. Data loss can occur for many reasons, including hardware failures, human error, or malicious events. Fortunately, data durability is one of the most useful features of Cloud computing. Thanks to redundant distribution of data across multiple geographical locations, true loss of data is vanishingly rare in Cloud systems. Amazon S3, for instance, boasts that its data durability is better than 99.999999999%, meaning that less than 10 files in a trillion (10^{12}) could be irreversibly lost in any given year. At this level, data loss is much more likely to be caused by user error or malicious attacks. Crypto viruses in particular have become an increasing threat in recent years, crippling research and hospital IT systems. Cloud systems are better protected against such threats than individual institutions or hospitals installations because they allocate significant amounts of expert resources to securing the system.

Another metric to consider is data availability. With so many health care processes critically dependent on data, even brief outages can severely disrupt patient care. These outages can be due to connectivity failure, which prevents systems from accessing data in the Cloud, or Cloud service disruption. Amazon S3 is supposed to be 99.99% available meaning that system downtime should not exceed 1 h per year. However, recent outages have shown this metric to be overly optimistic. In one instance, Amazon S3's datacenter on the East coast was shut down for over 5 h, causing many websites and web services in the United States to crash. Given this, it is critical for health care institutions to have robust contingency plans to handle such outages.

5.3.3 REGULATORY ISSUES IN HEALTH CARE CLOUDS

The security of patient-related information is a major concern in the adoption of Cloud computing by health care providers. In the United States, protected health information (PHI) is regulated according to the Health Insurance Portability and Accountability Act (HIPAA) of 1996. This extensive set of regulations applies to covered entities (i.e., hospitals) and business associates (e.g., commercial vendors) to assure patients that their PHI is confidential and protected. Under these rules, Cloud service providers are considered business associates. A business associate is defined as an entity that requires access to PHI to provide services either to a covered entity or to another business associate. In order for a Cloud provider to process PHI, it must enter in a HIPAA-compliant business associate agreement with a covered entity. A service level agreement can be established to ensure that the Cloud provider meets HIPAA regulatory obligations. This agreement covers topics such as system availability and reliability, backup and data recovery, security, and information disclosure.

Furthermore, HIPAA requires PHI to be encrypted both during its transmission and storage. Many Cloud providers such as Amazon and Box have developed HIPAA-compliant configurations for transferring and storing PHI on the Cloud. However, it should be noted that the use of encryption does not exempt Cloud vendors from a HIPAA-compliant business agreement. Data transmission between a health care institution and the Cloud can be encrypted using Transfer Layer Security (TLS) or a similar data transfer protocol. Data transfers between VMs within the Cloud must also be encrypted. One way to achieve this is to use Amazon Virtual Cloud, a service that automatically encrypts the transmission of data inside the Cloud, which can be extended to include the institution local IT resources. Amazon S3 also presents several options for encrypting stored data at the level required by HIPAA.

Finally, Cloud providers have a duty to disclose any potential breach of PHI as soon as it is detected. In some cases, these regulatory reporting requirements can be automated using a tool called Amazon Artifact.

5.4 OPPORTUNITIES FOR CLOUD COMPUTING IN RADIATION ONCOLOGY

With the increase in imaging and genomic testing, the quantity of medical data generated by cancer patients as part of their treatment is growing rapidly, creating new challenges for processing and storing of these data. Cloud computing technology is destined to play an increasing role in tackling this challenge.

5.4.1 INFORMATION SYSTEMS IN RADIATION ONCOLOGY

Physicians rely on computers and software routinely but few medical specialties use software as intensively as radiation oncology. With the move to intensity-modulated treatments, software and data have become essential to the practice of modern radiation oncology (Baumann et al. 2016). Historically, radiation oncology software (e.g., treatment planning systems, quality assurance, and image registration) was installed on individual workstations and connected to centralized databases that ran on multiple dedicated hardware servers. This rigid architecture is slowly being replaced by flexible systems that use virtualization to provision virtual servers from a central pool of resources. Newer radiation oncology IT systems are composed of a main chassis on which server blades are mounted and connected. These blades use a virtualization operating system to run virtual servers such as an image server, a radiation therapy database, web servers, billing servers, and application delivery servers. Virtualization has many advantages. First, applications are easily maintained and always kept up to date because they are housed in a central location. Servers can be upgraded with minimal interruption because it is possible to run multiple versions of the virtual server on the same hardware. Furthermore, applications can be accessed from any location using a light client application. Software is also better secured and less susceptible to viruses, computer glitches, and data corruption, resulting in less downtime for individual users. Finally, because the servers are virtualized, they are more resilient to disasters. Virtualized servers running on a failing blade can be migrated to other blades within the same chassis. Servers can be mirrored in real time to one or more off-site locations, allowing services to resume promptly following the catastrophic failure of an entire rack.

Radiation therapy vendors are now offering several solutions that implement this architecture. For instance, Varian provides Fullscale, a customizable solution that virtualizes the main IT systems used by radiation oncology. Varian's virtual servers can run either on existing servers at the customer site, on a managed appliance installed on site by Varian, or in Varian's Cloud. Varian also provides Cloud-based applications for quality assurance (Qumulate) as well as an application marketplace. In general, Cloud-based IT systems are two to three times cheaper than a dedicated on-site installation. Users of Fullscale can access their treatment planning system using either the thick-client Eclipse application or a thin Citrix client that loads the application remotely from the Cloud or local datacenter. Generally, the thick client is suitable for heavy users of the treatment planning system (dosimetrists and physicists), whereas the thin client is good for physicians. Overall, software virtualization provides more robust performance than older solutions that ran servers on dedicated hardware because redundancy is built into the infrastructure layer.

5.4.2 CLOUD STORAGE OF MEDICAL INFORMATION

Since its inception nearly a century ago, radiation therapy has continuously moved toward ever more accurate treatments. This paradigm places data at the center of the clinical workflow. Data are needed to delineate the contours of the tumor and prescribe the radiation dose that should be given to the tumor, to verify that the patient is correctly positioned during treatment and that the tumor is responding as expected, and to ensure that the linear accelerator functions correctly and delivers the planned dose to the tumor. As treatment of cancer has turned into a data-centric enterprise, information technology has played an increasingly important role in orchestrating the flow of data among different systems. However, the proliferation of software systems has had a negative impact on the efficiency of clinical workflows, increasing the costs and causing bottlenecks.

Cloud computing offers a unique opportunity for redesigning clinical IT systems in a way that places the patient right in the center. Rather than connecting inefficient and disparate systems, data should be moved to a common Cloud platform, and applications should be built on top using the layered architecture of Cloud systems to implement clinical applications using low-footprint SaaS platforms. This new Cloud paradigm could dramatically streamline radiation oncology workflows, improve the efficiency of radiation oncology personnel, and increase the quality of patient care.

Cancer patient data are extremely varied. At the simplest level, the data include physician notes, lab results, and patient biometrics. These data are traditionally stored in EHR systems. In addition, patient data often include radiological scans acquired for diagnosis, staging, and treatment planning purposes. Metadata are frequently associated with these scans, including radiologist reports, image annotations, and quantitative image biomarkers. These images are archived in large PACS system typically managed by the radiology department. Digital histopathological images are also starting to be used in the clinical workflow and are stored in laboratory information systems (LIS). An increasing number of cancer patients are also getting their genome sequenced as part of their treatment. Finally, radiation oncology generates a number of specialized data such as organ contours, dose prescriptions, dose maps, patient-specific quality assurance measurements, and radiation treatment plans. These data are also incorporated into patients' files.

Cloud storage represents an attractive strategy for storing the large files associated with individual patients. It is estimated that each patient generates 10 GB of data, most of which is genomic information. Given this, a medium-sized cancer center could accumulate nearly 1 petabyte of data over 10 years. This amount of data represents 3000 hard drives assuming 1 TB/HD and a redundancy factor of 3. Of note, research data sets can be even larger. For instance, The Cancer Genome Atlas (TCGA) comprises over 2 petabytes of data for 11,000 patients, or 200 GB per patient.

The portion of this data specific to radiation oncology is approximately 0.5–1 GB/patient. This includes simulation CT scans, imaging scans used for treatment planning and treatment monitoring (PET/CT, MRI), portal images, on-board cone-beam CTs, four-dimensional (4D) CT used for motion management, and perfusion CT scans. Given this, a medium-scale radiation oncology department could generate 1–2 TB of data per year. Given the requirement to store data a minimum of 10 years, storage costs can be significant, especially due to the requirements for disaster contingency plans. Cloud computing, with its efficient implementation model, can yield significant savings and can simplify the implementation of large-scale storage.

One example of this is for implementation of PACS. Image archiving is one of the drivers of big data in medicine, and Cloud-based PACS have been proposed (Kagadis et al. 2013, Moore et al. 2014). One of the functions of a PACS system is to render a 3D radiological data set into an image that can be displayed on a diagnostic workstation. For large 3D data sets, it is not efficient to transfer an entire data set to a remote workstation. Rather, an emerging approach is to push the heavy task of 3D image rendering to the Cloud and stream the resulting output to the radiologist workstation. Because the output of the rendering task is much smaller than the full 3D data set, it can be visualized with low latency even on low-power mobile devices such as tablets and cell phones.

Different client architectures can be employed to implement a Cloud-based PACS. One approach is to develop a thin-client application that can run either on a mobile operating system (e.g., iOS) or in a PC desktop environment. It is also possible to build a remote PACS viewer within a web browser either using a rich-Internet framework such as JAVA, Flash, or Silverlight or directly using standard HTML5 (this latter approach is called a *zero-footprint client*). Finally, an even more radical approach is to run the radiologist workstation entirely in the Cloud and use a remote desktop tool to forward the workstation screen to the radiologist's location. This approach moves nearly all of the computation to the Cloud and only requires the radiologist device to have the capability of remote desktop. It facilitates software upgrades because all the software is located remotely and minimizes cybersecurity risks because all the sensitive data is kept in one place.

The advantages of Cloud storage of medical data extend beyond cost effectiveness. For one, putting all the data in one place facilitates the necessary collaboration between clinical departments. It also facilitates continuity of care, for example, in the case where a patient is referred by a different institution. Previous studies performed at another institution are more likely to be easily available, making new scans unnecessary. Furthermore, specific cases can be easily shared with specialty consultants for a second opinion. Finally, some of the data can be made available to patients for their own information via web portals. This would go along with the current trend in medicine in which patients are given an increasing amount of medical information regarding their condition and the planned treatment.

5.4.3 MEDICAL COMPUTATION IN THE CLOUD

One of the areas that will greatly benefit from Cloud computing is radiation treatment planning (Na et al. 2013). Dose calculation is an essential part of treatment planning, but accurate dose calculation can be time consuming when an accurate physical model is used to represent dose deposition in the tissues. In the current workflow, optimal radiation therapy plans are obtained by trial and error: a sequence of plans is generated by the physics personnel until a good trade-off can be achieved between dose to target and organs at risk. Given this interactive workflow, it is critical for the treatment optimization to be completed as quickly as possible. Cloud computing, with its model of elastic resource scaling, is ideally suited for "bursty" tasks that are run with unpredictable frequency. Upon submission of a new set of optimization parameters, an array of VMs can be requisitioned to execute the new treatment optimization.

Given the massive computation throughput achievable with Cloud computing, it has even been suggested that dose computation could be performed routinely using Monte Carlo simulation. This type of dose computation is more accurate than analytical methods but more computationally demanding. A previous study demonstrated that Monte Carlo simulation of photodynamic therapy could be accelerated 1200-fold using MapReduce to distribute computation onto a Cloud of 240 nodes (Pratx and Xing 2011). A similar study used the same approach to distribute the execution of GATE in the Cloud (Liu et al. 2017). Other groups have used clusters of virtual Cloud computers to parallelize Monte Carlo simulation (Poole et al. 2012; Wang et al. 2011). Another group presented a multi-GPU Cloud-based server (MGCS) framework to accelerate the dose convolution-superposition method (Neylon et al. 2017).

Other operations can also be performed using Cloud computing. For instance, deformable image registration (DIR) of the planning CT to the treatment CBCT was implemented using Cloud computing (Zaki et al. 2016). Image reconstruction for CBCT is another process that can be accomplished efficiently using MapReduce in the Cloud (Meng et al. 2011). Here, the parallelization is efficiently handled by parallel Map tasks, and the accumulation of the different projections is performed by parallel Reduce tasks.

Based on its many advantages, an attractive use of Cloud-based treatment optimization is for adaptive radiotherapy (Xing et al. 2007). This type of radiation treatment incorporates online information obtained from on-board imaging to refine the treatment plan right before treatment. It can be used to account for shrinking of the target tumor, for differences in patient anatomy due to inter-fractional motion, or for individual therapy response revealed by mid-treatment molecular imaging. One of the critical requirements for online adaptive radiotherapy is that treatment planning optimization be performed quickly using all the information available at the time of treatment. This could be implemented using Cloud computing. Briefly, patient data would be uploaded onto the Cloud as they are being acquired. This could include mid-treatment scans, CBCT positioning scans, machine logs from radiation treatments, and so forth. The information could be used to automatically update the patient's treatment plan, which would then be reviewed and loaded onto the treatment machine. In practice, adaptive radiation therapy is still difficult to implement due to the requirement for patient-specific quality assurance, but in theory these challenges could be overcome.

Finally, the massive amount of computation available via Cloud computing creates opportunities for increasing the use of artificial intelligence and machine learning in radiation oncology. One can imagine that the burden associated with repetitive tasks (dose prescription, organ contouring, and treatment planning) will be slowly shifted to the machines. The World Economic Forum recently forecasted the advent of a fourth industrial revolution, in which artificial intelligence will be a key technology. New technologies emerging from this revolution will shift information processing from expert humans to more automated systems. The availability of Cloud-based frameworks for big data machine learning will likely facilitate the development of new intelligent software tools. One exciting prospect is that clinical databases from myriad institutions could be mined automatically to train artificial intelligence tools on tasks such as treatment planning, quality assurance, error detection, and dose prescription. It is hoped that, in a not too distant future, machine intelligence can be harnessed to derive initial organ contours, dose prescription, and treatment plan based on historical knowledge from tens of thousands of previously treated patients. These decisions will still undergo verification and validation by a team of humans, but the time saved in the process will enable human efforts to be allocated toward other important goals.

5.4.4 CLINICAL TRIALS

Cloud computing provides a convenient platform for sharing large medical data among different health care institutions, collaboration, and standardization of methods (D'Haese et al. 2015). Individualized precision radiation therapy relies on accurate models of tumor and normal tissue response that can only be built through large-scale controlled clinical trials. Large-scale cooperative clinical trials such as Radiation Therapy Oncology Group (RTOG) can leverage Cloud infrastructure to organize and summarize clinical data from multiple sources, including genomics, treatment plans, imaging data, dose–volume histograms, and clinical outcomes. Standardization of data formats is of critical importance for sharing of data to occur on a large scale. Many interoperable formats such as DICOM, DICOM-RT, ICD-9 and CPT codes, and LOINC can be used to encode and share clinical information.

Another advantage of storing clinical trial data in the Cloud is that the raw clinical data can be de-identified and opened to authorized research groups for hypothesis generation or for extracting information that is beyond the scope of the original trial. Such sharing of data accelerates the scientific discoveries and leverages existing investments in clinical research.

Federal sponsors of biomedical research often mandate the dissemination of research results. The National Institutes of Health (NIH) is piloting a new initiative called Big Data to Knowledge in which they are deploying Cloud resources to allow investigators to store, manage, disseminate, and interact with large data sets. The main goal of the program is to make these data findable, accessible, interoperable, and reusable. This effort follows many years of investment by the NIH in biomedical informatics, which has resulted in popular resources such as TCGA, the National Biomedical Imaging Archive, the Cancer Imaging Archive, and the Visible Human Project. One of the advantages of a Cloud platform over older archives is that it allows researchers to share their analysis methods so that others can interact with the data through SaaS platforms. These software programs can be run directly on the NIH Cloud without requiring these large data to be downloaded locally.

Other comparable services have been launched to facilitate the collection and sharing of cancer data. CancerLinQ was started by the American Society of Clinical Oncology to give oncologists personalized guidance to help optimize clinical decisions (Shah et al. 2016). As of 2017, the program was used by over 2000 oncologists and contained over 750,000 cancer patient records. The goal of this program is to collect information from the large majority of patients that are not enrolled in clinical trials. Another tool that was developed specifically for radiation oncology is Oncospace (Robertson et al. 2015). This database platform aggregates treatment planning and clinical information to facilitate data mining for improving evidence-based medicine and decision support. The database is structured according to disease site and it includes information such as target and organs at risk, dose distributions, diagnoses, toxicities, clinical outcomes, medication, laboratory results, and patient characteristics.

5.4.5 CHALLENGES OF CLOUD COMPUTING IN RADIATION ONCOLOGY

One of the concerns with moving data to a Cloud platform is the risk of lock-in. Once large data sets and processing tools are deployed on a given Cloud, it is difficult and costly to move them to a different provider. As a result, Cloud vendors have leverage on their customers and little incentive to reduce cost.

Another issue with Cloud computing is that it can lead to unnecessary duplication of image data stored in different locations. In current practice, it is common to use different systems for different applications (e.g., image archiving, treatment planning, and image processing). Treating a patient with radiation may require a CT scan, organ and tumor contours from an image segmentation package, and a dose distribution from a treatment planning system. Each time a different system is used, the data has to be imported and duplicated into the new system. As a result, a given CT scan may be stored three or more times (not accounting for additional copies made for data redundancy). An emerging solution to this problem is the use of vendor-neutral archives (Langer et al. 2013). In this paradigm, images are stored in a standard format in a central repository, and they can be cross-referenced by other peripheral systems. Such a system would also facilitate the sharing of image data with external institutions.

Customers have little control on technical aspects of the Cloud; they may not be able to guarantee that deleted data are truly deleted because copies of the data may persist in multiple locations. They also have little control over the security of the Cloud. In particular, a concern is that unpublished data from clinical trials have monetary value and could constitute a target for cybercriminals. Another potential concern is that the Cloud industry as a whole remains volatile and Cloud technology evolves at a significantly faster pace than health care technology in general.

5.5 FINAL THOUGHTS

Cloud computing has become a ubiquitous term that one can see placarded on airport walls or in magazine ads. The excitement about Cloud computing is real and justified. This model based on sharing of computational resources over the Internet can dramatically reduce the cost of storing and processing big data, which will have profound implications in radiation oncology. Cloud computing will enable researchers to help extract new information from big medical data sets through advanced machine learning and data mining. It will help automate many low-level tasks that are now performed by physicians, freeing their time to interact more deeply with patients. It will provide greater treatment accuracy and safety for patients by decreasing the chance of human errors. The future of Cloud-based radiation therapy is bright.

ACKNOWLEDGMENTS

The authors would like to thank Dr. Karl Bush for useful discussions.

REFERENCES

Armbrust, M., A. Fox et al. (2010). A view of cloud computing. *Communications of the ACM*. 53(4): 50–58.

Baumann, M., M. Krause et al. (2016). Radiation oncology in the era of precision medicine. *Nature Reviews Cancer*. 16(4): 234–249.

CFO Research. (2012). The Business Value of Cloud Computing: A Survey of Senior Finance Executives. http://lp.google-mkto.com/rs/google/images/CFO%2520Research-Google_research%2520report_061512.pdf.

D'Haese, P.F., P.E. Konrad et al. (2015). CranialCloud: A cloud-based architecture to support trans-institutional collaborative efforts in neurodegenerative disorders. *International Journal of Computer Assisted Radiology and Surgery*. 10(6): 815–823.

Dave, A., B. Patel et al. (2016). Load balancing in cloud computing using optimization techniques: A study. *International IEEE Conference on Communication and Electronics Systems (ICCES)*, pp. 1–6.

Dean, J., S. Ghemawat. (2008). *MapReduce: Simplified Data Processing on Large Clusters, Sixth Symposium on Operation*. San Francisco, CA, Vol. 51(1), pp. 107–113.

Foster, I., Y. Zhao, et al. (2008). Cloud computing and grid computing 360-degree compared. *IEEE In Grid Computing Environments Workshop, GCE'08*, pp. 1–10.

Kagadis, G.C., C. Kloukinas et al. (2013). Cloud computing in medical imaging. *Medical Physics*. 40(7).

Kephart, J.O., D.M. Chess. (2003). The vision of autonomic computing. *Computer*. 36(1): 41–50.

Langer, S.G., K. Persons et al. (2013). Towards a more cloud-friendly medical imaging applications architecture: A modest proposal. *Journal of Digital Imaging*. 26(1): 58–64.

Liu, Y., Y. Tang et al. (2017). GATE Monte Carlo simulation of dose distribution using MapReduce in a cloud computing environment. *Australasian Physical & Engineering Sciences in Medicine*. 40(4): 777–783.

Mell, P., T. Grance. (2011). The NIST definition of cloud computing. National Institute of Standards and Technology. Special Publication 800–145. doi:10.6028/NIST.SP.800–145.

Meng, B., G. Pratx et al. (2011). Ultrafast and scalable cone-beam CT reconstruction using MapReduce in a cloud computing environment. *Medical Physics*. 38(12): 6603–6609.

Moore, K.L., G.C. Kagadis et al. (2014). Vision 20/20: Automation and advanced computing in clinical radiation oncology. *Medical Physics*. 41(1): 010901.

Na, Y.H., T.S. Suh et al. (2013). Toward a web-based real-time radiation treatment planning system in a cloud computing environment. *Physics in Medicine and Biology*. 58(18): 6525.

Neylon, J., Y. Min et al. (2017). Analytical modeling and feasibility study of a multi-GPU cloud-based server (MGCS) framework for non-voxel-based dose calculations. *International Journal of Computer Assisted Radiology and Surgery* 12(4): 669–680.

Poole, C.M., I. Cornelius et al. (2012). Radiotherapy Monte Carlo simulation using cloud computing technology. *Australian Physical & Engineering Sciences in Medicine*. 35(4): 497–502.

Pratx, G., L. Xing. (2011). Monte Carlo simulation of photon migration in a cloud computing environment with MapReduce. *Journal of Biomedical Optics*. 16(12): 125003–1250039.

Reyes-Ortiz, J.L., L. Oneto et al. (2015). Big data analytics in the cloud: Spark on Hadoop vs MPI/OpenMP on Beowulf. *Procedia Computer Science*. 53: 121–130.

Robertson, S.P., H. Quon et al. (2015). A data-mining framework for large scale analysis of dose-outcome relationships in a database of irradiated head and neck cancer patients. *Medical Physics*. 42(7): 4329–4337.

Shah, A., A.K. Stewart et al. (2016). Building a rapid learning health care system for oncology: Why CancerLinQ collects identifiable health information to achieve its vision. *Journal of Clinical Oncology*. 34(7): 756–763.

Subashini, S., V. Kavitha. (2011). A survey on security issues in service delivery models of cloud computing. *Journal of Network and Computer Applications*. 34(1): 1–11.

Wang, H., Y. Ma et al. (2011). Toward real-time Monte Carlo simulation using a commercial cloud computing infrastructure. *Physics in Medicine and Biology*. 56(17): N175.

Xin, R.S., J. Rosen et al. (2013). Shark: SQL and rich analytics at scale. In *Proceedings of the 2013 ACM SIGMOD International Conference on Management of Data*, New York, pp. 13–24.

Xing, L., J. Siebers et al. (2007). Computational challenges for image-guided radiation therapy: Framework and current research. *Seminars in Radiation Oncology.* 17(4): 245–257.

Zaharia, M., M. Chowdhury et al. (2010). Spark: Cluster computing with working sets. *HotCloud.* 10(10–10): 95.

Zaki, G., W. Plishker et al. (2016). The utility of cloud computing in analyzing GPU-accelerated deformable image registration of CT and CBCT images in head and neck cancer radiation therapy. *IEEE Journal of Translational Engineering in Health and Medicine.* 4: 1–11.

Zhang, Q., L. Cheng et al. (2010). Cloud computing: State-of-the-art and research challenges. *Journal of Internet Services and Applications.* 1(1): 7–18.

Zou, L., Z. Xie et al. (2016). EP-1926: Hybrid of cloud computing and workstations for radiotherapy planning. *Radiotherapy and Oncology.* 119: S914.

6 Big data statistical methods for radiation oncology

Yu Jiang, Vojtech Huser, and Shuangge Ma

Contents

6.1 Background 79
    6.1.1 Big Data for Biomedicine 79
    6.1.2 The Small-Data Era of Radiation Oncology 80
    6.1.3 The Big Data Era of Radiation Oncology 80
6.2 Integrating Multidimensional Data 81
    6.2.1 Multidimensional Studies 81
    6.2.2 Statistical Methods 81
        6.2.2.1 A Marginal Screening-Based Approach 82
        6.2.2.2 Joint Analysis Based on the Additive Models 83
        6.2.2.3 More Advanced Joint Analysis Methods 84
        6.2.2.4 Functional-Group–Based Analysis 86
    6.2.3 The Past, Present, and Future 87
6.3 Integrating Multiple Independent Data Sets 88
    6.3.1 Collection of Multiple Independent Data Sets 88
    6.3.2 Statistical Methods 88
        6.3.2.1 "Classic" Meta-analysis 88
        6.3.2.2 Integrative Analysis Methods 88
        6.3.2.3 Accommodating Finer Data Structures 89
    6.3.3 The Past, Present, and Future 90
6.4 Computation: A Unique Challenge Faced by Big Data Analysis 91
6.5 Concluding Remarks 92
Acknowledgments 92
References 92

6.1 BACKGROUND

6.1.1 BIG DATA FOR BIOMEDICINE

"Big data" is no longer a new concept. From highly advanced scientific research to ordinary daily life, big data are now playing an increasingly important role. In biomedical research and practice, big data have attracted extensive attention. The definition of big data in biomedical science is discussed in other chapters of this book. It should be noted that the definition and contents are still evolving, with researchers in different subfields emphasizing different aspects. Loosely speaking, big data refer to data that have a high volume and significant complexity, demand real-time fast computation, and cannot be accommodated using "classic" small-data analytics (Benedict et al., 2016; Bibault et al., 2016; Huilgol, 2016).

The collection, analysis, and applications of big data have had instant and deep impact on biomedicine. For complex diseases such as cancer, cardiovascular diseases, and mental disorders, genetic profiling and imaging studies have been extensively conducted (Gillies et al., 2015; Benedict et al., 2016). Data generated in such studies have "small sample size, high data dimensionality" characteristics. If thinking in terms of a data matrix where each row corresponds to one sample and each column corresponds to one variable, this type of data is "short, but wide." The development of novel high-dimensional statistical techniques has generated disease markers that have led to a deeper understanding of disease etiology, therapeutic targets that have led to new venues for treating diseases, and accurate disease outcome models that can assist better decision making (Cancer Genome Atlas Research, 2008, 2014; Trifiletti and Showalter, 2015). In "classic" epidemiologic and clinical studies, one "annoying" but reoccurring phenomenon is that the findings made in a single study often lack reproducibility in independent studies. This can be partly or even largely explained by the limited sample sizes of individual studies and across–data set heterogeneity. With the decreasing cost of data collection and collective effort on data sharing, for many diseases there are now large consortiums and databases that host data on a huge number of heterogeneous samples. Thinking in terms of a data matrix, this type of data is "long." Representative examples of this type of data include electronic health record (EHR) data collected by, for example, large insurance companies or Medicare and Medicaid, as well as databases such as Gene Expression Omnibus (GEO; www.ncbi.nlm.nih.gov/geo/), Genomic Data Commons (GDC; https://portal.gdc.cancer.gov/), and database of Genotypes and Phenotypes (dbGaP; www.ncbi.nlm.nih.gov/gap). There have been extensive recent developments in novel statistical methods that can effectively conduct multi–data sets analysis and accommodate the across–data set heterogeneity. For example, data analysis has led to identifying more reliable risk factors that have been validated in independent studies (Ma et al., 2011a; Zhao et al., 2015a; Huang et al., 2017a, 2017b). Technical developments in mobile devices, user-friendly apps, and telecommunication have fundamentally changed the way that patients and physicians interact. With portable mobile devices, a large amount of data can be collected on patients and transferred back to clinics/physicians instantly. Such data need to be processed and analyzed on a real-time basis, reflecting the high-velocity feature of big data. The development in such medical devices, analytic methods, and decision-making strategies has greatly advanced the fields of cardiovascular diseases, diabetes, and other diseases. The literature on big data in biomedicine is too vast to be reviewed here. See Bender (2015), Lustberg et al. (2017), Kessel and Combs (2016), as well as other chapters of this book for relevant discussions.

6.1.2 THE SMALL-DATA ERA OF RADIATION ONCOLOGY

Analytics has always been an important component of radiation oncology (Stefani et al., 1979; Van Dye et al., 2013; Nieder and Gaspar, 2014). Before the big data era, there had been a long history of small-data analytics in radiation oncology. It should be acknowledged that the specific scientific questions investigated in radiation oncology are significantly different from the other fields of biomedicine. However, from a statistical perspective, the underlying questions and techniques are often similar. For example, a problem of critical interest is the tolerance of normal tissues to radiation (i.e., the toxicity of radiation) (Lyman and Wolbarst, 1987; Emami et al., 1991). This has been cast as a regression problem and various statistical models—including, for example, the liner regression model, multivariate logistic regression model, principal component analysis (PCA), and others—have been implemented (Burman et al., 1991; Skwarchuk et al., 2000; Dawson et al., 2005; Barnett et al., 2009). Although there are significant differences in study cohorts and treatment strategies, the analytics are largely standard and can be conducted using well-developed methods and software. For textbook-style references, see Brahme (2014). Another important scientific problem is to identify biomarkers that can predict radiotherapy outcomes (Taylor and Kim, 1993; Mazumdar and Glassman, 2000). Although scientifically challenging, in terms of statistical analysis, this is again a rather standard problem. Common techniques under the regression analysis framework include selecting important variables based on magnitude and significance level (p-value) (Mazumdar and Glassman, 2000). Such an analysis can reveal association but not causation, for which more complex causal inference analysis needs to be conducted (Frangakis and Rubin, 2002).

6.1.3 THE BIG DATA ERA OF RADIATION ONCOLOGY

With technological developments, the research and practice of radiation oncology has entered the big data era. With the other chapters of this book devoted to specific applications, in this chapter, we will focus on the statistical methodologies that have been recently developed for big data and are broadly applicable to a large number of applications. Unfortunately, big data statistical methods are often significantly more complicated than their small data counterparts, and it is impossible to provide all necessary details in a single chapter. Thus, only brief introductions are provided, and the original publications should be consulted for thorough discussions.

6.2 INTEGRATING MULTIDIMENSIONAL DATA

6.2.1 MULTIDIMENSIONAL STUDIES

The clinical outcomes and phenotypes of interest in radiation oncology, for example, response to radiation therapy, involve complex biological processes. It has been recognized that the statistical models built on a small number of clinical and pathological variables often have only moderate success. As in the other fields of oncology, more comprehensive data collection has been pursued with the expectation that more data (variables) can bring additional information and hence improve model fitting. Beyond the traditional low-dimensional clinical and pathological variables, the most prominent new data types include omics data (Medicine et al., 2012; Starkschall and Siochi, 2013) and imaging data (Lei, 2011), both of which are highly dimensional. Omics data are collected using advanced profiling techniques and describe molecular changes. Examples include genetic data (e.g., DNA-level mutations), genomic data (e.g., mRNA gene expressions), epigenetic data (e.g., DNA methylation), proteomic data (protein expressions), and others. For detailed discussions on omics data in radiation oncology, see Medicine et al. (2012). Imaging data, for example, obtained using a CT scan or MRI, describe the physical properties of tumors and normal tissues. For detailed discussions on imaging techniques in radiation oncology, see Brock (2016), Gaffney et al. (2012) and Naqa et al. (2015).

The collection of samples, technical aspects of data generation, and raw data processing are by no means trivial. Luckily, with the extensive effort of the past decades, such techniques are now relatively mature. For example, for the processing of raw omics data, the existing statistical methods and software include Minfi and quantile normalization for DNA methylation data (Marabita et al., 2013; Aryee et al., 2014), remove unwanted variation (RUV) for RNA sequencing data (Risso et al., 2014), quantile normalization, and variance stabilization normalization (VSN) for proteomic data (Kohl et al., 2012), and many others. In what follows, we will assume that data have been properly preprocessed and normalized and focus on the high-end modeling techniques.

6.2.2 STATISTICAL METHODS

In the low-dimensional small data era, the conceptual model is

$$Outcome \sim Clinical/Environmental/Pathological\ Variables \tag{6.1}$$

where the outcome variable can be continuous, categorical, or (censored) survival. Examples of variables on the right side of the regression are available in the literature (Gunderson et al., 2015) as well as other chapters of this book. In the big data era, with the consideration that omics variables and variables extracted from imaging data may have independent predictive power for the outcome variables, the conceptual model is now:

$$Outcome \sim Clinical/Environmental/Pathological\ Variables + Omics + Imaging \tag{6.2}$$

It is noted that most if not all of the existing studies have "incomplete" data collection, in the sense that not all relevant variables are measured. Prior to data analysis, properly managing and consolidating data also posed a demanding task. In Figure 6.1, we show the data management flowchart for the analysis of The Cancer Genome Atlas (TCGA; cancergenome.nih.gov) breast cancer (BRCA) data. For details, see Zhao et al. (2015b).

The estimation of Model 2 faces multiple challenges. The first and most prominent challenge is the high dimensionality. For example, in a typical genomewide gene expression study, the data dimensionality has

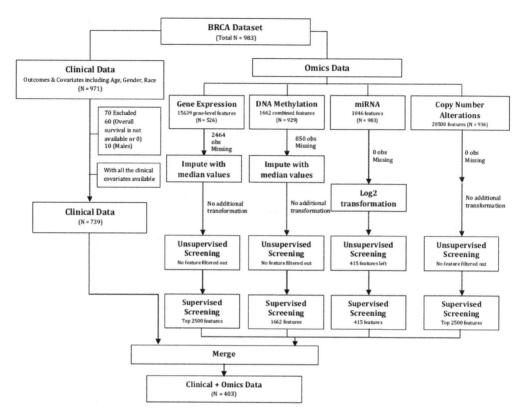

Figure 6.1 Data management for the analysis of TCGA BRCA data.

the order of 10^4; in a typical imaging study, the dimensionality has the order of 10^5 (Mwangi et al., 2014). In contrast, the sample size is on the order of 10^{1-3}. Statistically speaking, this poses a dimension reduction problem. The second challenge is the overlap in information contained in different types of variables. For example, both imaging and pathological variables can describe the physical characteristics of tumors. Statistically speaking, this poses a collinearity problem. The third challenge is that most of the high-dimensional measurements are "noises." For example, in a genomewide study, most of the genes are not expected to be relevant to response to radiation. Statistically speaking, this poses a variable/model selection problem. To address the aforementioned and other challenges, the following statistical techniques have been developed. Four such techniques are described below.

6.2.2.1 A marginal screening-based approach

To simplify notation and without loss of generality, consider the model:

$$Outcome \sim Clinical\ Variables + mRNA\ Gene\ Expression + Methylation \qquad (6.3)$$

In this model, there are two sets of high-dimensional omics variables. Models with more omics variables and/or imaging data can be analyzed in a similar manner. This approach proceeds as follows:

1. For each gene expression, fit a regression model: *outcome ~ gene expression*. Note that this is a univariate model and can be fit using many existing software packages. Depending on the type of outcome, the model can be linear (for a continuous outcome), logistic (binary), Cox (survival), or others.
2. For each methylation locus, fit a regression model: *outcome ~ methylation locus*.
3. Based on the model fitting results from Steps 1 and 2, select a small number of the most significant gene expressions and methylation loci. This selection can be based on the marginal *p*-values (which can be easily obtained from all statistical software) or likelihood.

4. Fit the outcome model with clinical variables and gene expressions/methylation loci selected in Step 3. With a small number of selected gene expressions and methylation loci, this step of model fitting is a standard low-dimensional problem and, again, can be accomplished using many existing software systems.

The biggest advantage of this approach is its computational and conceptual simplicity—no special software is needed for carrying out this analysis. As such, this approach was adopted in quite a few studies in the "early days." The underlying statistics, however, turns out to be highly nontrivial. The validity of this approach needs to be built on the highly challenging "sure independence screening" theory (Fan and Lv, 2008), which demands that genes (methylation loci) associated with the outcome are only weakly correlated with genes not associated with the outcome. Biologically speaking, such a condition is unlikely to be true in practice.

6.2.2.2 Joint analysis based on the additive models

Continuing the consideration of the conceptual model in Model 3, approaches more advanced than the marginal screening-based method described above have been developed. They do not need to make the weak correlation assumption and hence may be more practical.

The first family of approaches conducts *dimension reduction*, and a representative approach is the PCA. With the high dimensionality of the original (e.g., omics, imaging) measurements, this family of approaches searches for a small numbers of linear combinations and uses them in model fitting. Beyond PCA, other members of this family include partial least squares (PLS), independent component analysis (ICA), and others. Consider the PCA-based analysis as an example, which proceeds as follows:

1. Conduct PCA with gene expression and methylation data separately. This can be achieved using existing software, for example, R functions *prcomp* and *princomp* and SAS procedure *PRINCOMP*.
2. Select the top principal components (PCs) of gene expression and methylation data separately.
3. Fit a low-dimensional regression model using the clinical variables and selected PCs.

For applications of PCA and other dimension reduction approaches in oncology studies with high-dimensional variables, see Ma and Kosorok (2009), Ma and Dai (2011), and Ma et al. (2011b). Comparatively, this family of approaches is simple. The popular dimension reduction techniques have a long history and have been well developed in classic multivariate analysis. Software for conducting such analysis is widely available and easy to use. On the negative side, the results generated by an analysis that employs dimension reduction are usually difficult to interpret. For example, a PC is a linear combination of *all* genes (methylation loci). In published studies, such linear combinations have been referred to as "mega genes," "super genes," "eigen genes," "latent genes," and others. However, the biological interpretations have never been fully clear. In addition, for any oncology outcome of interest, it is expected that only a subset of genes (methylation loci) are relevant. In contrast, dimension reduction approaches do not have a selection component and build models using all genes.

The second family conducts *variable selection*, where the goal is to select a subset of relevant genes and methylation loci and use only them to build a model. Unlike the marginal screening-based approach, which achieves selection and model building in distinct consecutive steps, these two goals are realized *simultaneously*. In the literature, variable selection methods have been developed based on the following techniques:

1. Penalization. This family of approaches seeks for a model and parameter estimates that balance between goodness-of-fit (e.g., likelihood) and model complexity (which is usually defined as a function of the regression parameters). Representative methods include Lasso, elastic net, bridge, smoothly clipped absolute deviation (SCAD), and others (Tibshirani, 1996; Fan and Li, 2001; Zou and Hastie, 2005; Huang et al., 2009). Applications of the penalization techniques to omics and imaging studies in oncology include Llacer et al. (2001), Ma et al. (2007), Huang et al. (2008) and Ma and Huang (2008) et al.
2. Bayesian analysis is a classic statistical technique but has found new applications in high-dimensional studies (George and McCulloch, 1993). With omics and imaging data, the basic principles of Bayesian analysis are similar to those for classic low-dimensional data. However, the specification of prior distribution, which in some sense is the most critical step in Bayesian analysis, is fundamentally different. Priors that have certain sparsity property (that is, can select a small subset of relevant variables) are usually needed. In addition, with the high dimensionality of data, computation is often much more challenging. For applications of the Bayesian technique to omics and imaging oncology studies, see Lee et al. (2003), Sha et al. (2006), and Smith and Fahrmeir (2007).

3. Boosting, which is a popular machine learning technique and assembles multiple "weak learners" (individual genes and methylation loci that may have weak predictive power for the outcome) into a "strong linear" (a model with strong predictive power) (Freund and Schapire, 1995; Friedman, 2001). It has been extensively applied in other scientific fields such as signal recognition and image processing and, more recently, in biomedical omics and imaging studies (Tieu and Viola, 2000; Sun et al., 2007; He et al., 2015).

4. There are other possibly less popular variable selection techniques, such as thresholding, that have also been applied in omics and imaging studies. For statistical descriptions of such methods, see Beck and Teboulle (2009), Wehrens and Franceschi (2012), for example.

The development of variable selection techniques with applications to oncology and other biomedical domains has been one of the hottest topics in statistics in recent decades. Compared with dimension reduction and some other techniques, advantages of variable selection techniques include interpretable results (because only a small subset of genes are included in the model) and well-developed statistical theories (with the consistency properties rigorously established under high-dimensional settings). On the negative side, variable selection techniques are not as user-friendly—they have been developed more recently and demand high-level statistics. In addition, computation is challenging—usually complicated optimization is involved, which may demand special algorithms and new coding. Luckily, most recently, some software packages, such as the R packages *glmnet* and *ncvreg*, are becoming increasingly popular and have been used by investigators outside of the statistics community.

6.2.2.3 More advanced joint analysis methods

The most prominent limitation of the aforementioned approaches is that they ignore the interconnections among different types of variables. Consider, for example, studies like TCGA that collect multiple types of genetic, genomic, epigenetic, and proteomic measurements on the same subjects. As shown in Figure 6.2, DNA-level changes (e.g., mutations) and epigenetic changes (e.g., methylation) regulate gene expression levels and, therefore, protein levels. With such regulation relationships, intuitively, different types of omics measurements have overlapping information, and thus, their effects are not as simple as additive. It is noted that this kind of regulation relationship (or hierarchical structure) may also exist under the classic low-dimensional settings. However, the problem becomes more prominent for high-dimensional omics and imaging data, which have a more serious "lack of information" problem.

A *conditioning–integration strategy* has been recently developed (Zhao et al., 2015b). Continuing the consideration of Model 3, this approach proceeds as follows.

1. Consider the model: *outcome ~ clinical variables + gene expressions*. With a single type of high-dimensional omics measurements, dimension reduction, and variable selection techniques as described above can be applied for model fitting. Note that, after this step of estimation, the effects of clinical variables and gene expressions are "reduced" to a single term.

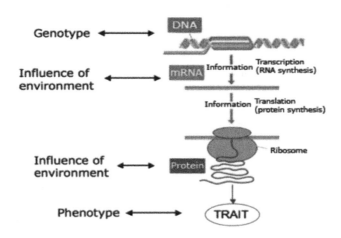

Figure 6.2 Regulations among different types of omics measurements.

2. Fit an outcome model with the single term obtained in Step 1 and methylation data. Note that in this model, there is just one type of high-dimensional measurements, and the variable selection and dimension reduction approaches adopted in Step 1 can be adopted again.

3. If there are other types of epigenetic or DNA-level change data, Step 2 can be repeated multiple times. For more details, see Zhao et al. (2015b).

The strategy is to start with the most downstream measurements, in this case, gene expressions as can be seen in Figure 6.2. Such measurements are "closest" to the outcome. Then *conditional on* that one type of measurements has been included in the model, the upstream measurements are integrated. In a series of recent studies (Zhao et al., 2015b; Jiang et al., 2016), it was shown that by accounting for the "order" of high-dimensional measurements, this seemingly simple approach can lead to models with superior predictive power for multiple cancer types. Another advantage of this approach is its computational "simplicity": It basically utilizes the same analysis technique multiple times and does not demand new computational algorithms or software. In Table 6.1, we present partial analysis results for TCGA data on multiple cancer types. The outcome variable of interest is overall survival, and a larger C-statistic suggests a better predictive model. Table 6.1 suggests that gene expression data have independent predictive power beyond clinical variables. Integrating other omics measurements may further improve model performance; however, this improvement

Table 6.1 **Analysis of TCGA data using the conditioning–integration approach**

METHOD	DATA TYPE	ESTIMATE OF C-STATISTIC (STANDARD ERROR)			
		BRCA	GBM	AML	LUSC
PCA	Clinical	0.54 (0.07)	0.65 (0.01)	0.65 (0.03)	0.56 (0.07)
	Clc + expression	0.73 (0.06)	0.64 (0.01)	0.68 (0.03)	0.74 (0.06)
	Clc + expr + methylation	0.72 (0.06)	0.64 (0.01)	0.76 (0.03)	0.70 (0.07)
	Clc + expr + miRNA	0.71 (0.06)	0.63 (0.01)	—	0.70 (0.07)
	Clc + expr + CNA	0.70 (0.07)	0.65 (0.01)	0.65 (0.03)	0.71 (0.07)
	Clc + expr + methyl + CNA	—	—	0.75 (0.03)	—
	Clc + expr + methyl + miRNA + CNA	0.73 (0.07)	0.63 (0.01)	—	0.68 (0.07)
PLS	Clinical	0.54 (0.07)	0.65 (0.01)	0.65 (0.03)	0.56 (0.07)
	Clc + expression	0.92 (0.04)	0.66 (0.01)	0.75 (0.03)	0.86 (0.04)
	Clc + expr + methylation	0.83 (0.05)	0.64 (0.01)	0.83 (0.02)	0.76 (0.05)
	Clc + expr + miRNA	0.82 (0.06)	0.65 (0.01)	—	0.69 (0.07)
	Clc + expr + CNA	0.91 (0.04)	0.66 (0.01)	0.74 (0.03)	0.82 (0.04)
	Clc + expr + methyl + CNA	—	—	0.83 (0.03)	—
	Clc + expr + methyl + miRNA + CNA	0.72 (0.06)	0.64 (0.01)	—	0.69 (0.07)
LASSO	Clinical	0.54 (0.07)	0.65 (0.01)	0.65 (0.03)	0.56 (0.07)
	Clc + expression	0.57 (0.08)	0.64 (0.01)	0.64 (0.03)	0.53 (0.08)
	Clc + expr + methylation	0.57 (0.08)	0.62 (0.02)	0.65 (0.04)	0.52 (0.08)
	Clc + expr + miRNA	0.65 (0.07)	0.61 (0.02)	—	0.55 (0.08)
	Clc + expr + CNA	0.52 (0.08)	0.65 (0.01)	0.63 (0.03)	057 (0.08)
	Clc + expr + methyl + CNA	—	—	0.65 (0.04)	—
	Clc + expr + methyl + miRNA + CNA	0.60 (0.08)	0.61 (0.02)	—	0.56 (0.08)

Big data in radiation oncology

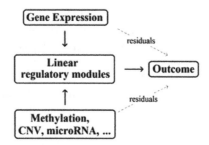

Figure 6.3 Scheme of the decomposition–integration approach in Zhu et al.(From Zhu, R. et al., *Biostatistics*, 17, 605–618, 2016.)

depends both on cancer types and on statistical techniques (highlighting the necessity of developing more effective and tailored methods). For more detailed results and discussions, see Zhao et al. (2015b).

A *decomposition–integration strategy,* which is more complicated but also "closer" to biology, has been developed in recent studies. In particular, one approach referred to as iBAG was developed based on the Bayesian technique (Wang et al., 2013), and an alternative was later developed based the penalization technique (Zhu et al., 2016). A schematic presentation of this approach is provided in Figure 6.3. Continuing the consideration of Model 3, this approach proceeds as follows:

1. Conduct the regression analysis of gene expressions on methylation. Gene expression levels are determined by methylation, other regulation mechanisms, as well as "random errors." This step is designed to decompose gene expressions into a component regulated by methylation (referred to as "linear regulatory modules" in Figure 6.3) and its complement ("residuals" in Figure 6.3). Note that, both the left and right hand sides of the regression are high dimensional, and hence advanced techniques are needed. For more details, see Zhu et al. (2016).

2. Based on the regression analysis in Step 1, decompose the effects of "*gene expressions + methylation*" into three components. The first is the component of gene expressions regulated by methylation; the second is the complement of the first component; and the third component represents the effects of methylation on outcome that are not reflected in gene expressions (also referred to as "residuals" in Figure 6.3).

3. Finally, an additive regression model is built using the three components obtained in Step 2.

By directly modeling the regulation relationship, this approach can be biologically more interpretable. In Zhu et al. (2016), the analysis of TCGA data on lung cancer and melanoma showed that this strategy has biologically sensible marker identification and constructs models with superior predictive power. On the negative side, this strategy is technically very challenging. In Zhu et al. (2016), the employed statistical techniques included high-dimensional regularized estimation, sparse singular value decomposition, linear space projection via matrix operations, and others, all of which demand advanced statistical training to comprehend and cannot be easily realized using existing software.

6.2.2.4 Functional-group–based analysis

In all of the aforementioned analyses, individual variables (e.g., genes, methylation loci, imaging features) are the basic units. A higher level of analysis is functional-group based (Huang et al., 2012a). Consider, for example, omics data, here a functional group can be a pathway, network module, and others, which consist of genes (e.g., methylation loci) that have related biological functions. With imaging data, a functional group consists of features that describe, for example, the same or related aspects of a tumor. By taking a functionality perspective and having a lower dimensionality (because the number of functional groups is usually much smaller than the number of variables), results generated from the functional-group–based analysis can be more interpretable and more reliable and reproducible (Ma et al., 2007; Liu et al., 2012). In this type of analysis, the functional groups are usually defined prior to the analysis, for example, biologically based on extensive data collected in previous studies. For omics data, the functional groups can be constructed using the Kyoto Encyclopedia of Genes and Genomes (KEGG), Gene Ontology (GO), and other online-curated databases. For imaging data, they can be constructed statistically, using, for example, clustering methods such as k-means. Multiple types of analysis with different goals have been conducted using the functional groups.

The first is to conduct marginal analysis and identify functional groups that are marginally associated with the outcome and can serve as markers. With omics data, the most popular functional-group–based analysis is perhaps the gene set enrichment analysis (GSEA) (Subramanian et al., 2005), which is needed to identify sets of genes (e.g., methylation loci) that are enriched with (i.e., they have unproportionally high percentages of) genes that are associated with the outcome. A large number of alternative methods are also available (Kim and Volsky, 2005; Backes et al., 2007). Unfortunately, the statistics involved is complicated and cannot be easily described. For comprehensive reviews and discussions, see Irizarry et al. (2009) and Hung et al. (2011). Luckily, for the purpose of routine data analysis, there are some well-developed software tools. For example, the GSEA software is freely available at software.broadinstitute.org/gsea/.

Analysis has also been conducted building cancer outcome models based on the functional groups. The expectation is that such models can be more interpretable than those based on individual genes (e.g., methylation loci). In general, such analysis has a two-level dimensionality problem. The first is that some functional groups may have large sizes, and in order to use them in model building, it is necessary to reduce the within-group high dimensionality. The second problem is that the number of functional groups can be comparable to (or larger than) the sample size. It is thus necessary to reduce dimensionality at the group level also. Following the same rationale, if it is desirable to discriminate important variables from random noises, it is necessary to conduct within-group and group-wise two-level selections. Examples of the existing approaches include that in Ma and Kosorok (2009), where the functional groups are KEGG pathways or network modules. The high dimensionality within-groups is accommodated using the PCA technique. At the group level, regularized selection is conducted using the thresholding and penalization techniques. A family of approaches that conduct two-level selection have been developed especially based on the penalization, thresholding, and Bayesian techniques (Breheny and Huang, 2009; Stingo et al., 2011; Simon et al., 2013; Jiang et al., 2017). Extensive methodological, theoretical, and computational developments have been conducted, and the analysis of a large number of data sets has demonstrated the superiority of functional-group–based analysis over individual-variable–based analysis.

6.2.3 THE PAST, PRESENT, AND FUTURE

Although it has been long acknowledged that the outcomes and phenotypes in cancer studies involve multiple types and levels of changes, the systematic collection and analysis of multidimensional data is very recent. As partly described above, quite a few advanced statistical methods have been developed. From a methodological perspective, many of these methods are highly successful in the sense that the desired statistical properties have been established under mild data and model conditions. Research efforts on computation and implementation are still ongoing. Some research software packages have been developed. However, they may not be user-friendly enough and cannot be easily used by practitioners with limited statistical training. Results from a series of recent studies show that the advanced methods can lead to biologically more meaningful markers and models with superior predictive power, which justify the merit of methodological development.

Despite great successes, compared with the real cancer biology, the existing models and methods are still "too naïve," and do not match in their complexity the real cancer biology. For example, for copy number variations, the proper way of modeling *trans*-acting effects is still being explored. More importantly, the current way of integrating multidimensional measurements is still too simplified, without sufficiently reflecting biology. A question of critical interest that has not been fully addressed so far is: How much additional value do multidimensional data bring (beyond clinical/pathological measurements or a single type of omics/imaging data)? There is no doubt that each new type of measurement brings additional information. However, noisy signals are also introduced. Statistically speaking, if the signal-to-noise ratio is not high enough, then more comprehensive data collection is in fact not beneficial. Practically speaking, there is also a cost-effectiveness problem. Although the cost of data collection has been decreasing, it can still pose a considerable burden. In a few recent studies (Weinstein et al., 2013), it has been suggested that for some cancers, the additional predictive power brought by comprehensive data collection may be too small to justify the cost of collecting it.

6.3 INTEGRATING MULTIPLE INDEPENDENT DATA SETS

6.3.1 COLLECTION OF MULTIPLE INDEPENDENT DATA SETS

For many scientific problems of common interest, multiple studies may have been independently conducted. Pooling and jointly analyzing multiple studies is by no means new. Meta-analysis and pooled analysis have been conducted for many years in radiation oncology as well as in other biomedical fields. There are statistical textbooks, tutorials, and software packages for conducting such analysis (Wallace et al., 2009; Borenstein et al., 2011; Schwarzer et al., 2015).

In the big data era, new developments include but are not limited to the following. First, with a strong push for reproducible research and to save resources for the community as a whole, there have been strong efforts in constructing well-curated databases and making the raw data publicly available. In contrast, in classic meta-analysis, only the summary statistics (e.g., effect sizes, p-values) are available. An example of sharing raw cancer data from clinical trials is Project Data Sphere. Second, in the small data era, different studies usually collected overlapping but different (and sometimes quite different) sets of variables, and the variables were low dimensional. With the "unification" and popularity of omics and imaging techniques, the sets of variables now collected in different studies are often highly similar or even identical and are high dimensional. Third, with the accumulation of data, the across–data set heterogeneity problem becomes more prominent. With a small number of carefully selected data sets, it can be reasonable to expect a high level of similarity across data sets. However, with a large number of data sets collected by different investigators in different regions/countries on different cohorts, the level of heterogeneity is expected to be much higher. The recent statistical developments have been largely motivated by the aforementioned new developments.

6.3.2 STATISTICAL METHODS

6.3.2.1 "Classic" meta-analysis

The first possibility is to adopt the existing meta-analysis techniques. In such an analysis, each data set is first separately analyzed. To achieve consistency, usually the same statistical technique or model is applied. If the data are low dimensional, standard regression techniques can be adopted. If the data are high dimensional, then some of the analysis techniques previously described can be applied. Then the summary statistics can be pooled across data sets. For effect sizes (e.g., regression parameters, hazard ratios, odds ratios), the final estimates are usually the averages (or weighted by sample sizes) of those from individual data sets, and the p-values can be obtained using, for example, the Fisher method. Here a "hidden" assumption is that different data sets share the same model, which is why parameter averaging is conducted.

Because meta-analysis has been well developed in the literature, here we will not discuss the details excessively. The most significant advantage of meta-analysis is its conceptual and computational simplicity. However, meta-analysis is challenged in big data analysis. With the increasing heterogeneity across data sets, the assumption of equal effect sizes can be too stringent. The heterogeneity should be better acknowledged and accounted for. There is also a special concern when the data sets have high-dimensional measurements, for example, omics and imaging. When the sample sizes of individual data sets are small, the analysis results can be unreliable. Averaging a set of unsatisfactory results is unlikely to lead to satisfactory results.

6.3.2.2 Integrative analysis methods

Integrative analysis has been developed in very recent literature and motivated by the limitations of meta-analysis (Huang et al., 2012b). In integrative analysis, the *raw data* of multiple data sets are jointly analyzed. That is, "information borrowing" starts at the very first step, as opposed to post-estimation as done by the traditional meta-analysis approach. In addition, one key model assumption, which differs significantly from that of meta-analysis, is that with the practically unknown across–data set heterogeneity, the regression parameters (and other effect sizes) are not necessarily equal across data sets. That is, different data sets can have their own regression models. Taking the relevance of variables into consideration, two conceptual models have been proposed. A schematic presentation is shown in Figure 6.4.

	Data 1	Data 2	Data 3
Variable 1	1.07	0.96	1.13
2	-0.18	-0.23	-0.07
3	0.33	0.42	0.23
4	0	0	0
5	0	0	0
6	0	0	0
...			
...			

	Data 1	Data 2	Data 3
Variable 1	1.07	0	1.13
2	-0.18	-0.23	0
3	0.33	0.42	0.23
4	0	0	0
5	0	0	0
6	0	0	0
...			
...			

Figure 6.4 Scheme of integrative analysis. Homogeneity (left) and heterogeneity (right) models.

1. The first model, as shown in the left panel, is the *homogeneity model*, under which different data sets have the same set of relevant variables. In this specific example, the first six regression coefficients of three data sets are shown (and "…" represents more variables). The first three variables are associated with the response variables and have nonzero regression coefficients in all three data sets, whereas Variables 4–6 have zero regression coefficients in all three data sets. This model is reasonable when multiple data sets are "sufficiently close," for example, as measured using metadata.

2. The second model, as shown in the right panel of Figure 6.4, is the *heterogeneity model*, under which different data sets can have possibly different sets of relevant variables. For example, Variable 1 has regression coefficients 1.07, 0, and 1.13 in the three data sets, respectively. That is, it is associated with the outcome variable for subjects in data sets 1 and 3, but not 2. The differences in quantitative results across data sets can be easy to comprehend. However, the differences in qualitative results are not commonly encountered in classic meta-analysis. The first scenario under which the heterogeneity model is sensible is that the differences in study subjects, their environmental and other exposures, and experimental settings are so great such that the homogeneity model may no long hold. In addition, in some recent publications (e.g., Shi et al., 2015), the interest is on the similarity and differences among different diseases (e.g., the shared and disease-specific markers for breast and ovarian cancers). In this case, the homogeneity model is insensible.

At first glance, it may not be clear why multiple data sets need to be jointly analyzed if they have different regression models or even different sets of relevant variables. In a series of recent studies (Ma et al., 2011a; Huang et al., 2012b; Liu et al., 2013, 2014a, 2014b; Shi et al., 2014; Huang et al., 2017a, 2017b), it was shown that as long as multiple data sets share some common relevant variables, with information borrowing in estimation, integrative analysis can outperform individual data set analysis and meta-analysis.

Statistical methods for conducting integrative analysis have been recently developed and are highly nontrivial, especially when the data dimensionality is high and/or the selection of relevant variables is needed. A relatively comprehensive review has been provided by Zhao et al. (2015a). Under the homogeneity model, in order to select a given variable, it only needs to be determined whether a variable is relevant. In contrast, under the heterogeneity model, the first step is to determine whether a variable is relevant at all, and then, for a relevant variable, the second step is to decide in which data set(s) it is relevant. Motivated by such considerations, one-level and two-level regularized selection and estimation methods have been developed using the thresholding, penalization, and boosting techniques (Huang et al., 2012b; Liu et al., 2013, 2014a; Shi et al., 2014; Huang et al., 2017b). Moreover, the statistical properties have been rigorously established. The computational aspect of integrative analysis is highly challenging, and almost every single approach demands a separate development of optimization algorithm. Data on the etiology (risk) and prognosis of lung cancer, breast cancer, liver cancer, and others have been analyzed (Ma et al., 2011a; Huang et al., 2012b; Liu et al., 2014b; Huang et al., 2017a). It is shown that the markers identified using integrative analysis methods are biologically sensible and have a better reproducibility. In addition, integrative analysis has led to outcome models with superior predictive power.

6.3.2.3 Accommodating finer data structures

In classic meta-analysis, the estimation of one data set is independent of that for another data set. Moreover, for a specific data set, usually additive covariate effects are assumed. In the early integrative analyses, because multiple data sets are independent, the goodness-of-fit measures (usually log-likelihood functions) are usually directly added. In addition, for a specific data set, the covariate effect formats are the same as in meta-analysis. Such a strategy, despite considerable successes, still has room for improvement. The first consideration is that although multiple data sets are independent and the regression models are allowed to differ, there should still be some similarity across data sets, which is the basis of conducting joint analysis. If it is reasonable to expect some similarity in the resulted regression models, then the statistical methods should *encourage* it. This may seem to have a Bayesian flavor and is a further development of the "information borrowing" strategy for parameter estimation. The second consideration is that variables in the same data set are more or less interconnected. When the number of variables is high (i.e., big data), there almost always exist highly correlated variables. From a biological perspective, as suggested by the functional-group–based analysis, some variables can be functionally interconnected. It has been suggested that statistically or biologically interconnected variables are likely to have similar regression coefficients in models. The strategy is thus if two variables are interconnected, then the statistical methods should encourage similar regression coefficients. Note that similar considerations may also hold for small data. With the significantly higher volume and complexity of big data, the need for accommodating finer data structures becomes more prominent.

Motivated by the aforementioned considerations, a *contrasted analysis* strategy has been recently developed (Shi et al., 2014) and consists of the following two main components. (1) For a specific variable's estimates in multiple data sets, the contrasted integrative analysis approach uses the penalization technique to shrink the differences in estimates across data sets so as to encourage similarity. With the degree of shrinkage data-dependently adjusted, it allows for different degrees of similarity and is able to accommodate the worst scenario with a lack of similarity. (2) For a specific data set, the interconnections among variables are described using a network structure, where the network is statistically constructed prior to regression analysis. The penalization technique is adopted again (although with different penalties) to shrink the estimates for parameters in the same data sets, where the degree of shrinkage is adjusted using the network "closeness" measure (connectivity). In Table 6.2, we reproduce partial simulation results from one of our recent studies (Huang et al., 2017b) and refer the reader to the original publication for more details. This simulation considers a simplified scenario where the three data sets share the same regression model (the same true regression coefficients). It is evident that the contrasted analysis (contrasted gBridge) has more accurate estimation than the benchmark integrative analysis (gBridge).

6.3.3 THE PAST, PRESENT, AND FUTURE

Pooling and jointly analyzing multiple independent data sets has been commonly accepted as an effective strategy in radiation oncology as well as in other fields. With the unique characteristics of big data, the old strategy has faced new challenges and led to new developments in statistical methodology. Despite promising successes, there are still multiple problems that need to be resolved. First, "bigger" is not necessarily better. With the increase in the number of data sets (sample size), heterogeneity inevitably

Table 6.2 **A simulation study of the contrasted penalized integrative analysis approach**

TRUE COEF.	GBRIDGE			CONTRASTED GBRIDGE		
	D1	D2	D3	D1	D2	D3
0.4	0.186	0.391	0.112	0.302	0.292	0.317
0.5	0.349	0.400	0.465	0.411	0.428	0.537
0.6	0.587	0.244	0.392	0.553	0.461	0.587
0.7	0.592	0.746	0.553	0.637	0.659	0.695
0.8	0.683	0.769	0.698	0.617	0.661	0.732
-0.4	-0.302	-0.312	-0.187	-0.309	-0.253	-0.287
-0.5	-0.627	-0.519	-0.482	-0.599	-0.575	-0.502
-0.6	-0.558	-0.742	-0.514	-0.583	-0.568	-0.599
-0.7	-0.571	-0.576	-0.556	-0.557	-0.612	-0.600
-0.8	-0.635	-0.622	-0.495	-0.704	-0.624	-0.730

increases. This leads to the classic bias–variance trade-off (possibly smaller bias caused by increased sample size, but larger variance caused by increased heterogeneity). Unfortunately, in the literature, there are no studies that would comprehensively evaluate the potential detrimental effects of pooling more data sets. Second, many of the meta-analysis and integrative analysis methods for big data are straightforward extensions of their counterparts for small data. More tailored developments are still needed. Third, for many of the recently developed methods, only research-grade statistical packages or libraries exist and these are not well tested and not yet fully user-friendly. To make a broader practical impact, more developments in efficient computational algorithms and usability improvements are needed.

6.4 COMPUTATION: A UNIQUE CHALLENGE FACED BY BIG DATA ANALYSIS

For any statistical analysis, computation cost can be a concern. The challenge becomes more serious with big data, which have a much higher volume and demand much more complicated analysis methods. For example, for large electronic medical record (EMR) data sets, such as those from Medicare (with millions of subjects and many more records), even storage, input, and output can pose a problem. Statistical analysis cannot be accomplished using popular statistical software such as SAS and R. In principle, there are two approaches that solve the large computation problem. Either more computing resources are used with existing, unmodified methods, or the methods are improved with additional computational shortcuts (or alteration or heuristics) that reduce the computing costs. In the recent years, there have been significant developments in computer hardware and algorithms, such as the use of graphics processing units (GPUs) and Hadoop-based techniques employing the MapReduce approach, that make large-scale computation more feasible. In addition, many statistical techniques have been developed, which can at least partly tackle the computational challenge. Some representative examples are as follows.

- *Screening*: In principle, marginal analysis (which analyzes one or a small number of variables at a time) is computationally more feasible than joint analysis (which simultaneously considers a large number of variables). The strategy of screening is to conduct simple marginal analysis and remove variables that are highly unlikely to be relevant prior to complicated analysis (Fan, 2007; Ray et al., 2016). There are two types of screening. The first is *unsupervised*, which does not utilize any information about the outcome variable. For example, variables with a high rate of missingness are removed from analysis; genes with a low level of expression or variation are removed from analysis. This type of analysis is more "mechanical" and mainly concerned with the quality of data. The second type is *supervised* and it uses the outcome variable information to select variables important in a marginal sense for further analysis. The supervised approach has similar operations as the marginal screening-based approach described above. The difference is that as the goal of screening is to remove nonrelevant variables (as opposed to accurately selecting relevant variables), a relatively large number of variables will be kept for downstream analysis to avoid false negatives.

- *Divide and conquer*: This is possibly the most popular computational technique for big data analysis (Jordan, 2012). The basic strategy is to first cut a big data set into small pieces, with each piece having a much smaller number of subjects and/or a much smaller number of variables. Then each piece is analyzed separately, which can be achieved in a highly parallel manner to reduce the overall computation time. Then the results from all pieces are assembled together. This strategy has roots in data mining methods such as bagging, assembling, random forests, and others. Despite the intuitive formulation, the underlying statistics are actually quite challenging: Each piece is not necessarily representative of the whole data set (which is especially true when each piece contains only a subset of the variables), and so the analysis results from a single piece may deviate significantly from those of the whole data set. Special statistical techniques are needed for the cutting and assembling steps. A recently developed strategy is to use clustering to assist cutting and weighted averaging and stability selection for assembling (Saigal et al., 2017).

- *Progressive analysis*: In a sense, this approach (Stolper et al., 2014) shares some common ground with the classic stepwise estimation approach as well as the conditioning–integration approach described above. As opposed to simultaneously considering a large number of variables, this approach first considers a small number of most important variables. Assuming that these variables are in the model, it then searches for other important variables. The searching and updating procedure is iterated multiple times until the model stabilizes. The key difference from the classic stepwise approach is that the estimate update is conducted in a stage-wise manner, similar to that in the boosting approach. Because the working model has a size much smaller than the number of variables, the computational cost can be significantly reduced.

- *Remarks*: Compared with methodology and theory, less attention has been paid to the implementation of big data statistical methods. For some practical applications—for example, the utilization of portable medical devices that transfer data on a real-time basis and demand instant analysis—the efficiency of computation can be of special concern. It is expected that with more applications of the aforementioned and other statistical techniques, we will see more developments in computation in the coming years. One issue that also has not received enough attention is the stability of computational algorithms and techniques. Practical applications demand stable computing. With the complexity of big data, the stability of computation is expected to decrease; however, there is still a lack of studies that carefully evaluate the stability issue.

6.5 CONCLUDING REMARKS

In this chapter, we have provided a very brief introduction of some big data statistical methods. Some of these methods have been applied to radiation oncology studies, whereas other have found applications in other biomedical studies but not yet in radiation oncology. With the special audience in mind, we have described the methods in a nontechnical way and refer the reader to the original publications for more details. The field of statistics for big data is moving fast, and this survey may need to be updated in the near future. In quite a few publications, it has been shown that novel statistical methods have led to important findings that were missed by the traditional analyses, and we expect to see more developments and applications of these methods in the near future.

ACKNOWLEDGMENTS

The authors' research has been partly supported by awards 2016LD01 from the National Bureau of Statistics of China, R01CA204120, R21CA191383 and the Intramural Research Program of the National Institutes of Health (NIH).

REFERENCES

Aryee MJ, Jaffe AE, Corrada-Bravo H, Ladd-Acosta C, Feinberg AP, Hansen KD, Irizarry RA (2014) Minfi: A flexible and comprehensive bioconductor package for the analysis of Infinium DNA methylation microarrays. *Bioinformatics* 30:1363–1369.
Backes C, Keller A, Kuentzer J, Kneissl B, Comtesse N, Elnakady YA, Müller R, Meese E, Lenhof H-P (2007) GeneTrail—Advanced gene set enrichment analysis. *Nucleic Acids Research* 35:W186–W192.
Barnett GC, West CM, Dunning AM, Elliott RM, Coles CE, Pharoah PD, Burnet NG (2009) Normal tissue reactions to radiotherapy: Towards tailoring treatment dose by genotype. *Nature Reviews Cancer* 9:134–142.
Beck A, Teboulle M (2009) A fast iterative shrinkage-thresholding algorithm for linear inverse problems. *SIAM Journal on Imaging Sciences* 2:183–202.
Bender E (2015) Big data in biomedicine: 4 big questions. *Nature* 527:S19.
Benedict SH, El Naqa I, Klein EE (2016) Introduction to big data in radiation oncology: Exploring opportunities for research, quality assessment, and clinical care. *International Journal of Radiation Oncology Biology Physics* 95:871–872.
Bibault JE, Giraud P, Burgun A (2016) Big data and machine learning in radiation oncology: State of the art and future prospects. *Cancer Letters* 382:110–117.
Borenstein M, Hedges LV, Higgins JPT, Rothstein HR (2011) *Introduction to Meta-Analysis*. Chichester, UK: Wiley.
Brahme A (2014) *Biologically Optimized Radiation Therapy*. London, UK: World Scientific Publishing Company.
Breheny P, Huang J (2009) Penalized methods for bi-level variable selection. *Statistics and its Interface* 2:369.
Brock KK (2016) *Image Processing in Radiation Therapy*. Boca Raton, FL: CRC Press.
Burman C, Kutcher G, Emami B, Goitein M (1991) Fitting of normal tissue tolerance data to an analytic function. *International Journal of Radiation Oncology* Biology* Physics* 21:123–135.

Cancer Genome Atlas Research N (2008) Comprehensive genomic characterization defines human glioblastoma genes and core pathways. *Nature* 455:1061–1068.

Cancer Genome Atlas Research N (2014) Comprehensive molecular characterization of gastric adenocarcinoma. *Nature* 513:202–209.

Dawson LA, Biersack M, Lockwood G, Eisbruch A, Lawrence TS, Ten Haken RK (2005) Use of principal component analysis to evaluate the partial organ tolerance of normal tissues to radiation. *International Journal of Radiation Oncology* Biology* Physics* 62:829–837.

Emami B, Lyman J, Brown A, Cola L, Goitein M, Munzenrider J, Shank B, Solin L, Wesson M (1991) Tolerance of normal tissue to therapeutic irradiation. *International Journal of Radiation Oncology* Biology* Physics* 21:109–122.

Fan J (2007) Variable screening in high-dimensional feature space. In: *Proceedings of the 4th International Congress of Chinese Mathematicians*. Somerville, MA: International Press, pp. 735–747.

Fan J, Li R (2001) Variable selection via nonconcave penalized likelihood and its oracle properties. *Journal of the American Statistical Association* 96:1348–1360.

Fan J, Lv J (2008) Sure independence screening for ultrahigh dimensional feature space. *Journal of the Royal Statistical Society: Series B (Statistical Methodology)* 70:849–911.

Frangakis CE, Rubin DB (2002) Principal stratification in causal inference. *Biometrics* 58:21–29.

Freund Y, Schapire RE (1995) A decision-theoretic generalization of on-line learning and an application to boosting. In: *European Conference on Computational Learning Theory*, pp. 23–37. Heidelberg, Germany: Springer.

Friedman JH (2001) Greedy function approximation: A gradient boosting machine. *Annals of Statistics* 29(5):1189–1232.

Gaffney DK, Anker CJ, Shrieve DC (2012) *Radiation Oncology: Imaging and Treatment*. Salt Lake City, UT: Amirsys.

George EI, McCulloch RE (1993) Variable selection via Gibbs sampling. *Journal of the American Statistical Association* 88:881–889.

Gillies RJ, Kinahan PE, Hricak H (2015) Radiomics: Images are more than pictures, they are data. *Radiology* 278:563–577.

Gunderson LL, Tepper JE, Bogart JA (2015) *Clinical Radiation Oncology*. Philadelphia, PA: Elsevier.

He S, Chen H, Zhu Z, Ward DG, Cooper HJ, Viant MR, Heath JK, Yao X (2015) Robust twin boosting for feature selection from high-dimensional omics data with label noise. *Information Sciences* 291:1–18.

Huang J, Breheny P, Ma S (2012a) A selective review of group selection in high-dimensional models. *Statistical Science: A Review Journal of the Institute of Mathematical Statistics* 27.

Huang J, Ma S, XIE H, Zhang C-H (2009) A group bridge approach for variable selection. *Biometrika*. 96(2):339–355.

Huang J, Ma S, Zhang C-H (2008) Adaptive Lasso for sparse high-dimensional regression models. *Statistica Sinica* 1603–1618.

Huang Y, Huang J, Shia BC, Ma S (2012b) Identification of cancer genomic markers via integrative sparse boosting. *Biostatistics* 13:509–522.

Huang Y, Liu J, Yi H, Shia BC, Ma S (2017a) Promoting similarity of model sparsity structures in integrative analysis of cancer genetic data. *Statistics in Medicine* 36:509–559.

Huang Y, Zhang Q, Zhang S, Huang J, Ma S (2017b) Promoting similarity of sparsity structures in integrative analysis with penalization. *Journal of the American Statistical Association* 112:342–350.

Huilgol N (2016) Big data in radiation oncology. *Journal of Cancer Research and Therapeutics* 12:1107–1108.

Hung JH, Yang TH, Hu Z, Weng Z, DeLisi C (2011) Gene set enrichment analysis: Performance evaluation and usage guidelines. *Briefings in Bioinformatics* 13(3):281–291.

Irizarry RA, Wang C, Zhou Y, Speed TP (2009) Gene set enrichment analysis made simple. *Statistical Methods in Medical Research* 18:565–575.

Jiang Y, Huang Y, Du Y, Zhao Y, Ren J, Ma S, Wu C (2017) Identification of prognostic genes and pathways in lung adenocarcinoma using a Bayesian approach. *Cancer Informatics* 1:7.

Jiang Y, Shi X, Zhao Q, Krauthammer M, Rothberg BE, Ma S (2016) Integrated analysis of multidimensional omics data on cutaneous melanoma prognosis. *Genomics* 107:223–230.

Jordan MI (2012) Divide-and-conquer and statistical inference for big data. In: *Proceedings of the 18th ACM SIGKDD International Conference on Knowledge Discovery and Data Mining*, pp. 4–4. New York: ACM.

Kessel KA, Combs SE (2016) Review of developments in electronic, clinical data collection, and documentation systems over the last decade—Are we ready for big data in routine health care? *Frontiers in Oncology* 6:75.

Kim SY, Volsky DJ (2005) PAGE: Parametric analysis of gene set enrichment. *BMC Bioinformatics* 6:144.

Kohl SM, Klein MS, Hochrein J, Oefner PJ, Spang R, Gronwald W (2012) State-of-the art data normalization methods improve NMR-based metabolomic analysis. *Metabolomics* 8:146–160.

Lee KE, Sha N, Dougherty ER, Vannucci M, Mallick BK (2003) Gene selection: A Bayesian variable selection approach. *Bioinformatics* 19:90–97.

Lei T (2011) *Statistics of Medical Imaging*. Boca Raton, FL: CRC Press.

Liu J, Huang J, Ma S (2013) Integrative analysis of multiple cancer genomic datasets under the heterogeneity model. *Statistics in Medicine* 32:3509–3521.

Liu J, Huang J, Ma S (2014a) Integrative analysis of cancer diagnosis studies with composite penalization. *Scandinavian Journal of Statistics, Theory and Applications* 41:87–103.

Liu J, Huang J, Ma S, Wang K (2012) Incorporating group correlations in genome-wide association studies using smoothed group Lasso. *Biostatistics* 14(2):205–219.

Liu J, Huang J, Zhang Y, Lan Q, Rothman N, Zheng T, Ma S (2014b) Integrative analysis of prognosis data on multiple cancer subtypes. *Biometrics* 70:480–488.

Llacer J, Solberg TD, Promberger C (2001) Comparative behaviour of the dynamically penalized likelihood algorithm in inverse radiation therapy planning. *Physics in Medicine and Biology* 46:2637–2663.

Lustberg T, van Soest J, Jochems A, Deist T, van Wijk Y, Walsh S, Lambin P, Dekker A (2017) Big Data in radiation therapy: Challenges and opportunities. *British Journal of Radiology* 90:20160689.

Lyman JT, Wolbarst AB (1987) Optimization of radiation therapy, III: A method of assessing complication probabilities from dose-volume histograms. *International Journal of Radiation Oncology* Biology* Physics* 13:103–109.

Ma S, Dai Y (2011) Principal component analysis based methods in bioinformatics studies. *Brief Bioinform* 12:714–722.

Ma S, Huang J (2008) Penalized feature selection and classification in bioinformatics. *Briefings in Bioinformatics* 9:392–403.

Ma S, Huang J, Song X (2011a) Integrative analysis and variable selection with multiple high-dimensional data sets. *Biostatistics* 12:763–775.

Ma S, Kosorok MR (2009) Identification of differential gene pathways with principal component analysis. *Bioinformatics* 25:882–889.

Ma S, Kosorok MR, Huang J, Dai Y (2011b) Incorporating higher-order representative features improves prediction in network-based cancer prognosis analysis. *BMC Medical Genomics* 4:5.

Ma S, Song X, Huang J (2007) Supervised group Lasso with applications to microarray data analysis. *BMC Bioinformatics* 8:60.

Marabita F, Almgren M, Lindholm ME, Ruhrmann S, Fagerstrom-Billai F, Jagodic M, Sundberg CJ et al. (2013) An evaluation of analysis pipelines for DNA methylation profiling using the Illumina HumanMethylation450 BeadChip platform. *Epigenetics* 8:333–346.

Mazumdar M, Glassman JR (2000) Categorizing a prognostic variable: Review of methods, code for easy implementation and applications to decision-making about cancer treatments. *Statistics in Medicine* 19:113–132.

Medicine I, Policy BHS, Services BHC, Trials CROBTPPOC, Omenn GS, Nass SJ, Micheel CM (2012) *Evolution of Translational Omics: Lessons Learned and the Path Forward*. Washington, DC: National Academies Press.

Mwangi B, Tian TS, Soares JC (2014) A review of feature reduction techniques in neuroimaging. *Neuroinformatics* 12:229–244.

Naqa IE, Li R, Murphy MJ (2015) *Machine Learning in Radiation Oncology: Theory and Applications*. New York: Springer International Publishing.

Nieder C, Gaspar LE (2014) *Decision Tools for Radiation Oncology: Prognosis, Treatment Response and Toxicity*. Berlin, Germany: Springer.

Ray MA, Tong X, Lockett GA, Zhang H, Karmaus WJ (2016) An efficient approach to screening epigenome-wide data. *BioMed Research International* 2016:2615348.

Risso D, Ngai J, Speed TP, Dudoit S (2014) Normalization of RNA-seq data using factor analysis of control genes or samples. *Nature Biotechnology* 32:896–902.

Saigal P, Khanna V, Rastogi R (2017) Divide and conquer approach for semi-supervised multi-category classification through localized kernel spectral clustering. *Neurocomputing* 238:296–306.

Schwarzer G, Carpenter JR, Rücker G (2015) *Meta-Analysis with R*. Cham, Switzerland: Springer International Publishing.

Sha N, Tadesse MG, Vannucci M (2006) Bayesian variable selection for the analysis of microarray data with censored outcomes. *Bioinformatics* 22:2262–2268.

Shi X, Liu J, Huang J, Zhou Y, Shia B, Ma S (2014) Integrative analysis of high-throughput cancer studies with contrasted penalization. *Genetic Epidemiology* 38:144–151.

Shi X, Yi H, Ma S (2015) Measures for the degree of overlap of gene signatures and applications to TCGA. *Briefings in Bioinformatics* 16:735–744.

Simon N, Friedman J, Hastie T, Tibshirani R (2013) A sparse-group lasso. *Journal of Computational and Graphical Statistics* 22:231–245.

Skwarchuk MW, Jackson A, Zelefsky MJ, Venkatraman ES, Cowen DM, Levegrün S, Burman CM, Fuks Z, Leibel SA, Ling CC (2000) Late rectal toxicity after conformal radiotherapy of prostate cancer (I): Multivariate analysis and dose–response. *International Journal of Radiation Oncology* Biology* Physics* 47:103–113.

Smith M, Fahrmeir L (2007) Spatial Bayesian variable selection with application to functional magnetic resonance imaging. *Journal of the American Statistical Association* 102:417–431.

Starkschall G, Siochi RAC (2013) *Informatics in Radiation Oncology*. Boca Raton, FL: Taylor & Francis Group.

Stefani SS, Hubbard LB, Sanders ES (1979) Mathematics for technologists in radiology, nuclear medicine, and radiation therapy. St. Louis, MO: C. V. Mosby Company.

Stingo FC, Chen YA, Tadesse MG, Vannucci M (2011) Incorporating biological information into linear models: A Bayesian approach to the selection of pathways and genes. *The Annals of Applied Statistics* 5.

Stolper CD, Perer A, Gotz D (2014) Progressive visual analytics: User-driven visual exploration of in-progress analytics. *IEEE Transactions on Visualization and Computer Graphics* 20:1653–1662.

Subramanian A, Tamayo P, Mootha VK, Mukherjee S, Ebert BL, Gillette MA, Paulovich A, Pomeroy SL, Golub TR, Lander ES (2005) Gene set enrichment analysis: A knowledge-based approach for interpreting genome-wide expression profiles. *Proceedings of the National Academy of Sciences* 102:15545–15550.

Sun Y, Liu Z, Todorovic S, Li J (2007) Adaptive boosting for SAR automatic target recognition. *IEEE Transactions on Aerospace and Electronic Systems* 43.

Taylor JM, Kim DK (1993) Statistical models for analysing time-to-occurrence data in radiobiology and radiation oncology. *International Journal of Radiation Biology* 64:627–640.

Tibshirani R (1996) Regression shrinkage and selection via the lasso. *Journal of the Royal Statistical Society Series B (Methodological)*:267–288.

Tieu K, Viola P (2000) Boosting image retrieval. In: *Computer Vision and Pattern Recognition, 2000. Proceedings. IEEE Conference on*, pp. 228–235: IEEE.

Trifiletti DM, Showalter TN (2015) Big data and comparative effectiveness research in radiation oncology: Synergy and accelerated discovery. *Frontiers in Oncology* 5:274.

Van Dye J, Batista J, Bauman GS (2013) Accuracy and uncertainty considerations in modern radiation oncology. *The Modern Technology of Radiation Oncology* 3:361–412.

Wallace BC, Schmid CH, Lau J, Trikalinos TA (2009) Meta-Analyst: software for meta-analysis of binary, continuous and diagnostic data. *BMC Medical Research Methodology* 9:80.

Wang W, Baladandayuthapani V, Morris JS, Broom BM, Manyam G, Do KA (2013) iBAG: Integrative Bayesian analysis of high-dimensional multiplatform genomics data. *Bioinformatics* 29:149–159.

Wehrens R, Franceschi P (2012) Thresholding for biomarker selection in multivariate data using Higher Criticism. *Molecular BioSystems* 8:2339–2346.

Weinstein JN, Collisson EA, Mills GB, Shaw KRM, Ozenberger BA, Ellrott K, Shmulevich I, Sander C, Stuart JM, Network CGAR (2013) The Cancer Genome Atlas Pan-Cancer analysis project. *Nature Genetics* 45:1113–1120.

Zhao Q, Shi X, Huang J, Liu J, Li Y, Ma S (2015a) Integrative analysis of "-omics" data using penalty functions. *Wiley Interdisciplinary Reviews: Computational Statistics* 7:99–108.

Zhao Q, Shi X, Xie Y, Huang J, Shia B, Ma S (2015b) Combining multidimensional genomic measurements for predicting cancer prognosis: Observations from TCGA. *Briefings in Bioinformatics* 16:291–303.

Zhu R, Zhao Q, Zhao H, Ma S (2016) Integrating multidimensional omics data for cancer outcome. *Biostatistics* 17:605–618.

Zou H, Hastie T (2005) Regularization and variable selection via the elastic net. *Journal of the Royal Statistical Society: Series B (Statistical Methodology)* 67:301–320.

7 From model-driven to knowledge- and data-based treatment planning

Morteza Mardani, Yong Yang, Yinyi Ye, Stephen Boyd, and Lei Xing

Contents

7.1 Treatment Planning   97
7.2 Clinical Domain Knowledge for Inverse Plan Optimization   97
7.3 Class Solution to Warm Start Inverse Planning   99
7.4 Improving the Convergence and Computational Efficiency of Plan Optimization   99
7.5 Automated Determination of Model Parameters   100
7.6 Knowledge-Based Planning Facilitated by Learning   103
    7.6.1 DVH Prediction   103
       7.6.1.1 Handcrafted Features   103
       7.6.1.2 Auto-extracted Features via Deep Learning   104
    7.6.2 Voxel-Wise Three-Dimensional Dose Prediction   105
    7.6.3 Learning in the Iso-dose Feature-Preserving Domain   106
7.7 Summary   107
References   107

7.1 TREATMENT PLANNING

Treatment planning is the gateway connecting various pretreatment and therapeutic information and is critical in determining the success of radiation therapy. Treatment planning aims to combine various relevant clinical domain knowledge and *patient-specific* information to provide the best possible dose distribution that delivers a maximum tumoricidal dose to the tumor target while minimizing the normal tissue toxicity. Inverse planning is a widely used approach to derive a patient-specific treatment plan by iteratively optimizing an *intricate* objective function, whose role is to mathematically rank candidate solutions. This approach has successfully led to clinical implementation of intensity modulated radiation therapy (IMRT) and Volumetric-modulated arc therapy (VMAT)[1–6] and the development of several other promising modalities such as station parameter optimized radiation therapy (SPORT). However, the planning process, routinely used in the clinical practice, is rather tedious and labor intensive, yet has no guarantees of generating truly *optimal* treatment plans.[7] This mainly emanates from the involvement of several model tuning parameters (e.g., the weighting importance factors and target prescription in the objective function) in treatment planning.[7–10] In general, these parameters are manually tuned on a trial-and-error and population-average basis because their influence on the final dose distribution is not known until performing the optimization. As a result, treatment planning remains one of the most labor-intensive and time-consuming tasks in current radiation therapy practice.

7.2 CLINICAL DOMAIN KNOWLEDGE FOR INVERSE PLAN OPTIMIZATION

Inverse planning typically starts with a model for dose optimization, where the construction of model or metric function relies heavily on our clinical knowledge of radiation therapy. For reliable guidance of the solution search process, the model should be as realistic as possible to reflect the clinical decision-making process. In practice, however, it is very challenging to objectively evaluate the figures of merit for different treatment plans. Despite intense research efforts in modeling the clinical decision-making strategies,[1–6] the optimal form of the objective function for clinical planning remains elusive. Two types of objective functions are being widely used: dose- and dose volume histogram (DVH)–based (physical objective functions)[7–16] and biological objective functions.[17–23] The major difference between these models lies in the endpoints that are used to evaluate the treatment and/or the fundamental quantities used to define the optimality. Dose–volume constraints are often introduced into the physical approaches to select a solution that meets a set of requirements for the target and sensitive structures. In practice, the construction of DVH constraints is *a priori* in nature.

Moreover, the objective in biological model-based inverse planning is usually stated as the maximization of the tumor control probability (TCP) while maintaining the normal tissue complication probability (NTCP) within a certain acceptable range.[5,17–21,23] Albeit the biologically based models are most relevant for radiotherapy plan ranking,[24–29] the dose–response function of various structures is not sufficiently understood, and there is still considerable controversy about the models for computing dose–response indices. The gap between the models and clinical decision-making entails tuning a large number of model parameters (such as the weighting factors), which often leads to undesirable inaccuracy and inefficiency in clinical treatment planning. Effective incorporation of additional prior domain knowledge represents a viable strategy to improve the situation and may greatly facilitate the computational and/or decision-making processes of clinical treatment planning. Figure 7.1 schematically shows the roles that prior clinical knowledge can play in the treatment planning process. These roles are discussed subsequently where the emphasis is placed on the recent development in (deep) machine learning–based treatment planning.

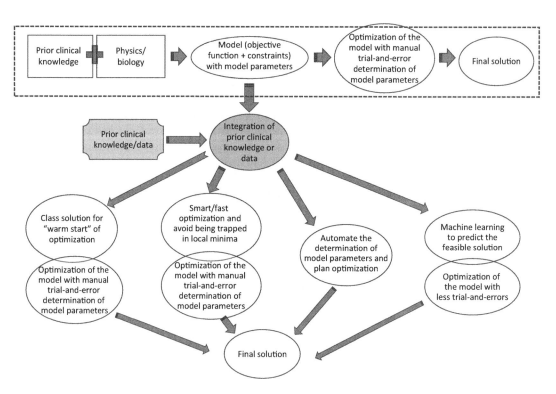

Figure 7.1 Roles of prior clinical knowledge and data in treatment planning. The items in the dashed box represent the conventional inverse planning procedure.

7.3 CLASS SOLUTION TO WARM START INVERSE PLANNING

Plan optimization typically proceeds in an iterative fashion with an *arbitrary* starting point. A carefully chosen starting point with the use of prior clinical knowledge would significantly reduce the number of iterations needed to find a convergent solution for the optimization. The warm startup idea has thus been integrated in many inverse planning systems. Along this line, class solution–based approaches have been investigated to facilitate the IMRT beam configuration selection process since the early days of inverse planning research.[11,12] The crux of this approach is to construct a representative beam configuration based on previous experience for each disease site, and then use this "class solution" for subsequent patient treatment planning of the same disease. The work by Reinstein et al. and Schreibmann and Xing[11,12] systematically investigated the issue and proposed a set of initial beam orientations for IMRT prostate irradiation. The effectiveness of class solutions by Reinstein et al. and Schreibmann and Xing[11,12] has also been well recognized by treatment planning system (TPS) vendors. Most, if not all, systems have the functionality to define and use class solutions. The challenge of identifying optimal configuration is, however, left to the user.

7.4 IMPROVING THE CONVERGENCE AND COMPUTATIONAL EFFICIENCY OF PLAN OPTIMIZATION

Prior knowledge is also valuable in guiding the optimizer concerning the optimal search path to avoid being trapped into local minima. Indeed, this can significantly accelerate the computations when adopting *nonconvex* objective functions that involve multiple groups of optimization variables. Imagine, for instance, the scenario with the beam orientations or apertures as the design variables. In this case, the objective function depends upon the beam orientations in addition to beamlet weights. Conceptually, this does not pose any challenge and the inverse planning framework can still be generalized to deal with the enlarged system. The main obstacle for a full optimization of the system is, however, the excessive computing time caused by the greatly enlarged search space.

Apparently, the optimal beam profiles and beam configurations are interdependent.[13,14] As a result, the beam intensity profiles need to be optimized for every trial configuration because the influence of gantry angles on the dose distribution cannot be known until the profile optimization is executed. Furthermore, a stochastic search algorithm is needed to optimize the beam angles due to the complicated dependence of the objective function on these variables. Thus, the angular variables are sampled by simulated annealing, and the beamlet weights are optimized using an iterative algorithm.[13] Although useful in proving feasibility, this approach incurs prohibitive computational complexity. This increases the need for more efficient optimization techniques to make it clinically acceptable.

Prior knowledge, such as the data derived from single beam scoring (e.g., beam-eye's view, BEV, or beam-eye's view dosimetrics, BEVD[15]), can be used to facilitate the computer search for the optimal beam configuration. During calculations, a trial beam configuration will be introduced by assigning a random angular variation to a randomly selected beam. The strategy is to assign a higher probability, $P(\theta)$, to accept a trial beam if it has a high BEVD score, and *vice versa*. That is, the value of $P(\theta)$ determines the probability for a trial to be rejected right away or to be further verified by the simulated annealing acceptance criterion. (For this verification, the beam profiles of the trial beam configuration must be optimized.)

Because of $P(\theta)$ prescreening, those trials that are less likely to be the final solution are discarded. We found that adopting this strategy in beam orientation optimization not only speeds up the computations but also improves the convergence behavior of the conventional simulated annealing calculation.[15] Figure 7.2 depicts the evolution of objective function as the iterations go by for the algorithms with and without prior knowledge guidance. It is evident that the BEVD-guided simulated annealing significantly outperforms the standard simulated annealing technique in terms of better fulfilling the planning objectives. In addition, all five independent BEVD-guided calculations converge within 300 iterations, whereas the standard simulated annealing takes around 3,000–5,000 iterations to approach the ground truth.

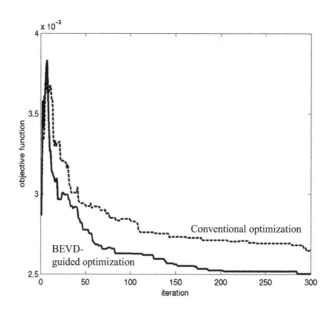

Figure 7.2 The objective function versus the iteration index during the optimization process. Dashed and solid lines correspond to the simulated annealing performed with standard sampling and the BEVD-guided sampling, respectively.

7.5 AUTOMATED DETERMINATION OF MODEL PARAMETERS

A central issue in treatment plan optimization pertains to dealing with conflicting requirements of different structures[7] or even different voxels.[16,17] Typically one constructs an objective function as a weighted sum of all the involved structural objectives for optimizing the treatment plan.[1,18] The weights emphasize the importance of different organs. It has long been recognized that this type of modeling is far from perfect, and determination of a large number of model parameters in the metric function is quite labor intensive.

In essence, one can utilize the existing domain knowledge and prior planning data to *automate* the parameter selection and to improve the efficiency and quality of treatment planning. Prior knowledge appears in various forms, ranging from modeling results, formulas derived from clinical experience, or simply prior treatment plans. The use of prior knowledge for automation of treatment planning dates back to the early days of IMRT research. Xing et al.[7,8] pioneered the autonomous parameter selection two decades ago by using a plan evaluation function constructed based on the prior clinical experience. Recent attempts have been reported by Zarepisheh et al., McIntosh et al., and Zarepisheh et al.[10,19,20] to replace the planning objective for each individual patient, which was constructed based on empirically known DVHs, with prior treatment plans from other patients with similar anatomy. With the effective use of an ensemble of prior treatment plans, a two-loop search strategy is invoked where the inner loop executes the standard treatment planning and the outer loop evaluates the plan generated by the inner loop optimization and finds the final plan with clinically sensible dosimetric distribution automatically.[21]

In contrast with the simplified inverse treatment planning platform,[22–25] the two-loop planning scheme, depicted in Figure 7.3, uses an application programming interface (API) that is capable of interacting with a commercial TPS. The outer loop optimization mimics a planner's interactive planning and decision-making process in searching for a sensible solution. With this adoption, Xing et al.[7,8] validates that IMRT/VMAT treatment planning can be readily automated. In particular, Figure 7.4 shows the ensemble of DVHs for the spinal cord and brainstem and the planning target volume (PTV) of the reference plans for a head-and-neck (HN) case study. Figure 7.5 also compares the empirical DVH for the clinical and autopiloted planning, where the resultant iso-dose distributions for two representative plans are depicted in Figure 7.6. Only minor discrepancies are observed between the autopiloted plan and the clinical plan generated independently by a human planner.

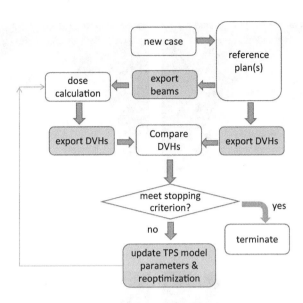

Figure 7.3 An architectural overview of the automated VMAT/IMRT plan optimization scheme.

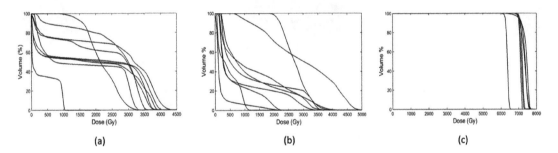

Figure 7.4 Reference DVH curves of (a) spinal cord, (b) brainstem, and (c) PTV for the head-and-neck case under planning. The ensemble of DVHs presents the preferred range of the resultant DVH curve.

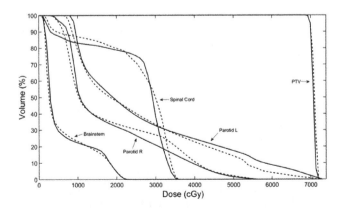

Figure 7.5 A comparison of DVHs of the clinical and autopiloted plans for a two-arc head-and-neck VMAT case. The dashed and solid curves represent the DVHs of clinical and autopiloted plans, respectively.

Big data in radiation oncology

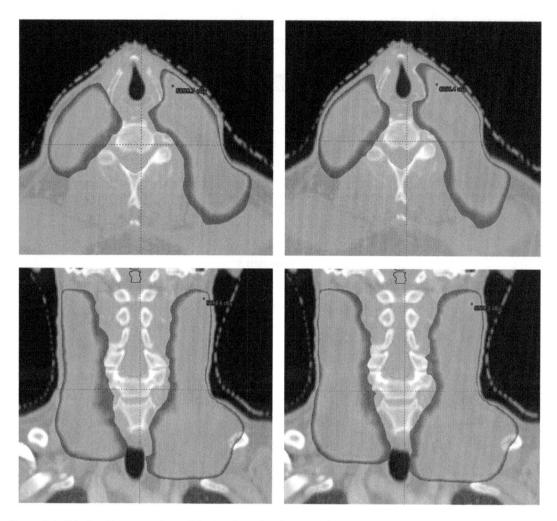

Figure 7.6 Side-by-side comparison of the iso-dose distributions of the autopiloted (right) and clinical (left) plans for the head-and-neck case.

In addition to automating the model parameter determination, significant efforts have been made to improve modeling with the aim of replacing the nonintuitive modeling parameters with more physically plausible quantities. This in turn facilitates the collection and utilization of prior planning data. Along this line, Liu et al.[26] recently developed an inverse planning framework that parameterizes the dosimetric tradeoff among the structures using physical quantities that are more meaningful than the traditional weighting factors. To this end, the inverse planning is cast as a convex feasibility program, the so-termed dosimetric variation-controlled model (DVCM), whose goal is to generate plans with dosimetric or DVH variations of all involved structures consistent with a set of prescribed values. In this framework, the prescribed dose and its permissible variation range for each structure reflects the importance of the structure. The variation for a structure is extracted from a library of previous cases that possessed similar anatomy and prescription. The proposed technique[26] is applied to planning for two prostate and two head-and-neck patients. The results are compared with those obtained using a conventional approach and with a moment-based optimization scheme.[20] In all four cases, the TPS generated a very competitive plan as compared with those obtained using the alternative approaches. With a physically more meaningful modeling of the interstructural trade-offs, the technique enables us to substantially reduce the need for trial-and-error adjustment of the model parameters and opens new opportunities of incorporating prior knowledge to facilitate the treatment planning process.

It is useful to reiterate that treatment planning based on prior knowledge, or historical patient plans, has been discussed previously.[20,27–30] These approaches are generally classified into three categories, namely (1) those that predict the weighting factors for different structures using machine learning algorithms, (2) those that generate initial solutions for the "warm startup" of the subsequent optimization, and (3) those that estimate the desired DVH curves to be achieved by the subsequent optimization. It is worth noting that estimating the permissible range of the final plan is only part of the inverse planning problem. Ultimately, the beam parameters that produce the best possible dose distribution are needed for patient treatment. The subsequent search process after knowledge-based plan prediction would still demand manual trial and error.

For the sake of completeness, note that the heuristic optimization techniques described by, Purdie et al., Lian et al., Liu and Wu, Shiraishi et al., Amit et al., and Schreibmann and Fox[31–37] and the multi-objective optimization schemes by Cotrutz et al., Kamran et al., Unkelbach et al.[38–40] have also been developed to facilitate the inverse planning process. In essence, one can improve the effectiveness of these algorithms with incorporation of population-based plan data by means of recent advances in machine learning and neural networks as elaborated in the ensuing sections.

7.6 KNOWLEDGE-BASED PLANNING FACILITATED BY LEARNING

Learning-based planning can be performed in various domains. In each situation, a machine learning model is first deployed with the use of a disease site–specific library and a handcrafted or auto-extracted (in the case of deep learning[41,42]) set of features. These characterizing features are extracted based on the treatment simulation CT image, segmentations, and the relative relationship of the geometric structure of the case. The model-predicted planning parameters such as the DVHs of the involved organs are then used as input to drive the subsequent inverse planning of a new patient case[33,41,43–46] and the references therein. The ensuing sections summarize in some detail the major aspects of machine learning–based planning in different domains. In particular, the focus will be placed on the predictive analytics of the DVH and the volumetric dose distribution.

7.6.1 DVH PREDICTION

A recent development in knowledge based planning (KBP) is to use machine learning to leverage the historical DVHs from previous treatment plan data to provide the best estimate of the final DVHs for a new patient ([29,33,35,41,44–46]). Subsequently, the predicted DVHs can be used as input parameters to start an IMRT/VMAT inverse planning.

7.6.1.1 Handcrafted features

Yuan et al.[46] studied organ at risk (OAR) dose-sparing variations among patients with IMRT-based plans and their correlation with anatomical features. It trains a stepwise multiple regression method that matches the *handcrafted* features to the principal components of the DVH. The handcrafted anatomical features encompass two major groups that encode the spatial and volumetric information. Besides the PTV and OAR volumes, the geometry of OAR relative to PTV is captured through the distance-to-target histogram (DTH), where, in particular, a few principal components of DVH are extracted as representative features. Upon training on a cohort of 64 prostate, and 84 head-and-neck historical plans, the significant anatomical factors contributing to OAR sparing are identified as the median distance between PTV and OAR; the OAR portion within an OAR-specific distance range, and the volumetric overlap of PTV and OAR. The predicted DVHs complied reasonably well with the optimized plans.

The DVH-prediction analytics by Yuan et al.[46] have appeared in commercial treatment planning systems such as Eclipse. However, it is still useful to recognize that all knowledge-based schemes use only prior knowledge-derived parameters (i.e., weighting either factors or prescription dose) to "warm start" the inverse planning, instead of using them to guide the plan search throughout the optimization process. Moreover, the predicted DVHs prescription by machine learning may not always be physically realizable or optimal. Although clinically valuable, it seems to be fair to state that the current implementation of knowledge-based planning is not yet an ideal approach, certainly not for all cases and all disease sites. Notably, manual interventions are required in the following situations: (1) establishment of a disease-specific library to set up the model based on machine learning or alike, (2) planning of special cases that have failed in knowledge-based planning, and (3) planning of special cases that do not belong to any group of prior classification (i.e., they cannot be described by modeling).

A main reason for the breakdown of the machine learning approaches comes from the suboptimum handcrafted-based feature engineering, which cannot truly capture the geometrical details in a three-dimensional (3D) anatomy. Deep learning[47] seems to be a promising approach to circumvent this difficulty as discussed next. Before moving on the next section, it is worth commenting that the method has also been applied to predict the weighting factors[28,48,49] (only for prostate IMRT) needed for driving an inverse plan optimization. In treatment plan optimization, a central issue is how to deal with the conflicting criteria and goals between the different structures. The conventional radiation therapy inverse planning relies on the use of weighting factors to balance the conflicting requirements of the PTV and OARs for a given patient. The objective function with the weighted sum of all the involved structural objectives is then optimized to derive a treatment plan for the patient.[1,18] The determination of the optimal weighting factors that lead to a clinically sensible solution is not a transparent procedure and requires manual trial-and-error effort, which has long been recognized as a "black art" and represents the most time-consuming part of clinical treatment planning.[1,7,8,50] Efforts of using advanced machine learning for weighting factors determination may thus be useful.

7.6.1.2 Auto-extracted features via deep learning

Recently, Mardani and Xing[41] attempted to cope with the aforementioned robustness and generalizability challenges by developing a deep-learning-based pipeline that directly maps the CT scan to the DVH. In this regard, the work of Mardani et al. cast the DVH-prediction problem as a multitask linear regression that directly maps the input contour to the principal components of the DVH through a deep convolutional neural network (CNN). CNNs are widely known as powerful architecture for deep learning with a great record of success in image recognition tasks.[47,51,52] The technique has been applied to many medical applications,[53–55] such as automated cancer diagnosis, pathology slice analysis, image processing, and treatment outcome prediction. In principal, CNN works by passing the data through multiple serial layers of encoding, where the output of each layer is fed into the next layer to create a higher-level set of features.[47,51,52] After several layers of encoding, a group of high-level features are created that best describe the considered task. CNNs particularly benefit from the simple convolution operation that renders them computationally very efficient, with desirable deformation-invariant properties leading to a high generalization capability.

The input contours for PTV and OARs (rectum and bladder) are rearranged in the polar coordinate (r, φ), which tend to be more interpretable for radiation therapy, and can better preserve the rotation invariance. The contoured CT scans for PTV and OARs are considered as two different input channels to an eight-layer CNN, which outputs principal components of the DVH; see the CNN network architecture in Figure 7.7. Three convolution layers (CL) are followed by average pooling, and, finally, two fully connected layers (FL) to arrive at the output. We trained the network with 110 patients and tested it on five independent patients who were treated at the Stanford Cancer Center, adopting the cross-entropy loss as the training cost. The optimized and predicted DVHs are plotted for a representative patient in Figure 7.8, where it is evident that the CNN-based prediction closely complies with the optimized DVH. In terms of mean-squared-error criterion, the predicted DVH for the rectum and bladder cancer deviates by only 10% and 20%, respectively, from the optimized plan.

Mardani and Xing also compared the prediction performance of the CNN-based scheme[41] with the RapidPlan[46] for a computer-simulated phantom that randomly generated 5,000 optimized plans. After testing the trained model on 1,000 independent plans, the CNN-based scheme achieved a 9% average mean-squared-error, whereas the average error for RapidPlan exceeded 19%. This indeed corroborates the superior feature extraction power of CNNs relative to handcrafting. Returning to the manual intervention situations listed under Situations 1–3, after training with sufficient data, multilayer neural networks have the potentiality to generalize across different organs and tumor sites with minimal tweaking by clinicians;

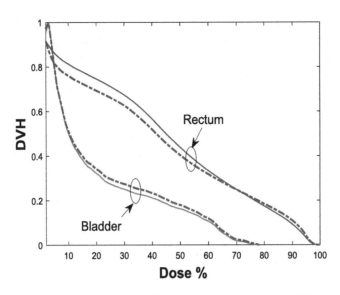

Figure 7.7 Architecture of the adopted eight-layer CNN for mapping the contoured CT scan to DVH. Thirty neurons were used for the convolutional layers (CONV), and 50 neurons for fully connected (FC) layers.

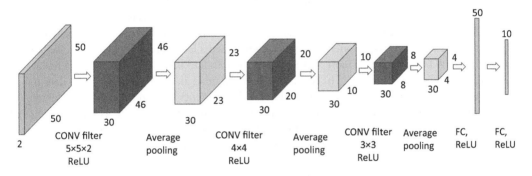

Figure 7.8 Ground-truth (solid line) and predicted (dashed line) DVHs for a prostate cancer patient. ReLU, Rectified Linear Unit.

similar observations already exist in the context of transfer learning.[56,57] Therefore, CNN-based prediction could be the prime candidate for DVH prediction in clinical practice.

7.6.2 VOXEL-WISE THREE-DIMENSIONAL DOSE PREDICTION

DVH is commonly employed as a means to statistically summarize the iso-dose distribution for a treatment plan. In reality, the DVH only provides first-order dosimetry information, that is, the volume (or fractional volume) of an organ receiving a certain dose without the spatial information of the doses. Note also that the DVH can be obtained *only after* an iso-dose plan has been crafted, prohibiting its usage during the process of dose optimization. Furthermore, in order to enable *real-time* image-guided radiotherapy, one may need more detailed dosimetry information. In essence, for real-time therapy the plan needs to be re-optimized frequently in an online fashion to track the dynamics of the tumor site over time.[58,59] Recall that the current treatment planning technology, adopted, for example, by Eclipse and RapidPlan,[46] relies solely on DVH and maximum/minimum dose volumes for OARs, which require a lot of tuning and replanning to assure the resulting plan meets the prescribed DVH. This can incur significant computational overhead (e.g., for VMAT) and hinders real-time planning. In principle, a detailed estimate of the dose profile serves as a valuable prior for plan optimization, which can confine the search space for the optimal plan and thus greatly accelerate the planning.

First efforts to develop 3D dose prediction analytics have been reported by.[42,60,61] Mardani and Xing[42] applied a deep learning–based dose prediction model for VMAT/IMRT treatment to develop predictive analytics for the *achievable* dose distribution based on a data set of quality plans designed for past patients. Specifically, Mardani et al.[42] predicted the 3D dose volume for a new patient based on the geometrical features of the contoured anatomical scan. The following two steps are involved with the process: (1) *Feature extraction via convolutional auto-encoders*: To extract a rich set of features resilient to patient and tumor-shape variations, two multilayer convolutional auto-encoders (CAE)[53] are applied separately to the contour and dose data sets to extract latent features that are invariant to possible image deformations; (2) *Learning the relation map:* Based on the historical dose and contour features, a predictive model based on the so-termed multitask linear regression is trained, where dose features are linearly correlated with a subset of contour features that are selected via sparsity regularization. Preliminary tests are performed on a cohort of 125 prostate patients (110 for training and 15 for test). Dose and contour volumes are rearranged to form a 3D polar grid with (r, φ, z) coordinates. The z-coordinate is divided into 22 overlapping patches, and from each patch the 100 most informative features are extracted. Our preliminary results, which were based on a two-layer linear auto-encoder, achieved 65% overlap volume between the actual and predicted iso-dose surfaces associated with V60%, V75%, and V90% dose volumes.

Shiraishi and Moore [60] also studied the correlation of dose at a given voxel to a number of geometric and plan parameters using an artificial neural network. For each plan in the study, PTV and OARs contours, dose matrix, and field arrangements were exported from the Eclipse treatment planning platform for 23 prostate and 43 stereotactic radiosurgery patients. Per each voxel, a few features (~10) concerning the relative geometry were chosen, and then fed into a simple two-layer fully connected neural network, which was trained to output the dose per voxel. Two separate networks were trained for the PTV voxels and the outside ones. Validations with independent patients demonstrated a less than 10% dose prediction error, and the resulting DVH seemed to be more accurate than the standard knowledge-based DVH prediction schemes as described by Shiraishi et al.[35] In contrast with Mardani et al.'s work,[42] this scheme does not fully leverage the feature extraction power of neural networks and relies solely on a small set of handcrafted features. In addition, the voxel-by-voxel modeling tends to be computationally inefficient for training and testing where each patient's volume contains around a few thousand voxels. The voxel-by-voxel training also leads to discontinuous dose distributions that typically necessitate further refinement.[60]

Before moving on, it is worth noting that McIntosh and Purdie[61] also recently developed a multi-patient atlas-based dose-prediction approach that learns to predict the dose-per-voxel for a novel patient directly from the CT scan. This method learns to automatically select the most effective atlas, and then map the dose from those atlases onto the patient. A conditional random field model is adopted for the optimization of a joint distribution prior that matches the complementary goals of a spatially distributed dose volume while adhering to the desired DVHs. The resulting distribution can then be used for inverse planning with a new spatial dose objective or it can be used to create typical dose volume objectives for the canonical optimization pipeline. Six treatment sites were evaluated with promising results.

7.6.3 LEARNING IN THE ISO-DOSE FEATURE-PRESERVING DOMAIN

To reduce the scale of computation, and improve characterization of iso-dose plans, an iso-dose feature-preserving voxelization (IFPV) framework was introduced recently by Liu and Xing[26] for machine learning–based applications in radiation therapy. An implementation of inverse planning in the more intelligently down-sampled IFPV domain was demonstrated. Dose distribution in IFPV scheme was characterized by partitioning the voxels into subgroups according to their geometric and dosimetry values. Computationally, the iso-dose feature-preserving (IFP) clustering combines the conventional voxels that are spatially and dosimetrically close into physically meaningful clusters. k-Means clustering and, subsequently, the support vector machine (SVM) classification algorithms are applied sequentially to group the voxels into IFP clusters. To illustrate the utility of the formalism, an inverse planning framework in the IFPV domain was implemented, and the resultant plans were compared quantitatively with those obtained using the conventional inverse planning technique. Overall, the IFPV generates models with significant dimensionality reduction without compromising the spatial resolution seen in the conventional downsampling schemes. In addition to the improved computational efficiency, it was found that the IFPV-domain inverse planning yields better treatment plans than that of DVH-based planning primarily because of the more effective use of both geometric and dose information by the system during plan optimization.

By removing the redundant dosimetry information of voxels within homogeneous volumes, the IFPV enables the use of a significantly reduced number of spatial parameters to describe the dose distribution, enabling a more compact representation of iso-dose plan with a slight loss of information. It has also been noticed that inverse planning in IFPV-domain seems to outperform existing DVH-based optimization techniques. Given the fundamental roles of the voxelization in therapeutic planning, the proposed IFPV framework should find useful applications in many other applications in radiation therapy. In particular, the formalism opens a new vista for IFPV-domain machine learning and/or other data-driven autonomous planning.

7.7 SUMMARY

An overriding paradigm of modern radiation therapy since the age of 3D CRT and IMRT has been the efficient and robust optimization of dose distribution. The goal of dose optimization is generally to find a solution that best conforms the PTV(s) while sparing the OARs. Clinical knowledge underlies many ongoing efforts in the development of new-generation treatment planning strategies. In this chapter, we summarized the developments of using prior knowledge and historical data to facilitate the inverse treatment planning. Recent advances in machine learning and deep neural networks to incorporate physical, biological, and clinical data into treatment planning systems were discussed in some detail. It is important to mention that, in our opinion, none of the existing DVH- or voxel domain (deep) machine learning–based schemes is ideal. We believe a new framework specially designed for deep learning–based planning is highly desirable to truly benefit from state-of-the-art technology. One of the fundamental deficiencies of the existing approaches lies in the implicit relationship, or the disconnection, between the weighting parameters and the resulting dose distribution. To navigate toward the right solution, it is desirable to relate the model parameters to the characteristic properties of the resultant plan. Furthermore, extracting the essential information of a treatment plan for concise description of the plan and other applications in radiation oncology is a fundamental research problem. A treatment plan is generally given in the form of voxelized dose distribution in the context of the 3D anatomy of the patient. The information is, from the perspective of practical usage, heavily oversampled in most regions because of the need for high spatial sampling in only a fraction of the anatomical regions. The newly proposed IFPV provides a low-dimensional and parametric representation of an iso-dose plan without sacrificing the essential characteristics of the plan, thus providing a practical, valuable framework for various applications in radiation therapy. In addition to the above discussion, the incorporation of artificial intelligence for advanced treatment planning tasks is emerging. Finally, we emphasized that the incorporation of newly available radiomics and other clinical data is an important area of research in KBP for the years to come.

REFERENCES

1. AAPM IMRT Sub-committee: Guidance document on delivery, treatment planning, and clinical implementation of IMRT: Report of the IMRT subcommittee of the AAPM radiation therapy committee. *Med Phy* 2003, 30:2089–115.
2. Bortfeld T: Optimized planning using physical objectives and constraints. *Semin Radiat Oncol* 1999, 9:20–34.
3. Xing L, Chen GTY: Iterative algorithms for inverse treatment planning. *Phys Med Biol* 1996, 41:2107–23.
4. Yu CX: Intensity-modulated arc therapy with dynamic multileaf collimation: An alternative to tomotherapy. *Phys Med Biol* 1995, 40:1435–49.
5. Otto K: Volumetric modulated arc therapy: IMRT in a single gantry arc. *Med Phys* 2008, 35:310–7.
6. Crooks SM, Wu X, Takita C, Matzich M, Xing L: Aperture modulated arc therapy. *Phys Med Biol* 2003, 48:1333–44.
7. Xing L, Li JG, Donaldson S, Le QT, Boyer AL: Optimization of importance factors in inverse planning. *Phys Med Biol* 1999, 44:2525–36.
8. Xing L, Li JG, Pugachev A, Le QT, Boyer AL: Estimation theory and model parameter selection for therapeutic treatment plan optimization. *Med Phys* 1999, 26:2348–58.

9. Yu Y, Zhang JB, Cheng G, Schell MC, Okunieff P: Multi-objective optimization in radiotherapy: Applications to stereotactic radiosurgery and prostate brachytherapy. *Artif Intell Med* 2000, 19:39–51.

10. Zarepisheh M, Long T, Li N, Tian Z, Romeijn HE, Jia X, Jiang SB: A DVH-guided IMRT optimization algorithm for automatic treatment planning and adaptive radiotherapy replanning. *Med Phys* 2014, 41:061711.

11. Reinstein LE, Hanley J, Meek AG: A feasibility study of automated inverse treatment planning for cancer of the prostate. *Phys Med Biol* 1996, 41:1045–58.

12. Schreibmann E, Xing L: Feasibility study of beam orientation class-solutions for prostate IMRT. *Med Phys* 2004, 31:2863–70.

13. Pugachev A, Li JG, Boyer AL, Hancock SL, Le QT, Donaldson SS, Xing L: Role of beam orientation optimization in intensity-modulated radiation therapy. *Int J Radiat Oncol Biol Phys* 2001, 50:551–60.

14. Pugachev A, Xing L, Boyer AL: Beam orientation optimization in IMRT: To optimize or not to optimize? *XII International Conference on the Use of Computers in Radiation Therapy*. Heidelberg, Germany, 2000.

15. Pugachev A, Xing L: Incorporating prior knowledge into beam orientation optimization. *Int J Radiat Oncol Biol Phys* 2002, 54:1565–74.

16. Cotrutz C, Xing L: Using voxel-dependent importance factors for interactive DVH-based dose optimization. *Phys Med Biol* 2002, 47:1659–69.

17. Yang Y, Xing L: Inverse treatment planning with adaptively evolving voxel-dependent penalty scheme. *Med Phys* 2004, 31:2839–44.

18. Bortfeld T: IMRT: A review and preview. *Phys Med Biol* 2006, 51:R363–79.

19. McIntosh C, Welch M, McNiven A, Jaffray DA, Purdie TG: Fully automated treatment planning for head and neck radiotherapy using a voxel-based dose prediction and dose mimicking method. *Phys Med Biol* 2017, 62:5926–44.

20. Zarepisheh M, Shakourifar M, Trigila G, Ghomi PS, GCouzens S, Abebe A, Norena L, Shang W, Jiang SB, Zinchenko Y: A moment-based approach for DVH-guided radiotherapy treatment plan optimization. *Phys Med Biol* 2013, 58:1869–87.

21. Wang H, Dong P, Liu H, Xing L: Development of an autonomous treatment planning strategy for radiation therapy with effective use of population-based prior data. *Med Phys* 2017, 44:389–96.

22. Deasy JO, Blanco AI, Clark VH: CERR: A computational environment for radiotherapy research. *Med Phys* 2003, 30:979–85.

23. Tewell MA, Adams R: The PLUNC 3D treatment planning system: A dynamic alternative to commercially available systems. *Med Dosim* 2004, 29:134–8.

24. Kim H, Li R, Lee R, Xing L: Beam's-eye-view dosimetrics (BEVD) guided rotational station parameter optimized radiation therapy (SPORT) planning based on reweighted total-variation minimization. *Phys Med Biol* 2015, 60:N71–82.

25. Dong P, Ungun B, Boyd S, Xing L: Optimization of rotational arc station parameter optimized radiation therapy. *Med Phys* 2016, 43:4973.

26. Liu H, Dong P, Xing L: Using measurable dosimetric quantities to characterize the inter-structural tradeoff in inverse planning. *Phys Med Biol* 2017, 62:6804–21.

27. Kang J, Schwartz R, Flickinger J, Beriwal S: Machine learning approaches for predicting radiation therapy outcomes: A clinician's perspective. *Int J Radiat Oncol Biol Phys* 2015, 93:1127–35.

28. Lee T, Hammad M, Chan TC, Craig T, Sharpe MB: Predicting objective function weights from patient anatomy in prostate IMRT treatment planning. *Med Phys* 2013, 40:121706.

29. Li T, Wu Q, Yang Y, Rodrigues A, Yin FF, Jackie Wu Q: Quality assurance for online adapted treatment plans: Benchmarking and delivery monitoring simulation. *Med Phys* 2015, 42:381–90.

30. Fogliata A, Nicolini G, Bourgier C, Clivio A, De Rose F, Fenoglietto P, Lobefalo F et al.: Performance of a knowledge-based model for optimization of volumetric modulated arc therapy plans for single and bilateral breat irradiation. *PLoS One* 2015, 10:e0145137.

31. Purdie TG, Dinniwell RE, Letourneau D, Hill C, Sharpe MB: Automated planning of tangential breast intensity-modulated radiotherapy using heuristic optimization. *Int J Radiat Oncol Biol Phys* 2011, 81:575–83.

32. Purdie TG, Dinniwell RE, Fyles A, Sharpe MB: Automation and intensity modulated radiation therapy for individualized high-quality tangent breast treatment plans. *Int J Radiat Oncol Biol Phys* 2014, 90:688–95.

33. Lian J, Yuan L, Ge Y, Chera BS, Yoo DP, Chang S, Yin F, Wu QJ: Modeling the dosimetry of organ-at-risk in head and neck IMRT planning: An intertechnique and interinstitutional study. *Med Phys* 2013, 40:121704.

34. Liu H, Wu Q: Evaluations of an adaptive planning technique incorporating dose feedback in image-guided radiotherapy of prostate cancer. *Med Phys* 2011, 38:6362–70.

35. Shiraishi S, Tan J, Olsen LA, Moore KL: Knowledge-based prediction of plan quality metrics in intracranial stereotactic radiosurgery. *Med Phys* 2015, 42:908.

36. Amit G, Purdie TG, Levinshtein A, Hope AJ, Lindsay P, Marshall A, Jaffray DA, Pekar V: Automatic learning-based beam angle selection for thoracic IMRT. *Med Phys* 2015, 42:1992–2005.

37. Schreibmann E, Fox T: Prior-knowledge treatment planning for volumetric arc therapy using feature-based database mining. *J Appl Clin Med Phys* 2014, 15:4596.
38. Cotrutz C, Lahanas M, Kappas C, Baltas D: A multiobjective gradient-based dose optimization algorithm for external beam conformal radiotherapy. *Phys Med Biol* 2001, 46:2161–75.
39. Kamran SC, Mueller BS, Paetzold P, Dunlap J, Niemierko A, Bortfeld T, Willers H, Craft D: Multi-criteria optimization achieves superior normal tissue sparing in a planning study of intensity-modulated radiation therapy for RTOG 1308-eligible non-small cell lung cancer patients. *Radiother Oncol* 2016, 118:515–20.
40. Unkelbach J, Bortfeld T, Craft D, Alber M, Bangert M, Bokrantz R, Chen D et al.: Optimization approaches to volumetric modulated arc therapy planning. *Med Phys* 2015, 42:1367–77.
41. Mardani M, Dong P, Xing L: Personalized dose prescription for treatment planning via deep convolutional neural networks. *Annual Meeting of American Society of Radiation Oncology*, Boston, MA, September 25–28, 2016.
42. Mardani M, Dong P, Xing L: Deep-learning based prediction of achievable dose for personalizing inverse treatment planning. *Annual Meeting of American Association of Physicists in Medicine*, Boston, MA, July 31–August 4 2015, 2016.
43. Zarepisheh M, Li R, Ye Y, Xing L: Simultaneous beam sampling and aperture shape optimization for SPORT. *Med Phys* 2015, 42:1012–22.
44. Appenzoller LM, Michalski JM, Thorstad WL, Mutic S, Moore KL: Predicting dose-volume histograms for organs-at-risk in IMRT planning. *Med Phys* 2012, 39:7446–61.
45. Wu Q, Yuan L, Li T, Ying F, Ge Y: Knowledge-based organ-at-risk sparing models in IMRT planning. *Pract Radiat Oncol* 2013, 3:S1–2.
46. Yuan L, Ge Y, Lee WR, Yin FF, Kirkpatrick JP, Wu QJ: Quantitative analysis of the factors which affect the interpatient organ-at-risk dose sparing variation in IMRT plans. *Med Phys* 2012, 39:6868–78.
47. Goodfellow I, Bengio Y, Courville A: *Deep Learning*. Cambridge, MA: MIT Press, 2016.
48. Good D, Lo J, Lee WR, Wu QJ, Yin FF, Das SK: A knowledge-based approach to improving and homogenizing intensity modulated radiation therapy planning quality among treatment centers: An example application to prostate cancer planning. *Int J Radiat Oncol Biol Phys* 2013, 87:176–81.
49. Boutilier JJ, Lee T, Craig T, Sharpe MB, Chan TC: Models for predicting objective function weights in prostate cancer IMRT. *Med Phys* 2015, 42:1586–95.
50. Wu Q, Djajaputra D, Wu Y, Zhou J, Liu HH, Mohan R: Intensity-modulated radiotherapy optimization with gEUD-guided dose-volume objectives. *Physics in Medicine & Biology* 2003, 48:279–91.
51. Krizhevsky A, Sutskever I, Hinton GE: Imagenet classification with deep convolutional neural networks. *Advances in Neural Information Processing Systems*, 2012. pp. 1097–105.
52. LeCun Y, Bengio Y, Hinton G: Deep learning. *Nature* 2015, 521:436–44.
53. Arik SO, Ibragimov B, Xing L: Fully automated quantitative cephalometry using convolutional neural networks. *J Med Imaging (Bellingham)* 2017, 4:014501.
54. Ibragimov B, Xing L: Segmentation of organs-at-risks in head and neck CT images using convolutional neural networks. *Med Phys* 2017, 44:547–57.
55. Ibragimov B, Toesca D, Chang D, Koong A, Xing L: From popution-average to deep learning-based toxicity rist prediction for the creation of individualized radiotherapy treatment plan. *Sci Transl Med* 2017:submitted.
56. Yosinski J, Clune J, Bengio Y, Lipson H: How transferable are features in deep neural networks? *Advances in Neural Information Processing Systems*, 2014. pp. 3320–8.
57. Sharif Razavian A, Azizpour H, Sullivan J, Carlsson S: CNN features off-the-shelf: an astounding baseline for recognition. *Proceedings of the IEEE Conference on Computer Vision and Pattern Recognition Workshops*, 2014. pp. 806–13.
58. Ten Haken R, Forman J, Heimburger D, Gerhardsson A, McShan D, Perez-Tamayo C, Schoeppel S, Lichter A: Treatment planning issues related to prostate movement in response to differential filling of the rectum and bladder. *Int J Radiat Oncol Biol Phys* 1991, 20:1317–24.
59. Zhu X, Ge Y, Li T, Thongphiew D, Yin FF, Wu QJ: A planning quality evaluation tool for prostate adaptive IMRT based on machine learning. *Med Phys* 2011, 38:719–26.
60. Shiraishi S, Moore KL: Knowledge-based prediction of three-dimensional dose distributions for external beam radiotherapy. *Med Phys* 2016, 43:378–87.
61. McIntosh C, Purdie TG: Voxel-based dose prediction with multi-patient atlas selection for automated radiotherapy treatment planning. *Phys Med Biol* 2017, 62:415–31.

8 Using big data to improve safety and quality in radiation oncology

Eric Ford, Alan Kalet, and Mark Phillips

Contents

8.1 Introduction: The Quality Gap in Radiation Oncology and the Need for New Approaches 111
8.2 Overview of Needs: Data Standards, Exchange, and Interoperability 112
8.3 Structure, Process, Outcomes: A Structure for Quality and Safety Data 113
    8.3.1 Outcomes 113
    8.3.2 Process 114
        8.3.2.1 Machine Learning and Big Data in Process Control: An Example Study 114
    8.3.3 Structure 117
        8.3.3.1 Event Reports from Incident Learning Systems 117
        8.3.3.2 Culture Metrics 117
        8.3.3.3 Human Performance Metrics 117
        8.3.3.4 Safety Profile Assessment 118
8.4 Validation of Quality and Safety Measures 118
8.5 Conclusions: Quality and Safety Data in Radiation Oncology 118
References 119

8.1 INTRODUCTION: THE QUALITY GAP IN RADIATION ONCOLOGY AND THE NEED FOR NEW APPROACHES

Outcome measures in oncology are notoriously challenging. Meaningful endpoints often take many years to acquire and are often confounded by disease progression and other effects. Contrast this with other areas of health care where patient safety and quality improvement efforts have shown an impact on outcomes. Examples include central line acquired bloodstream infection (CLABSI)[1] and postoperative morbidity and mortality.[2] In these contexts relatively acute endpoints are readily available. Nevertheless, there are data in the realm of radiation oncology that tie the parameters of treatment to quality and safety. Perhaps the best examples are dose–response data for toxicities of normal tissues. Many studies are available and summaries appear from the quantitative analysis of normal tissue effects in the clinic (QUANTEC) study,[3] the American Association of Physicists in Medicine (AAPM) Working Group on Biological Effects of Hypofractionated Radiotherapy/Stereotactic Body Radiation Therapy (SBRT),[4] and other studies. Although these results guide practice, they would not be considered as big data by most investigators. Typically only one dimension is considered (e.g., dose) and only a very few data points are evaluated per patient. Also problematic in these studies is the high variability in the quality of the radiation treatment that is actually delivered. Recent studies suggest that normal tissue doses that are achieved are highly variable between institutions and even within a single institution.[5,6]

As with normal tissue sparing, tumoricidal doses are also challenging to achieve in practice. Even though tumor-dependent dose–response data are available in many cases and decision making is guided by this experience, actual treatment plans often do not achieve the goal. The deficiencies that can arise are demonstrated by a reanalysis of the Radiation Therapy Oncology Group (RTOG) 9704 trial for pancreatic cancer where it was found that 48% of patients received a treatment that deviated in some way from the recommended protocol.[7] Furthermore, these deviations were highly predictive of overall survival in patients at 5 years. These data are not unique. Recent meta-analyses demonstrate that protocol deviations are relatively common and are linked with outcomes measures in almost all the cooperative group trials examined.[8,9] Part of the reason for this is that "errors of omission" (e.g., underdosing all or part of a tumor) are difficult to identify. Physician peer review of treatment plans is one method of addressing this, and studies suggest that such review would alter treatment plans in approximately 10% of cases (Brunskill et al.[10] and references therein) although the rate is >20% in some studies.[11]

Furthermore, a single number is likely not a good predictor of patient outcome. Take, for example, prescribed dose. Although this would seem to be a simple and reasonable data point to collect, it is often problematic. The delivered dose can vary by a substantial amount from the prescribed dose depending on the details of how the dose is prescribed (normalization, point vs. volume). In one large multi-institutional study, 63% of patients received a dose that was more than 10% below that prescribed.[12] Prescribed dose also indicates nothing about the quality of the plan or the delivery that clinical trials data indicate is important. All of this begs for a more comprehensive approach that includes reliable metrics of safety and quality and associate improvements.

To appreciate this, consider the example of CLABSI and the efforts to reduce life-threatening infections in the ICU. Here a checklist tool was developed and was shown to decrease the rate of infections by 70%, an effect which was durable over years.[1] The success of these programs was enormous and became the subject of several national bestselling books.[13,14] It was later realized though that a key to the success of these programs was not necessarily the checklist itself (although this was important), but equally important was the availability of a well-defined metric for success, in this case post-procedure infection rates.[15] This provided a goal and also a normative pressure. Feedback to staff is crucial. This can also be observed in the context of incident learning, where it has been shown that meaningful feedback to staff is crucial for the success of the program.[16] There is a clear need, therefore, for metric(s) and approaches to quality and safety in the oncology context. Although this is clearly challenging, big data offers a promising approach. More predictive power may be gained by moving beyond single dimensional parameters (e.g., prescribed dose) into larger dimensions of data and associated machine learning tools.

This chapter reviews new approaches to quality and safety improvement in radiation oncology with a specific focus on big data. Only a small handful of studies are available at this time on this topic. However, an understanding is beginning to emerge about what big data might mean in the context of quality and safety in radiation oncology. The reader is also referred to a thought-paper on this topic[17] that emerged from the 2015 conference "Big Data Workshop: Exploring Opportunities for Radiation Oncology in the Era of Big Data" sponsored by American Society of Radiation Oncology (ASTRO), National Cancer Institute (NCI), and AAPM.[18] Substantial advances can be expected in the coming years and it is hoped that the ideas here may inform that process.

8.2 OVERVIEW OF NEEDS: DATA STANDARDS, EXCHANGE, AND INTEROPERABILITY

There is currently an unmet need for data standardization in oncology and this is directly related to quality and safety. Many efforts are underway, examples include the development of a standard radiation prescription nomenclature by ASTRO[19] and the development of standards for structure names and dose–volume histogram data by AAPM Task Group 263.[20] In the United States, some of the interest will be driven by the need for Centers for Medicare and Medicaid Services (CMS)-approved Qualified Clinical Data Registries needed to qualify for the new reimbursement models.[21] Such standards can be expected to have a huge impact as they have in other areas of health care, for example, the widely-used Systematized Nomenclature of Medicine (SNOMED) system, which is the categorization schema for medical terminology.

In developing informatics standards, several key things are worth consideration. First, systems should be designed so that the data are collected in the standard process of care. This has been referred to as "data farming,"[22] that is, developing systems so that the data are naturally collected. This is to distinguish it from "data mining," which assumes that a complete and useful data set exists. Second, it must be realized that multiple vendors' information systems are at work, each of which has a different set of stakeholders. This includes the Oncology Information System (e.g., Aria or Mosaiq) and the hospital electronic medical record (EMR; e.g., Epic or Cerner). Information standards must be established so that there is an interoperability between these systems.

8.3 STRUCTURE, PROCESS, OUTCOMES: A STRUCTURE FOR QUALITY AND SAFETY DATA

One way to conceptualize big data's aspects of quality and safety is through the well-established quality model proposed by Avedis Donebedian and colleagues at the University of Michigan in the 1960s.[23] In this paradigm, quality and safety measures fall into one of three categories: Structure (e.g., health care infrastructure), Process (e.g., diagnosis, process of care), and Outcomes (e.g., mortality, quality of life, cost).

Figure 8.1 illustrates this paradigm. The Structure category is the overarching system within which care is delivered. Structure includes aspects such as equipment, training, and quality management systems. Within this structure, care is delivered. Figure 8.1 shows some example components of this care path. Finally, patient outcomes are shown in Figure 8.1 with some example endpoints. Figure 8.1 may be a useful representation because each of the components has measureable data associated with it. Because it is somewhat easier to conceptualize, we will begin the discussion with "Outcomes" (Section 8.3.1) then work backward through the Donabedian model in "Process" (Section 8.3.2) and "Structure" (Section 8.3.3). In Section 8.3.2 ("Process"), we present a specific example of a quality improvement approach being pursued by our group that is based on big data and machine learning.

8.3.1 OUTCOMES

In the context of big data and oncology, most of the outcomes studies to date have focused on new methods to provide prognostic accuracy using radiomics signatures and the like.[24] In the therapy context, most studies are of a few variables (e.g., dose or concurrent therapy) partly due to the difficulties in establishing meaningful oncologic endpoints because of the relatively late time points involved and the confounding factors of disease progression. Recent studies, however, have begun to consider expanded data sets relevant to big data. The study of Robertson et al.[25] considered 100 patients receiving radiotherapy for head-and-neck cancer for which data were available for both radiation dosimetry and outcomes. A total of 57 normal tissue structures and 97 outcomes measures were analyzed in combination. Of the various possible combinations, 17% resulted in significant odds ratios > 1.05 on logistic regression analysis. Many of these combinations were already well known (e.g., dysphagia and dose to larynx), but some were novel. Other studies have examined plan quality using "knowledge-based planning" algorithms (e.g., Moore et al.[5] and Wu et al.[6]) and these are discussed further in Section 8.3.2. There are several challenges with these approaches. First, it is sometimes challenging to assess patient outcomes and so these studies often rely on a surrogate such as a normal tissue complication probability model. Second, as discussed in other chapters, there is the challenge of overfitting the data, especially when there are few patients and many variables. In spite of such challenges, linking radiotherapy parameters to outcomes represents a key area for progress in the near future.

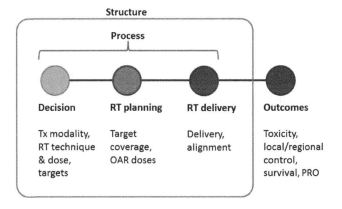

Figure 8.1 A model for quality and safety based on the Donabedian categories of Structure, Process, and Outcomes. The relevance to big data is that each component has measurable data connected with it. PRO, patient reported outcomes; RT, radiation therapy.

8.3.2 PROCESS

Process metrics are more straightforward to measure than outcomes in the oncology context. Many process components lend themselves to measurement (Figure 8.1) and many studies have been published in this realm. The first process component to consider is in the category of "medical decision making," the first component in Figure 8.1. This includes the decision to treat, what modality(ies) to use and the specifics of the radiation therapy (RT) approach (e.g., dose and fractionation, delineation of targets and normal tissues). There are many types of process metrics that can be collected. These include medical decision-making aspects, for example, Surveillance, Epidemiology, and End Results (SEER)–database studies on treatment use. Analysis of plan quality is another aspect and this is treated in more detail in other chapters, so it will not be considered further here. A final aspect is the actual delivery of treatment, that is, delivered versus intended, and there are data that show deficiencies in this realm. Data from phantom-based audits from the Imaging and Radiation Oncology Core-Houston (IROC-H) group, for example, show that 18% of institutions failed the head-and-neck phantom irradiation test with a criterion of 7%/7-mm plan versus delivered.[26] There are clearly very many process-related data that can be collected. In the remainder of this section, we present a concrete example of how these data can be used to influence quality and safety.

8.3.2.1 Machine learning and big data in process control: An example study

In this section, we present an effort by our group to develop a semiautomatic method to identify anomalies (errors, potentially) in a patient's EMR. Such a tool may be useful at various points in the workflow as a verification layer. As such, it falls firmly into the category of "process" under the structured outlined here.

8.3.2.1.1 The need for artificial intelligence–assisted error detection method

The development of software systems and artificial intelligence (AI) or machine learning tools to facilitate decision-making tasks in radiation oncology is a long-standing recognized need.[27] Much of what was formerly considered to be specialized adaptations of computers to medicine is now in routine clinical use and, along with that, the electronic codification of almost all radiation oncology processes from imaging to treatment plan creation, to radiation delivery. The complexity of knowledge and information in the process thus continues to drive a need for practical software solutions to assist in tasks such as plan quality evaluation.

Despite the explosion of electronic data capture and clinical software use, the current standard method for finding errors in treatment plans is human inspection. In radiation oncology, this happens at the point when a treatment plan is complete. Such checks can be performed by a medical physicist, a radiation therapist, and/or physicians in quality assurance (QA) peer-review chart rounds as recently advocated by ASTRO and other organizations.[28] Research has shown that these inspections can be particularly effective.[29] Despite being the most effective pretreatment check, research also suggests that a physicist's initial chart review identifies errors only 65% of the time.[30] With the growing number of plan variables and plan complexity in radiation oncology, human inspection becomes more and more challenging to complete in an accurate and efficient way.

8.3.2.1.2 A Bayesian network approach to error detection

There are a number of ways to approach the QA problem programmatically, the two most common being rules-based systems and probabilistic models. Recent research has shown that rules-based and Bayesian network verifications can catch errors not otherwise found by human inspection.[64] A rules approach involves comparison of values or sets of values to a predetermined range or set of specific values, the result of the comparison often being Boolean (for an example see Yang and Moore[31]). Establishment and use of the various rules does not necessitate (nor take advantage of) large sources of or dynamically changing data. In the case of the probabilistic reasoning approach, Bayesian network models have the ability to both leverage large data and adapt to dynamic changes in clinical practice. One of the other important advantages of such a model is that it incorporates the range of possibilities—quantified by mining existing treatment plans—that clinical realities impose. Rules-based systems have a difficult time instantiating all of the possible decision pathways.

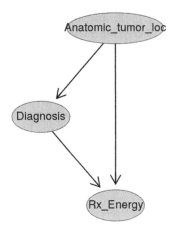

Figure 8.2 A simple Bayesian network for evaluating energy use in treatment plans, representing the mathematical probability P(Rx_Energy|Diagnosis|Anatomic_tumor_loc).

Two major components are required to define a Bayes net model. (1) Topology: the network graph that connects concepts together and defines the direction of dependent conditional variables. (2) Conditional probability tables: the sets of conditional probability values explicitly forming the relationships between any one variable and the other(s).

A simple Bayes Net (BN) structure might look like Figure 8.2. Here, the structure depicts that the prescription energy ("Rx_Energy") depends on both the patient's type of disease ("Diagnosis") as well as the location of the tumor ("Anatomic_tumor_loc"). There is no clearly defined rule set for what energy to use for any given tumor class or location, but there are "more-likely-than-not" cases that can be computed based on distributions built from historical clinical practice data.

Distributions among previous historic cases provide real-world clinical data to be compiled into a knowledge base via machine learning algorithms. Figure 8.3 shows the distribution of Modalities ("Rx_Energy") for all patients with a Diagnosis of "*scalp neoplasm*." This distribution along with the distributions of tumor location are used to calculate a *joint probability* which describes a new probability set for the downstream node ("Rx_Energy") in the presence of knowledge about the dependent variables

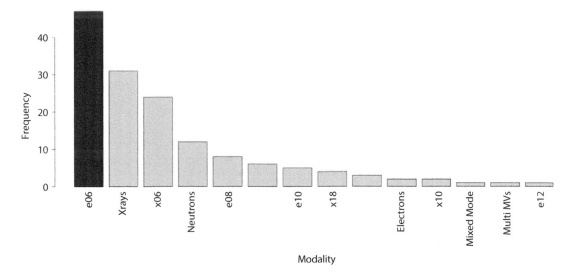

Figure 8.3 Distribution of energy modalities used in all "scalp neoplasm" treatments at one institution, extracted from a Mosaiq relational database via SQL. The green bar highlights the most probable value (un-normalized frequency) for the modality among all similar cases.

Big data in radiation oncology

("Diagnosis" and "Anatomic_tumor_loc"). In this paradigm, updating the distributions over timescales of practice change (2–5 years) can be done to keep the model current with standard care.

Computation of potential errors in the simple example above involves instantiating the known variables and computing the new probability for the distribution. We might find that the probability of using 15-MeV electrons for a superficial tumor depth is very low, for example, P(Rx_Energy = 15 MeV | Diagnosis = Spindle cell carcinoma | Anatomic_tumor_loc = scalp) = 0.021. If these variable values were found in a plan together, then the energy selected for the plan can be determined to be *most likely* in error.

Using these basic principles, BN models of higher complexity have been validated with simulated clinical errors.[32] Networks utilizing a million or more unique probabilities can be built from thousands of unique historical patient cases using machine learning algorithms. Figure 8.4 shows an example of a more complex network structure designed to evaluate multiple aspects of a treatment plan, given some initial staging and diagnostic information.

Increasing complexity comes at a computational cost, however. As models grow, employment of big data reduction/analysis methods will need to take on a larger role. A MapReduce type method could be employed for BN learning, for example, splitting a large network into d-separated subnets, learning CPTs, then recombining them back into the larger network in post processing. In addition, the complexity growth of networks may extend beyond the single database model. Information from hospital EMRs, treatment planning systems, Picture Archiving and Communication System (PACS), and other third-party vendor software systems may become required to form the most comprehensive and useful models. The drive for some form of semantic interoperability to connect disparate data sources and establish consistency and compatibility for model and data sharing purposes is recognized not just for probabilistic models but for a myriad of other applications.[33,34] It remains to be seen which overarching semantic structure will be adopted.

The BN approach demonstrates an adaptation of AI modeling to big data as applied to areas of high impact. Probabilistic models can leverage a variety of data and data types, and they show promising performance in radiotherapy error detection, but their full potential has yet to be realized. Larger complex networks representing more comprehensive descriptions of radiation oncology (RO) processes will be needed. Moreover, such models need to make their way into clinical workflows and software systems to achieve the goals of improving safety, reliability, and transparency in treatment planning.

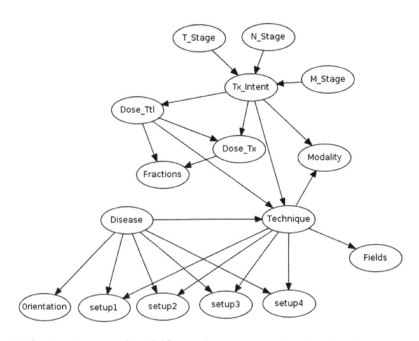

Figure 8.4 A complex Bayesian network model for simultaneous evaluation of multiple treatment plan components.

8.3.3 STRUCTURE

Structure metrics operate at an even broader level than the Process measures discussed earlier. Whereas Process focuses on the details of actions performed, Structure focuses on the system within which those actions are performed. Process metrics are most often patient-specific (e.g., a measure of the quality of a plan or a QA test for the validity of treatment delivery). Structure metrics, however, extend beyond the specific-patient realm (e.g., the quality management program in place at a facility or the system for enrolling patients in clinical trials). As with the Process and Outcomes dimensions discussed earlier, a wealth of data is emerging in the Structure dimension, making it a legitimate target for big data initiatives. Here we describe several examples of Structure metrics in radiation oncology that may fall into the classification of big data.

8.3.3.1 Event reports from incident learning systems

Incident learning systems (ILS) are widely used in health care and are specifically recommended by virtually every major stakeholder organization in radiation oncology.[35–39] Over the past 5 years, over 40 manuscripts have appeared on ILS in the radiation oncology literature on this topic, demonstrating the keen interest and the wealth of data that is emerging. Incident reports include not only major incidents but also near-miss events and unsafe conditions.[40,41] In each of these reports there is a rich data in the form of the event narrative and potentially hundreds of structured data elements (compare Ford et al.[42]). Altogether, this means that ILS data sets are potentially extremely large. ILS systems exist within most institutions and, more recently, as national and international systems (e.g., Safety in Radiation Oncology [SAFRON] from the International Atomic Energy Agency [IAEA][43]). As of this writing, the largest system is RO-ILS™: Radiation Oncology Incident Learning System, a system in the United States sponsored by ASTRO and AAPM with over 300 participating facilities and over 3,300 reports.[44] To date, there are almost no reports on data mining or machine learning as applied to ILS data. However, given the rich data sets on radiation oncology, this presents an attractive avenue for further research.

8.3.3.2 Culture metrics

Tightly coupled to incident learning systems is the concept of "Safety Culture." This phrase, first used in the context of the Chernobyl Nuclear Accident, refers to the awareness of high-risk activity and a determination to maintain safe operations. Culture is one of the best predictors of the quality of patient care and adverse events,[45] and so it represents an interesting target of study. Although culture may not seem a likely target for big data, there are ways to measure culture, such as the validated Hospital Survey of Patient Safety Culture™ from the U.S. Agency for Healthcare Research and Quality (AHRQ). Each year the AHRQ collects data from hundreds of thousands of providers through this instrument. Data on safety culture measures have begun to appear in the radiation oncology literature as well,[46–48] and this represents a future direction for big data efforts.

8.3.3.3 Human performance metrics

There are a variety of ways to assess and measure human performance, and although such measurements can be viewed as task-specific systems-level elements often influence human performance (e.g., the lighting levels in reading rooms affects the performance of radiologists[49]). As such, human performance metrics may best fall into the "Systems" category of the Donabedian model.

There is a rich history of the study of human performance in health care and the associated data set often comprise many thousands of performance data points for each subject. In diagnostic radiology, for example, eye-tracking systems have been used to study the performance of observers to various aspects of the system, for example, one versus two monitors.[49] Studies have begun to appear in the radiation oncology literature as well. Mazur and colleagues at the University of North Carolina have used performance measurement systems to study various tasks[50] and have also studied the impact of task demands and workload on performance.[51] Chan et al.[52] performed a study where they redesigned a popular radiation oncology EMR and then measured the task performance of staff to simulated errors. Task time was significantly reduced and error detection rates were significantly increased through the use of human factors design methodologies. This study also involved large volumes of human performance data.

8.3.3.4 Safety profile assessment

One valuable goal in assessing a care delivery system is to measure the performance with respect to safety-critical aspects. In 2010, a group within the AAPM set out to develop such a tool for radiation oncology. The result was the "Safety Profile Assessment," an online tool designed to provide an overall metric for performance in safety-critical areas based on recommendations for best practices from various definitive sources.[53] Launched in 2013 and provided free to the community, the tool consists of 92 indicator questions covering four broad topic areas in quality and safety. In 2015, results were reported from the first 114 respondents.[54] It was found that some indicators had strong compliance across practices, but others were more highly variable. As more data come in from practices across the world this database may represent a valuable source of information to guide quality and safety improvement.

8.4 VALIDATION OF QUALITY AND SAFETY MEASURES

The sections above outline the various quality and safety-related data and methods for analysis assisted by machine learning algorithms (e.g., the automated chart review tool described in Section 8.3.2.1). It is not enough to propose methods or measures, however. These methods or measures must be validated. This is one of the "Four Vs" commonly referred to in big data: volume, variety, veracity, and velocity. The "veracity" is what might be called validation, and it refers to the uncertainties and the "trustworthiness" of the information presented.

An illustration of the potential pitfalls in validation can be found in the experience with QA tests of intensity modulated radiation therapy (IMRT) treatments. Over the previous decade, numerous tests have been developed for verifying the quality of IMRT delivery.[55,56] However, validation was lacking for these tests. It was later realized that these tests have severe limitations; they are not a reliable predictor of plan quality within an institution[57,58] nor do they predict for plan quality as measured by external phantom-based audits from the IROC-H group.[59] The result is that there is now a wealth of data on plan quality whose meaning is questionable and the field is struggling to understand what the next steps might be to restructure the way QA is performed. This experience underscores the need for validation.

Recently, Carlone et al.[60] have pioneered an approach to measuring the performance of quality assurance measurements that may be useful in the context of big data. They introduced quality defects (or errors) into a simulated radiotherapy plan and measure the ability of the system to detect these defects using a receiver operating characteristic (ROC) curve methodology. This approach is widely used in radiology (e.g., Gong et al.,[61] Kalender et al.,[62] and Krupinski and Jiang[63]). Other studies have appeared in radiation oncology, for example an assessment of the detectability of various different types of errors using a new Electronic Portal Imaging Device (EPID)-based dosimetry measurement system.[64] This type of validation is particularly relevant to quality and safety algorithms and data sets because the goal is often to identify outliers.

In the era of "precision medicine," a conflict may arise between decisions to treat an individual patient's disease and the recommended approach, for example, ASTRO clinical practice statements or National Cancer Care Network (NCCN) guidelines. The variety between treatments may reflect actual differences in patients' values and disease states or it may be a result of individual practitioners' (or their institutions') preconceived ideas regarding treatment strategies. In general, the former is considered a positive approach, whereas the latter may result in substandard treatments due to historical, educational, and training prejudices.

8.5 CONCLUSIONS: QUALITY AND SAFETY DATA IN RADIATION ONCOLOGY

A wide variety of quality and safety data has been discussed here. Traditional metrics such as radiotherapy plan quality or patient outcomes play a central role, but the data sources extend well beyond these to include error reports, human performance metrics, and even measures of culture. Thus the inclusion of quality and safety data is supportive of the "variety" dimension of the commonly cited "four Vs" of big data (volume, variety, veracity, and velocity). Using this data, new algorithms are being developed to detect quality and safety problems in radiotherapy treatments, and the results appear to be very promising. At the same time, new validation approaches are being employed that may in the future prevent deficient

QA procedures that have plagued the field in recent years. There are many challenges ahead, not least of which is informatics standards for quality-related data that are only now beginning to emerge. It is clear, though, that as in all realms of human enterprise big data and associated techniques will play a major role in the future. It will be vital to explicitly consider quality and safety data as part of this, for without quality we have nothing.

REFERENCES

1. Pronovost P, Needham D, Berenholtz S et al. An intervention to decrease catheter-related bloodstream infections in the ICU. *N Engl J Med.* 2006;355(26):2725–2732.
2. Haynes AB, Weiser TG, Berry WR et al. A surgical safety checklist to reduce morbidity and mortality in a global population. *N Engl J Med.* 2009;360(5):491–499.
3. AQ. Quantitative Analysis of Normal Tissue Effects in the Clinic (QUANTEC). 2010; http://aapm.org/pubs/QUANTEC.asp.
4. Grimm J. Outcomes of hypofractionated treatements: Results of the WGSBRT. *Med Phys.* 2015;42(6):3685.
5. Moore KL, Schmidt R, Moiseenko V et al. Quantifying unnecessary normal tissue complication risks due to suboptimal planning: A Secondary study of RTOG 0126. *Int J Radiat Oncol Biol Phys.* 2015;92(2):228–235.
6. Wu B, Ricchetti F, Sanguineti G et al. Patient geometry-driven information retrieval for IMRT treatment plan quality control. *Med Phys.* 2009;36(12):5497–5505.
7. Abrams RA, Winter KA, Regine WF et al. Failure to adhere to protocol specified radiation therapy guidelines was associated with decreased survival in RTOG 9704—A phase III trial of adjuvant chemotherapy and chemoradiotherapy for patients with resected adenocarcinoma of the pancreas. *Int J Radiat Oncol Biol Phys.* 2012;82(2):809–816.
8. Fairchild A, Straube W, Laurie F, Followill D. Does quality of radiation therapy predict outcomes of multicenter cooperative group trials? A literature review. *Int J Radiat Oncol Biol Phys.* 2013;87(2):246–260.
9. Ohri N, Shen X, Dicker AP, Doyle LA, Harrison AS, Showalter TN. Radiotherapy protocol deviations and clinical outcomes: A meta-analysis of cooperative group clinical trials. *J Natl Cancer Inst.* 2013;105(6):387–393.
10. Brunskill K, Nguyen TK, Boldt RG et al. Does peer review of radiation plans affect clinical care? A systematic review of the literature. *Int J Radiat Oncol Biol Phys.* 2017;97(1):27–34.
11. Matuszak MM, Hadley SW, Feng M et al. Enhancing safety and quality through preplanning peer review for patients undergoing stereotactic body radiation therapy. *Pract Radiat Oncol.* 2016;6(2):e39–e46.
12. Das IJ, Cheng CW, Chopra KL, Mitra RK, Srivastava SP, Glatstein E. Intensity-modulated radiation therapy dose prescription, recording, and delivery: Patterns of variability among institutions and treatment planning systems. *J Natl Cancer Inst.* 2008;100(5):300–307.
13. Gawande A. The Checklist Manifesto. 2013; http://gawande.com/the-checklist-manifesto.
14. Pronovost PJ, Vohr E. *Safe Patients, Smart Hospitals: How One Doctor's Checklist Can Help Us Change Health Care from the Inside Out.* Hudson Street Press; 2010.
15. Dixon-Woods M, Bosk CL, Aveling EL, Goeschel CA, Pronovost PJ. Explaining Michigan: Developing an ex post theory of a quality improvement program. *Milbank Q.* 2011;89(2):167–205.
16. Burlison JD, Quillivan RR, Kath LM et al. A multilevel analysis of U.S. hospital patient safety culture relationships with perceptions of voluntary event reporting. *J Patient Saf.* 2016.
17. Potters L, Ford E, Evans S, Pawlicki T, Mutic S. A systems approach using big data to improve safety and quality in radiation oncology. *Int J Radiat Oncol Biol Phys.* 2016;95(3):885–889.
18. Benedict SH, Hoffman K, Martel MK et al. Overview of the American Society for Radiation Oncology–National Institutes of Health–American Association of Physicists in Medicine Workshop 2015: Exploring opportunities for radiation oncology in the era of big data. *Int J Radiat Oncol Biol Phys.* 2016;95(3):873–879.
19. Evans SB, Fraass BA, Berner P et al. Standardizing dose prescriptions: An ASTRO white paper. *Pract Radiat Oncol.* 2016;6(6):e369–e381.
20. Mayo C, Moran JM, Xiao Y et al. AAPM Task Group 263: Tackling standardization of nomenclature for radiation therapy. *Int J Radiat Oncol Biol Phys.* 2015;93(3):E383–E384.
21. ASTRO. Medicare Access and CHIP Reauthorization Act of 2015 (MACRA). 2016; www.astro.org/macra.
22. Mayo CS, Kessler ML, Eisbruch A et al. The big data effort in radiation oncology: Data mining or data farming? *Adv Radiat Oncol.* 2016;1(4):260–271.
23. Donabedian A. The quality of care—How can it be assessed? *J Am Med Assoc.* 1988;260(12):1743–1748.
24. Aerts HJWL, Velazquez ER, Leijenaar RTH et al. Decoding tumour phenotype by noninvasive imaging using a quantitative radiomics approach. *Nat Commun.* 2014;5:4006.
25. Robertson SP, Quon H, Kiess AP et al. A data-mining framework for large scale analysis of dose-outcome relationships in a database of irradiated head and neck cancer patients. *Med Phys.* 2015;42(7):4329–4337.

26. Molineu A, Hernandez N, Nguyen T, Ibbott G, Followill D. Credentialing results from IMRT irradiations of an anthropomorphic head and neck phantom. *Med Phys.* 2013;40(2):022101.
27. Zink S. The promise of a new technology—Knowledge-based systems in radiation oncology and diagnostic-radiology. *Comput Med Imag Grap.* 1989;13(3):281–293.
28. Marks LB, Adams RD, Pawlicki T et al. Enhancing the role of case-oriented peer review to improve quality and safety in radiation oncology: Executive summary. *Pract Radiat Oncol.* 2013;3(3):149–156.
29. Ford EC, Terezakis S, Souranis A, Harris K, Gay H, Mutic S. Quality control quantification (QCQ): A tool to measure the value of quality control checks in radiation oncology. *Int J Radiat Oncol Biol Phys.* 2012;84(3):e263–e269.
30. Gopan O, Zeng J, Novak A, Nyflot M, Ford E. The effectiveness of pretreatment physics plan review for detecting errors in radiation therapy. *Med Phys.* 2016;43(9):5181.
31. Yang D, Moore KL. Automated radiotherapy treatment plan integrity verification. *Med Phys.* 2012;39(3):1542–1551.
32. Kalet AM, Gennari JH, Ford EC, Phillips MH. Bayesian network models for error detection in radiotherapy plans. *Phys Med Biol.* 2015;60(7):2735–2749.
33. Ibrahim A, Bucur A, Dekker A et al. Analysis of the suitability of existing medical ontologies for building a scalable semantic interoperability solution supporting multi-site collaboration in oncology. *Paper presented at IEEE Bioinformatics and Bioengineering (BIBE)* 2014.
34. Kalet A, Doctor JN, Gennari JH, Phillips MH. Developing Bayesian networks from a dependency-layered ontology: A proof-of-concept in radiation oncology. *Med Phys.* 2017;44(8):4350–4359.
35. World Health Organization (WHO). *WHO Draft Guidelines for Adverse Event Reporting and Learning Systems: From Information to Action.* Geneva, Switzerland: World Health Organization Document Production Services; 2005.
36. Agency TNPS. *Seven Steps to Patient Safety.* London, UK; 2004.
37. Donaldson L. *Radiotherapy Risk Profile: Technical Manual.* Geneva, Switzerland: World Health Organization; 2008.
38. Ortiz Lopez P, Cossett JM, Dunscombe P et al. *ICRP Report No. 112: Preventing Accidental Exposures from New External Beam Radiation Therapy Technologies.* ICRP; 2009.
39. Zeitman A, Palta J, Steinberg M. Safety is no accident: A framework for quality radiation oncology and care. *Am Soc Radiat Oncol.* 2012.
40. Arnold A, Delaney GP, Cassapi L, Barton M. The use of categorized time-trend reporting of radiation oncology incidents a proactive analytical approach to improving quality and safety over time. *Int J Radiat Oncol Biol Phys.* 2010;78(5):1548–1554.
41. Mutic S, Brame RS, Oddiraju S et al. Event (error and near-miss) reporting and learning system for process improvement in radiation oncology. *Med Phys.* 2010;37(9):5027–5036.
42. Ford EC, Fong de Los Santos L, Pawlicki T, Sutlief S, Dunscombe P. Consensus recommendations for incident learning database structures in radiation oncology. *Med Phys.* 2012;39(12):7272–7290.
43. IAEA. Safety Reporting and Learning System for Radiotherapy (SAFRON). 2017; https://rpop.iaea.org/RPOP/RPoP/Modules/login/safron-register.htm.
44. ASTRO. Radiation Oncology Incident Learning System (RO-ILS). 2016; www.astro.org/roils.
45. Mardon RE, Khanna K, Sorra J, Dyer N, Famolaro T. Exploring relationships between hospital patient safety culture and adverse events. *J Patient Saf.* 2010;6(4):226–232.
46. Kusano AS, Nyflot MJ, Zeng J et al. Measurable improvement in patient safety culture: A departmental experience with incident learning. *Pract Radiat Oncol.* 2015;5(3):e229–e237.
47. Mazur L, Chera B, Mosaly P et al. The association between event learning and continuous quality improvement programs and culture of patient safety. *Pract Radiat Oncol.* 2015;5(5):286–294.
48. Woodhouse KD, Volz E, Bellerive M et al. The implementation and assessment of a quality and safety culture education program in a large radiation oncology department. *Pract Radiat Oncol.* 2016;6(4):e127–e134.
49. Krupinski EA, Kallergi M. Choosing a radiology workstation: Technical and clinical considerations. *Radiology.* 2007;242(3):671–682.
50. Mazur LM. TU-EF-BRD-03: Mental Workload and Performance. AAPM; 2015.
51. Mazur LM, Mosaly PR, Moore C et al. Toward a better understanding of task demands, workload, and performance during physician-computer interactions. *J Am Med Inform Assoc JAMIA.* 2016;23(6):1113–1120.
52. Chan AJ, Islam MK, Rosewall T, Jaffray DA, Easty AC, Cafazzo JA. The use of human factors methods to identify and mitigate safety issues in radiation therapy. *Radiother Oncol.* 2010;97(3):596–600.
53. Dunscombe P, Brown D, Donaldson H et al. Safety profile assessment: An online tool to gauge safety-critical performance in radiation oncology. *Pract Radiat Oncol.* 2015;5(2):127–134.
54. Ford EC, Brown D, Donaldson H et al. Patterns of practice for safety-critical processes in radiation oncology in the United States from the AAPM safety profile assessment survey. *Pract Radiat Oncol.* 2015;5(5):e423–e429.

55. Ezzell GA, Burmeister JW, Dogan N et al. IMRT commissioning: multiple institution planning and dosimetry comparisons, a report from AAPM Task Group 119. *Med Phys.* 2009;36(11):5359–5373.
56. Low DA, Moran JM, Dempsey JF, Dong L, Oldham M. Dosimetry tools and techniques for IMRT. *Med Phys.* 2011;38(3):1313–1338.
57. Kruse JJ. On the insensitivity of single field planar dosimetry to IMRT inaccuracies. *Med Phys.* 2010;37(6):2516–2524.
58. Nelms BE, Zhen HM, Tome WA. Per-beam, planar IMRT QA passing rates do not predict clinically relevant patient dose errors. *Med Phys.* 2011;38(2):1037–1044.
59. Kry SF, Molineu A, Kerns JR et al. Institutional patient-specific intensity-modulated radiation therapy quality assurance does not predict unacceptable plan delivery as measured by IROC Houston's head and neck phantom. *Int J Radiat Oncol Biol Phys.* 2014;90(5):1195–1201.
60. Carlone M, Cruje C, Rangel A, McCabe R, Nielsen M, Macpherson M. ROC analysis in patient specific quality assurance. *Med Phys.* 2013;40(4):042103.
61. Gong X, Glick SJ, Liu B, Vedula AA, Thacker S. A computer simulation study comparing lesion detection accuracy with digital mammography, breast tomosynthesis, and cone-beam CT breast imaging. *Med Phys.* 2006;33(4):1041–1052.
62. Kalender WA, Polacin A, Suss C. A comparison of conventional and spiral CT: An experimental-study on the detection of spherical lesions. *J Comput Assist Tomogr.* 1994;18(4):671–671.
63. Krupinski EA, Jiang Y. Anniversary paper: Evaluation of medical imaging systems. *Med Phys.* 2008;35(2):645–659.
64. Bojechko C, Ford EC. Quantifying the performance of in vivo portal dosimetry in detecting four types of treatment parameter variations. *Med Phys.* 2015;42(12):6912–6918.

9 Tracking organ doses for patient safety in radiation therapy

Wazir Muhammad, Ying Liang, Gregory R. Hart, Bradley J. Nartowt, David A. Roffman, and Jun Deng

Contents

9.1 Introduction 123
9.2 Various Contributions to Organ Doses 124
9.3 Personal Organ Dose Archive 127
    9.3.1 What Is PODA? 127
    9.3.2 Monte Carlo Dose Calculations 129
    9.3.3 A GPU-Based MC Engine for Super-Fast Dose Calculations 129
        9.3.3.1 GPU Basics 129
        9.3.3.2 GPU-Based MC Dose Engine 130
        9.3.3.3 GPU-Based Multiple Source Modeling for MC Simulations 131
        9.3.3.4 GPU-Based Patient-Specific MC Dose Calculations of Protons 132
        9.3.3.5 GPU-Based Patient-Specific MC Dose Calculations of Brachytherapy Sources 133
        9.3.3.6 GPU-Based Patient-Specific MC Dose Calculations of kV Photons 133
        9.3.3.7 GPU-Based Patient-Specific MC Dose Calculations of Special Procedures 133
    9.3.4 Dose Mapping with Deformable Image Registration 134
    9.3.5 A Self-Contained Database for Automatic Dose Accumulation 135
    9.3.6 Dose Report and Morbidity Analysis 136
    9.3.7 Early Warnings and Intervention 137
9.4 Normal Tissue Tolerances for Different Treatment Schema 137
9.5 Clinical Applications of the PODA for Patient Safety in Radiotherapy 139
    9.5.1 Organ Protection and Radiation Treatment Intervention 139
    9.5.2 Cancer Risk Monitoring and Control 139
    9.5.3 Integration into a Personal Health Data Archive 140
    9.5.4 Normal Tissue Tolerance Benchmark and Revision 140
9.6 Conclusion and Outlook 140
References 141

9.1 INTRODUCTION

Ionizing radiation kills tumors but may also damage normal tissues. The primary goal of radiation therapy is to deliver a lethal dose to the tumor tissues while sparing normal tissues. Efforts have been constantly made by the scientific community to maximize the therapeutic roles of radiation and to minimize its detrimental effects ever since the beginning of radiation oncology. Technology improvements in imaging and beam delivery have been changing the ways of cancer diagnosis and treatment in the clinic. The past two decades have witnessed the transition of modern radiotherapy from three-dimensional conformal radiation therapy (3D-CRT) to more advanced and complicated techniques such as intensity-modulated radiotherapy (IMRT), image-guided radiotherapy (IGRT), stereotactic body radiation therapy (SBRT), and volumetric modulated arc therapy (VMAT) [1–5]. Advanced beam delivery techniques have greatly improved dose conformity, making dose escalation possible for localized tumors. With the use of higher-than-conventional radiation doses, local and regional tumor control has been significantly improved over the years, although increased cases of late toxicity in normal tissues have been reported [6,7].

The practice of radiation oncology over the past 100 years has greatly advanced our understanding of the dosimetric criteria of normal tissue tolerance. However, it is still far from perfection. Pioneering attempts to address normal tissue toxicity to radiation were carried out by Rubin and Casarett [8]. Their work provided a foundation for the study of the late effects of radiation therapy and served as a raw reference for radiation oncologists. With the advent of 3D-CRT and 3D treatment planning in the 1980s, concurrent clinical needs for accurate and comprehensive knowledge of dose–time–volume relationships for normal tissues have become apparent. In 1991, Emami and colleagues published a landmark article that summarized the tolerance dose (TD) in terms of TD 5/5 (5% complication probability within 5 years from treatment) and TD 50/5 (50% complication probability within 5 years) of three volume categories (one third, two thirds, and the whole organ) for 28 critical sites of normal tissues based on literature search and their clinical experience [9].

Modern radiotherapy techniques deliver near-uniform doses to the target volume while dose distributions in the surrounding normal tissues are usually nonuniform and variable. This provides treatment planners with extra flexibility in determining which regions of, and to what extent, normal tissues will be irradiated. Accordingly, there is a pressing need for more accurate information on normal tissue tolerance and models to predict the risk of normal tissue injury to complete the 3D treatment plan that optimizes the therapeutic ratio. It was for this reason that the Quantitative Analyses of Normal Tissue Effects in the Clinic (QUANTEC) Steering Committee was formed with a long-standing interest in dose–volume modeling. Reviews from the QUANTEC group have updated the available dose–volume–outcome data and normal tissue complication probability (NTCP) models since the seminal paper by Emami [10].

In order to predict NTCP and optimize a treatment plan, one needs accurate and comprehensive information regarding organ dose. Although modern treatment planning systems (TPS) can calculate 3D dose distributions in patient anatomy for any given treatment plans, the actual delivered dose in an individual patient after fractionated radiotherapy is not well recorded in the current clinical procedures. This issue is of great clinical importance and has been listed as one of the research priorities in the QUANTEC group report [10]. Organ doses calculated by TPS may deviate from the actual delivered doses for the following four major reasons: (1) leakage and scatter doses associated with advanced beam delivery techniques such as IMRT and VMRT are often not accurately simulated by commercial TPS [11,12]; (2) TPS dose calculations are only performed for contoured organs within a patient's anatomy volume of interest while providing no dose information for non-contoured organs; (3) changes in organ volume, shape and location during the course of treatment may result in variations in the planned organ dose [13,14]; (4) kilo-voltage imaging doses, which can be considerable in some IGRT cases, are not included into the total dose accumulation due to inability of current commercial TPS in simulating kilo-voltage X-rays [15–17]. For these reasons, some patients may have accumulated dangerously high doses in some radiosensitive organs after treatment while the clinicians and physicists might be completely unaware of it. Notably, many recent studies have shown that non-negligible second cancer risks are associated with increased organ doses from scatter and leakage radiations that are not correctly accounted for by commercial TPS [11,18–24].

9.2 VARIOUS CONTRIBUTIONS TO ORGAN DOSES

Leakage through the accelerator head and multi-leaf collimator (MLC) and scattering of high energy photons within the patient and treatment room are thought to increase organ dose. Leakage and scatter dose is a function of several factors, including distance from the field edge, field size, photon energy, and beam-on time. There have been concerns that leakage and scatter dose would not be negligible for highly conformal radiotherapy techniques, such as IMRT, where a large number of monitor units (MU) are used. Athar et al. simulated the out-of-field photon doses to various organs in 6 MV IMRT in the head and neck and spine region for adult and pediatric patients. The reported out-of-field doses to organs near the field edge were in the range of 2, 5, and 10 mSv Gy^{-1} for 3-, 6-, and 9-cm diameter IMRT fields, respectively [25].

Simulation results also reveal a general decrease of out-of-field doses with increasing patient age, suggesting an increased risk of late effects ascribable to leakage and scatter for pediatric patients. Sharma and coworkers measured the peripheral dose for 6 MV X-rays using a 0.6-cm³ ionization chamber inserted into a plastic water phantom at 5-cm depth for different field sizes and strip field widths [26]. Under static MLC mode with 6 × 6 cm field size, peripheral doses (normalized to 100% on the central axis at depth of maximum dose of static open field) were 0.8% and 0.2% at 5 and 20 cm from the field edge. When shifted to dynamic MLC mode with the same field size, peripheral doses increased to 0.8%–1.4% and 0.3%–1.2% depending on different strip field widths at 5 and 20 cm. Strikingly, when field size was increased to 14 × 14 cm in dynamic MLC mode, peripheral doses climbed up to 3.2%–5.6% and 1%–2.3% with the same setup of strip field widths and distances.

The accurate delineation of the organ at risk (OAR) is crucial for NTCP estimation and treatment plan optimization. However, in practice, OAR delineation could be subjective. Although the Contouring Atlas project led by the Radiation Therapy Oncology Group (RTOG) has been trying to provide consensus and guidelines for OAR delineation, variation still exists at least in some cases [13]. The non-contoured organs and those outside of a patient's anatomy volume of interest are not taken into consideration by TPS dose calculation. Hence, radiation risks of these organs are underestimated. Shown in Figure 9.1 is a clinical treatment plan for a patient with a cranial-spinal lesion treated by VMAT with a total dose of 23.4 Gy in 13 fractions. As expected, a highly conformal dose distribution is around the tumor volume with sparing of adjacent normal tissues that are contoured. However, because the heart and liver are not contoured, all the 3D dose information of these two radiosensitive organs is missing. Essentially, the mean doses and dose ranges of 3.7 (1.8–7.9) Gy and 4.7 (1.7–20.4) Gy to the heart and liver are totally ignored, which is dangerous if future planning fails to include this part of dose burden into clinical consideration.

Over the course of radiation therapy, geometric variation in the patient's anatomy may occur due to factors such as tumor response, edema, weight loss, and posture change [14]. It is recognized that anatomical changes will lead to variation in the planned dose, which may undermine the therapeutic ratio of the treatment and bring about unexpected normal tissue toxicity. To quantify the differences between the planned and delivered doses in the target and peripheral normal tissues in IMRT, O'Daniel et al. performed in-house deformable image registration to map dose distributions to the original treatment plan and to calculate the delivered doses for 11 patients with head and neck cancer [27]. Using conventional radiopaque markers alignment method, they found mean doses delivered to the parotid gland were 5–7 Gy higher than the planned dose in 45% of the patients (median, 1.0 Gy contralateral, $p = 0.016$; median, 3.0 Gy ipsilateral, $p = 0.026$). Simulation using an imaginary bone alignment method predicted a reduction of parotid doses in most of the patients. However, it was still significantly higher than the planned doses (median, 1.0 Gy, $p = 0.007$). George et al. investigated the effects of respiratory motion and setup error during breast IMRT treatment planning [28]. Results showed that respiratory motion increased delivered doses in the lung and heart as well as Planning Target Volume (PTV) dose heterogeneity, and the differences increased with intra-fraction motion.

Currently, organ doses from image-guidance and diagnostic radiological procedures are not included in the TPS, but they can be considerable in some IGRT cases. In a recent study where 4832 cancer patients

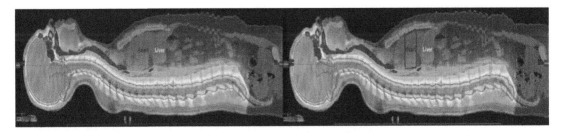

Figure 9.1 Dose distributions in a patient with cranial-spinal lesion when heart and liver are not contoured (left) and contoured (right). Because the heart and liver were not contoured in this case, the mean doses of 3.7 and 4.7 Gy, and the maximum doses of 7.9 and 20.4 Gy to the heart and liver were not accumulated, which is dangerous if future planning fails to include this dose burden into clinical consideration.

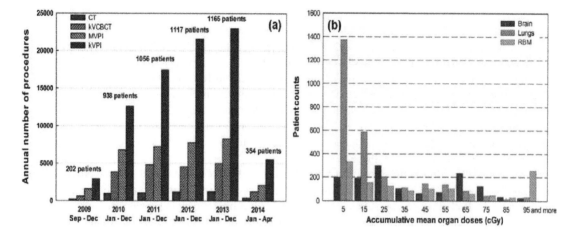

Figure 9.2 Imaging dose accumulation to the brain, lungs, and red bone marrow for 4832 cancer patients who received a total of 142,017 imaging procedures during image-guided radiotherapy at our institution from September 2009 to April 2014. (a) Year-wise number of procedures of the listed imaging techniques while year-wise number of patients are also given (b) Accumulative mean organ doses received by number of patients to their listed organs.

received a total of 142,017 imaging procedures during IGRT in our institution, it was found that the mean cumulative imaging doses and dose ranges to the brain, lungs, and the red bone marrow (RBM) were 38.0 (0.5–177.3) cGy, 18.8 (0.4–246.5) cGy, and 49.1 (0.4–274.4) cGy, respectively (Figure 9.2). While low for most cases, the cumulative imaging doses may be high and further worsen normal tissue toxicity. One early study we did was on the radiation dose to the testes during image-guided IMRT of prostate cancer [15]. As listed in Table 9.1, our Monte Carlo (MC) simulations revealed that the mean dose to the testes was 3.6 cGy, which was 400% more than the 0.7 cGy estimated by TPS. The cumulative doses to the testes may be clinically significant because impairment of spermatogenesis occurs at doses as low as 10 cGy. Our recent MC study on the kilovoltage (kV) imaging doses in the radiotherapy of pediatric cancer patients indicated that kV cone beam CT (CBCT) deposits much larger doses to critical structures in children than adults, usually by a factor of 2–3 [16,17]. On average, one kVCBCT scan operated at 125 kV in half-fan mode would deposit 2.9, 4.7, 7.7, 10.5, 8.8, 7.6, 7.7, 7.8, and 7.2 cGy to testes, liver, kidneys, femoral heads, spinal cord, brain, eyes, lens, and optical nerves, respectively. In addition, imaging doses from kVCBCT were highly size dependent, and decreased monotonically with increasing patient size (Figure 9.3). It is,

Table 9.1 Mean absorbed doses to the organs compared between whole pelvis irradiation (WPI), prostate only intensity-modulated radiation therapy (PO-IMRT), and IGRT, that is, PO-IMRT plus daily kilo-voltage cone beam computed tomography (kVCBCT) in half-fan pelvis protocol

ORGANS	WPI	PO-IMRT	IGRT (PO-IMRT + KVCBCT HALF-FAN PELVIS PROTOCOL)			
	10 MV	10 MV	10 MV + 60 KV	10 MV + 80 KV	10 MV + 100 KV	10 MV + 125 KV
Prostate	181.9	203.3	203.7	204.3	205.1	206.7
Rectum	169.0	117.3	117.8	118.5	119.4	121.1
Bladder	183.2	126.4	127.1	127.9	128.8	130.5
Testes	1.7	0.7	1.8	2.2	2.7	3.6
Left femoral head	111.0	69.1	69.9	71.1	72.4	74.9
Right femoral head	110.9	67.1	67.9	69.0	70.3	72.7

Note: All the doses in the table are in cGy per fraction.

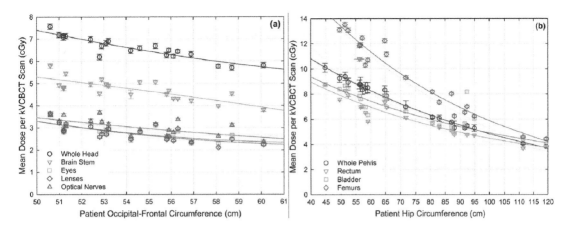

Figure 9.3 The kVCBCT-contributed mean doses in critical structures of the (a) head and (b) pelvis decreased monotonically with the increase in patient size.

therefore, essential to include the radiation doses from imaging procedures into treatment planning and choose an appropriate scanning protocol when kVCBCT is routinely applied to pediatric patients.

9.3 PERSONAL ORGAN DOSE ARCHIVE

In the previous sections, we have discussed four major factors that may lead to variations in the accumulated organ doses. In this section, we will introduce a big data approach termed personal organ dose archive (PODA) to track the organ organs for individual patients in order to improve patient safety in radiation therapy.

9.3.1 WHAT IS PODA?

PODA is a methodology that combines big data and radiation oncology to improve safety of individual patients undergoing radiation therapy through daily 3D organ dose tracking and mapping. From the discussions in the previous sections, it is clear that the delivered doses may be different from the planned doses. However, all the information we need to calculate the delivered doses is readily available. Specifically, we have treatment head mechanical structure to simulate leakage dose. We can access treatment parameters (beam energy, gantry, collimator, and couch angles, isocenter location, and field sizes) from TPS and MLC sequence and MU through the Dynalog file to account for leakage and scattered dose from MLC and within a patient's anatomy. Dose mapping can be performed using volumetric imaging such as kVCBCT on the treatment day to compensate for position error and geometrical changes in patient's anatomy. Parameters related to image guidance, including tube voltage, scan mode, and current, can be extracted to calculate imaging doses to the organs. Using the big data generated and recorded during treatment, we are able to calculate the delivered 3D organ doses on our MC engine and monitor the accumulated organ doses to improve the safety of the patients. Moreover, in combination with the patient's electronic medical record and other biomarkers and genomic information, we may be able to make a personalized treatment plan, analyze, and predict radiation risk at an individual level based on an artificial neural network algorithm that we have developed. Essentially, PODA takes us one step closer to precision radiation oncology.

The PODA system includes four major parts: (1) a Digital Imaging and Communications in Medicine (DICOM) library, (2) a database, (3) a graphics processing unit (GPU)–based MC dose engine, and (4) dose analysis and early warning. Figure 9.4 shows the schematic drawing of the PODA and the relationship between different parts. The DICOM library hierarchically stores all the raw data for each individual patient, including, for example, CT images, CT voxel phantom, radiotherapy (RT) plan, RT dose, and Dynalog files. The PODA database organizes and manages all the patients' data, including patients' general information, radiation events, organ doses, warning level, and so on. A GPU-based dose engine computes organ doses for each event involving ionizing radiation. The dose analysis and early warning subsystem

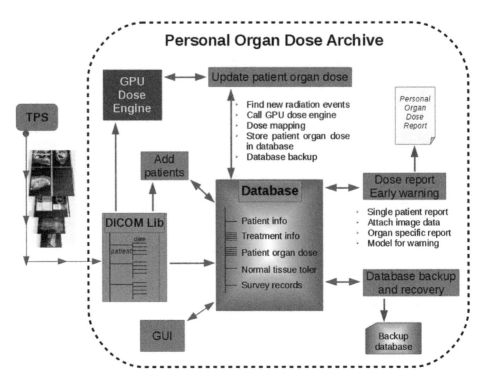

Figure 9.4 Schematic drawing of PODA and the relationship between different components.

includes modules such as Update Organ Dose, Report & Alert and Database Backup & Recovery. These modules help to connect all parts of the PODA and make the PODA workflow (Figure 9.5) run efficiently and smartly. Basically, the PODA works like this. First, at the end of each treatment day, data related to all kinds of radiation events are exported from the TPS and stored in the DICOM library. Second, the Update Organ Dose module scans the DICOM library to find new radiation events and calls the GPU-based MC

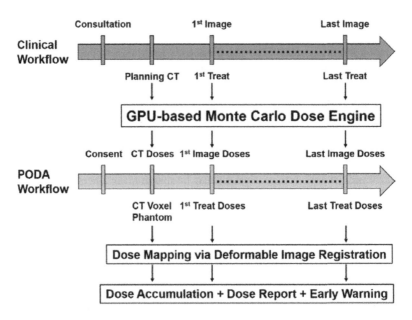

Figure 9.5 A schematic showing a regular clinical workflow in radiation therapy clinic and a PODA workflow where organ doses from each clinical event involving ionizing radiation are simulated, mapped, accumulated, and reported.

dose engine to perform the simulation. The GPU dose engine returns the calculated 3D dose distribution in the patient's anatomy. Event-specific organ doses and cumulative organ doses are extracted from the 3D dose and stored in the database. Dose mapping will be performed if necessary. This process repeats until all the new radiation events for all the patients are handled. Third, risk analysis will be performed based on organ doses accumulated during the previous treatments. Depending on risk level, a dose report and early warning will be issued if the accumulated organ dose will exceed the preset dose criteria by prediction.

9.3.2 MONTE CARLO DOSE CALCULATIONS

The accuracy in the prescribed radiation dose to the cancer patients are assessed by studying the dose distribution prior to the dose delivery. The accuracy in the dose delivery to the tumor while minimizing dose to the surrounding healthy tissues depends on the dose calculation methodology. The statistics shows that the tumor control probability is decreased by more than 15% with a 5% underdosage in the dose delivery to the tumor [29]. On the other hand, an increase of 5% in the dose delivery can result in intolerable side effects in the surrounding healthy tissues [29]. Among the dose calculation methods, MC simulation of radiation transport is considered the most accurate one.

The MC method for radiation transport is a numerical solution to a problem that is based on random number sampling combined with statistical methods. Through this method, one can predict many quantities of interest that are difficult to measure physically [29]. A number of general-purpose MC-based computer codes are currently available for the radiation transport (i.e., MCNP, EGSnrc, PENELOPE, GEANT4). In radiotherapy, the MC method has the capability to accurately describe the physical processes and to efficiently handle the complicated geometries. However, due to its stochastic nature, the simulation of a large number of particle histories is required to achieve a desired statistical accuracy.

The MC simulation for dose calculation is extensively used in radiation therapy owing to its well-accepted accuracy. However, computational efficiency has been a main concern because a large number of particle histories are required for achieving the desired level of precision for dose calculation, limiting its applications in the clinical environment [30]. Despite huge advances in computer architecture in recent years, the currently available MC engines for radiation dose calculation are still not efficient enough for routine clinical applications in radiotherapy. This issue has been somewhat mitigated through parallel computing. Through this technique, a significant computational speedup factor can be achieved. Extensive amount of research has been carried out regarding the implementation of various MC dose calculation packages on a variety of computing platforms. These include multicore CPU, CPU cluster, GPU, and Cloud computing [31]. Currently, the CPU cluster is the main platform for parallel dose calculations. In this parallel computation, task and data parallelization approach is applied. Initially, all available MC dose calculation packages such as MCNP, EGSnrc, and PENELOPE and the dose planning model (DPM) were developed on a single CPU platform. However, later on these packages were successfully transferred to CPU clusters by applying parallel processing interfaces such as message passing interface (MPI) to produce a higher computational efficiency [31]. The idea of parallel computing is achieved in such a way that the simulated particle histories are equally divided or distributed to all the processors. These histories are then executed simultaneously without interfering with each other, and finally the dose distributions are summed from all processors. Cloud computing is another technology that has been employed as a parallel processing for MC dose computing. In this technology, available parallel MC packages developed for CPU clusters are launched on the new Cloud platform provided by a third party (such as Amazon). The new technology is not much different from CPU cluster-based dose computing because here the virtual platform is essentially a CPU cluster. However, due to the initializing of the remote virtual cluster and data transfer, the efficiency is slightly improved [31].

9.3.3 A GPU-BASED MC ENGINE FOR SUPER-FAST DOSE CALCULATIONS

9.3.3.1 GPU basics

The GPU was designed to accelerate the processing of graphics information. It was a specialized hardware to write frame buffer that will ultimately be displayed on the screen. Originally, it was designed for computer gaming. Initially, these had a fixed pipeline with no control left for the programmers [31]. In 2001, the direct programming access to the parts of the hardware resources, rather than through a fixed pipeline, was introduced with the NV20 (from NVIDIA; Santa Clara, CA) and the R200 (from ATI Technologies now AMD; Sunnyvale, CA). With this access, the programmers could write and execute on the vertex and fragment processors. However, limited scientific computing with challenges was possible in this generation of GPUs. In 2006, NV80 and R600 were introduced which had a "unified shader architecture." This newly introduced feature placed vertex and fragment shaders into a single processor. This new feature was capable of handling both tasks and, soon after its introduction, computing programing languages/streams such as CUDA (by NVIDIA) and SDK (by AMD, formerly ATI) were also introduced. With this feature, a GPU could be directly accessed by application programming interfaces (API) through C-language, without needing computer graphics context (e.g., OpenGL or DirectX) [31].

A GPU has many processing elements termed CUDA cores, which are further split into several computing units. It follows a single instruction, multi data (SIMD) design. Together, these computing units are known as a streaming multiprocessor. The scalability in the number of streaming multiprocessors per device (i.e., the GPU, or the GPU and its graphics card) is achieved through this architecture. The system on which the device is installed is called the "host" [31].

In general, various memory levels are presented on a GPU and a graphics card. The streaming multiprocessors have a pool of local memory that is shared across the threads in the work group and has a set of registers (i.e., variable over brands and generations). A device-wide data cache is introduced on newer devices. The cache serves as a hardware-controlled data cache. The graphics card has a pool of global memory that is located outside a GPU. This memory is linked with the system and the GPU memories [31].

9.3.3.2 GPU-based MC dose engine

Recently, GPU drew the attention of relevant computational scientists working in the field of medical physics due to its effective scientific computing capability (i.e., rapid parallel processing capability). Its capability is really tested with graphic cards (e.g., NVIDIA's GeForce and Tesla series) in the field of medical physics and medical image processing [31]. It also includes the development of fast MC-based dose engines on a GPU platform. The aim is to improve the computational efficiency as compared with conventional CPU-based calculations [30]. However, due to the randomness nature of MC simulation and special architecture of the GPU, an inherent conflict exists and, therefore, a linear scalability is hardly achievable on GPU architectures [31].

By utilizing parallel computing architectures of a GPU, numerous advantages can be achieved for the dose calculation in radiotherapy. Computing power similar to that of a CPU cluster can be achieved at much lower cost. Its installation and maintenance is easier as compared with cluster. Moreover, GPU-based MC dose calculations do not need an initialization stage as compared with Cloud computing and the high bandwidth among the CPU and GPU memory, the overhead due to data transfer is usually negligible [31]. The GPU is a relatively new platform as compared with the CPU, and modifications are required to launch already developed MC dose calculation packages. To do so, a GPU-based MC dose engine is to be developed with a programming language usable on the GPU. Currently, the following three GPU-coupled electron–photon MC codes are reported in the literature and are developed for radiotherapy applications: GPUMCD [32], GPU-based DPM (gDPM) [33–35], and GMC [36]. The main challenge included in developing such types of MC dose calculation packages is that the GPU is suitable for data-parallel problems but not for task-parallel problems [31].

High efficiency is expected from the GPU-based platform for data parallelization tasks. However, when launching an MC dose engine on this platform, if the divergence between GPU threads is not treated carefully, the parallel nature of the computation will be lost and will result in poor computational efficiency [31]. To perform GPU-based MC dose computations, two types of GPU thread divergence may be encountered. Performing simultaneous simulations in different manners for the different types of particles (e.g., photons and electrons) will result in divergence between the threads. Many particles are simulated in parallel on the GPU platform and the number of processing elements of the GPU defines the parallelization efficiency. However, all threads in the same warp should follow the same instruction flow; thus, there is always a chance of a high degree of divergence while tracking different type of particles simultaneously. Moreover, a greater degree of divergence is added when all secondary particles are simulated at the same time. These factors would lead to a high degree of efficiency loss on the GPU platform. This type of divergence was successfully handled in an MC-based dose calculation engine, GPUMCD [32] and gDPM v2.0 [34], by separating the simulations of electrons and photons. Considerable acceleration in the simulation time has been achieved. In the second type, thread divergence can be produced even between the same type of particles due to the randomness of the particle interaction with matter during their transport through the medium. However, explicit control is difficult on this type of divergence.

9.3.3.3 GPU-based multiple source modeling for MC simulations

The accuracy of an MC dose engine depends upon how accurately the detailed information regarding characteristics (i.e., energy, angular, and spatial distributions of all particles) of the radiation beams are incorporated into the engine. This information is stored in a file commonly known as a phase–space file. However, these files require a large amount of disk space/memory and CPU time. The alternative to the phase–space approach is beam modeling. To do this, a phase–space file of a full MC simulation can be used to build a multiple-source model that will be a direct input to the dose calculation engine. These multiple-source models are characterizations of photon and electron beams from a radiotherapy treatment unit such as a medical linear accelerator (LINAC), as shown in Figures 9.6 and 9.7. It is evident that particles emitted from different LINACs have different energy, angular, and spatial distributions, whereas the same particles from a single LINAC have similar energy, angular, and spatial distributions. Therefore, each component of the LINAC is considered as a sub-source. A sub-source represents one or more components in the LINAC and has the same energy characteristics and its own energy spectrum and fluence distributions [29]. The advantages of multiple-source models include high efficiency, a significant decrease in particle simulation time, and data storage. However, accurate source modeling is critical [32,34–36].

GPU-based multiple source modeling for MC simulations is relatively new field. Recently, a phase–space-based source model has been developed for gDPM based on a phase–space-let (PSL) concept [34,35,37]. In doing so, the phase–space file was split into small parts. This was done based on energy, type, and location of the particles. The efficiency was considerably improved [38]. However, the PSL-based source model is still less optimal for the GPU platform because loading large amounts of particle data on a CPU memory from a hard drive and manipulating and transferring it to GPU memory take tens of seconds. Although this time is relatively small for CPU-based MC dose simulation, it is a large portion of the total computational time required for GPU-based dose calculations and limits further improvement in the computational efficiency [30]. The data preparation can be avoided by using an analytical source modeling

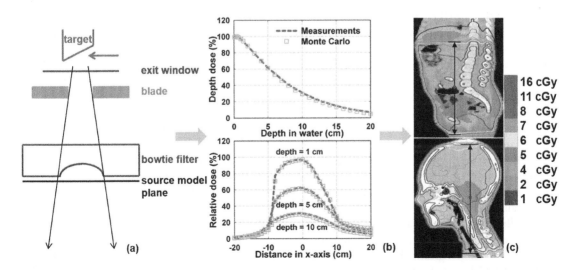

Figure 9.6 Monte Carlo treatment planning of kV photons consists of (a) multiple source modeling, (b) validation of a multiple-source model with measurements, and (c) 3D MC dose calculations in patient anatomy with a multiple-source model.

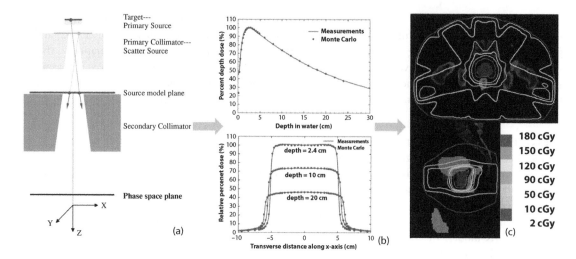

Figure 9.7 Monte Carlo treatment planning of MV photons consists of (a) multiple source modeling, (b) validation of a multiple-source model with measurements, and (c) 3D MC dose calculations in patient anatomy with a multiple-source model.

consisting of multiple beam components. In this type of modeling, analytical functions are designed to express direction distributions, energy, and location of the particles, making sampling tasks more light weighted and hopefully more efficient than phase–space file-based models [30].

The following are the main features of a good analytical source model [30]:

- It represents the actual particle distributions emitted from a clinical unit.
- It should make it easier to derive parameters.
- An easy sampling procedure of source particles should be implemented.
- It should be commissionable to an actual clinical beam with an automated commissioning process.

A thread-divergence may occur due to the running of the sampling processes on different GPU-threads into different sub-sources. This will affect the overall computational efficiency. To improve the efficiency, it is recommended that the source model coordinate well with the sampling processes among different GPU threads [30].

9.3.3.4 GPU-based patient-specific MC dose calculations of protons

Proton beam therapy (PBT) is considered superior due to its ability to accurately deliver radiation dose to the clinical target while sparing the surrounding healthy tissues. PBT can be delivered in uniform scanning, passive scattering, or pencil-beam scanning (PBS) mode. The most advanced is PBS mode because intensity-modulated PBT is possible through this mode [39]. Higher dose conformality is offered by PBT as compared with conventional radiation therapy. However, a poor delivery of the dose and uncertainties in dose calculation may significantly under- or even overdose the tumor and consequently, overdose the surrounding healthy tissues. Consequently, MC simulation can play a key role in proton therapy in calculating accurate dose and a significant reduction in treatment planning margins is expected. Although, pencil-beam–based algorithms are widely used for proton therapy in the clinics for their high computational efficiency, they produce an unsatisfactory dose calculation accuracy for tissues with a large degree of heterogeneity [40]. Currently, MC dose calculations in proton therapy are used for research purpose only by recalculating existing treatment plans because of its low computational efficiency for routine applications for all patients [40].

Regarding the GPU-based PBT, a track-repeating algorithm was developed on a GPU-platform for the proton dose calculations. Furthermore, a simplified GPU-based MC method has been developed and used clinically. These efforts greatly shorten proton dose calculation time. However, the development of a full GPU-based MC dose engine for PBT with satisfactory accuracy and efficiency is still highly desirable and a challenging task. To achieve a balance between accuracy and efficiency, much work is required to derive approximations for the interactions and transport processes (e.g., electromagnetic and nuclear interactions, scattering) for dose calculations that will cause less effect within the healthy human tissues. Another critical issue is the inherent conflict between the SIMD processing scheme of GPU and the statistical nature of an MC method. Due to this conflict, high performance in the GPU-based MC dose calculation engine for PBT is not straightforward to achieve [40].

9.3.3.5 GPU-based patient-specific MC dose calculations of brachytherapy sources

MC simulation is also a valuable tool in brachytherapy dose calculations and dosimetry. The brachytherapy sources, shielding, and applicators can be accurately characterized through this method. The MC simulations are most commonly used to investigate heterogeneity effects in mathematical phantoms [41]. The use of patient geometries constructed from CT images for such studies is limited [41]. The simulation of complex brachytherapy geometries, long calculation time, and laborious preprocessing steps needed to model the brachytherapy applicator and/or seed structures limit the use of MC methods for clinical patient-specific dosimetry [41].

Recently, some studies have reported the use of GPU-based patient-specific MC dose calculations [32,42]. These studies showed a potential reduction in computational times. Hissoiny et al. [32] reported a computational time of 70 ms on average for a single brachytherapy seed simulation by considering only water-based material and 106 particles. However, the simulations were performed for a single low-energy brachytherapy seed in water while neglecting the modeling of Rayleigh scattering and hence cannot be used in a clinical brachytherapy environment.

9.3.3.6 GPU-based patient-specific MC dose calculations of kV photons

The combination of X-rays and CT has played an important role in medical imaging, in diagnostic/preventive screenings, and in radiation therapy since its introduction, and its usage is dramatically increasing with time due to its immense benefits. However, the radiation doses received by the patients during potentially excessive use of X-rays are of serious concern. The MC simulation is considered the most accurate method as stated earlier in this section of this chapter. By utilizing this method, one can not only accurately model a CT system, such as the source location and scanner geometry, but can also accurately describe the interactions between radiation and matter. Many general-purpose MC simulation packages (such as MCNP, PENLELOPE, and EGSnrc) have been reported in numerous research articles to have been used to calculate CT dose [43]. However, the computational efficiency is the main concern here in addition to performing such patient-specific dose calculations in the clinical environment [43]. A fast and accurate MC-CT dose engine will not only be essential for patient-specific dose evaluation but will also facilitate radiotherapy treatment because in many radiotherapy procedures such as IGRT, CBCT and CT scans are performed on a daily basis before each fraction to verify the patient positioning. The doses received as the result of these scans are normally not counted toward the planned treatment dose. These doses may differ considerably from the planed dose received by a patient. The fast and accurate MC dose engine can easily calculate, track, and incorporate these doses into the treatment planning, and the total dose to the target and critical organs can be precisely predicted [43]. Recently, GPU has been increasingly used to speed up several intensive computational tasks in the field of medical physics. GPU-based MC dose calculation packages have been developed for radiotherapy and considerable calculation efficiency has been achieved [32,35]. Similarly, one such limited package for the kV energy range has been developed on a GPU platform. However, the package can be upgraded for dose calculation in CT dose calculations. Later, a kV photon dose simulation package was also made available for assessing the CT dose received by patients [40,43].

9.3.3.7 GPU-based patient-specific MC dose calculations of special procedures

Recently, specialized techniques (either for dose delivery or for target localization) are also used in conjunction with conventional radiotherapy techniques. These specialized techniques include
- Image guided radiotherapy (IGRT)
- Conformal radiotherapy and intensity modulated radiotherapy (IMRT)

- Total body irradiation (TBI) with photons
- Total skin electron irradiation (TSEI)
- Stereotactic irradiation
- Intraoperative radiotherapy (IORT)
- Endorectal irradiation
- Respiratory gated radiotherapy
- Positron emission tomography (PET)/computed tomography (CT) fused images

In these specialized techniques, equipment modifications and special quality assurance programs are required. Due to their complex nature, the precise dose calculation with high computational efficiency is much needed, and the GPU-based MC packages are the best candidate for this task. The previously developed GPU-based MC package (i.e., gDPM) is already used to calculate the plan dose for some of these specialized techniques such as IMRT or a VMAT [31]. These codes will be very useful to utilizing MC techniques for special procedures in the real clinical environment. However, to do so, much effort is required.

9.3.4 DOSE MAPPING WITH DEFORMABLE IMAGE REGISTRATION

Because patient anatomy may change during the treatment, the organ dose simulated on the treatment day may be quite different from those simulated with the planning CTs. We apply a deformable imaging registration (DIR) algorithm to map the 3D dose distribution on the treatment-day CTs back to the original planning CTs. Specifically, we use MIM package version 6.4 (MIM Software Inc., Cleveland, OH), which has been installed in our department for clinical applications in Gross Tumor Volume (GTV) contouring and atlas-based segmentation, as shown in Figure 9.8. For CT-CT registrations, the MIM package uses a constrained intensity-based free-form deformable registration algorithm. The MIM package allows deformable transformation of contours, adaptive recontouring for replanning, propagating contours across time points in gated studies, and monitoring therapy response. In addition, dose maps can be co-registered for dose accumulation and from prior treatments to aid in replanning for tumor recurrences. After fusion, deformed contours need to be reviewed by physicians. Special attention must be paid to co-registered volumes in the regions of particular concern because MIM performance is less robust when excessively noisy CT data or pairs of volumes with inconsistent Hounsfield units are present.

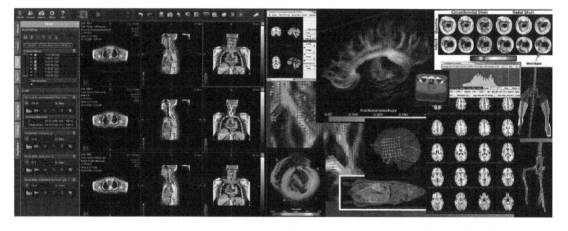

Figure 9.8 Two deformable image registration tools that will be used for 3D dose mapping from treatment-day CTs to reference planning CTs: MIM software package (left) and BioImage Suite (right).

9.3.5 A SELF-CONTAINED DATABASE FOR AUTOMATIC DOSE ACCUMULATION

Tracking organ doses for all patients and for all the radiation events over the entire course of treatment would produce a large amount of data, which would need to be carefully maintained by a database engine for data storage, update, and query. At the current stage, we choose to use the relational database management system SQLite as a demonstration implementation of the PODA database in our institution. The SQLite is a highly efficient, open-source database engine suitable for small-to-medium sized data (<2 TB). Moreover, it is an embedded database engine, which means the database could be contained in the PODA system itself and not need an independent thread running a database. The advantage of SQLite is that it adds convenience during PODA development and deployment, which compensates for its drawback in concurrency.

The design of the database is shown in Figure 9.9. It has a Main Table to manage the general information (e.g., identification number [ID], name, age, sex, weight, race, specific table names) of all the registered patients. The Main Table serves as an entrance to access each patient's specific data, which are stored in independent specific tables. There are three types of patient-specific tables. The Radiation Events Record Table is the largest table and is designed to store information related to all radiation events, including the event type, event date, directory of storage, dose update status, cancer type, RT technique, number of fraction, event-specific organ doses, and cumulative organ doses. Each time a new radiation event is imposed on the patient—regardless of whether it is diagnostic imaging, external beam treatment, or brachytherapy—a new entry will be automatically appended to this table to record the irradiation event. The Medical Survey Table records the patient's survey results. It could be used to make personalized predictions based on our artificial neural network algorithm. It also leaves room for follow-up research of any late effects of radiation. The Alert Level Table stores the accumulated doses, estimated treatment days to exceed dose criteria, and alert level for each of 32 organs. Accumulated doses could be maximum dose, mean dose, or volume, depending on the specific criteria for different organs. When we register a new patient on the PODA, an entry containing the patient's general information will be added to the Main Table and three patient-specific tables with unique names will be automatically created in the database. In addition, a link

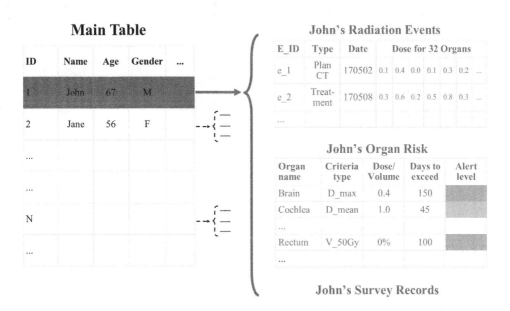

Figure 9.9 Layout of a PODA database. The Main Table records the general information (e.g., ID, name, age, sex, weight, race, specific table names) of all the registered patients. In addition, there are three specialized tables created for each patient to store radiation events, organ dose and risk level, and survey records.

between the entry and the newly built tables will be established to access and update data in the tables. The database will be updated after every treatment day to record organ doses delivered on that day and to accumulate organ doses.

9.3.6 DOSE REPORT AND MORBIDITY ANALYSIS

After organ doses are mapped back onto the planning CTs on treatment day, dose accumulation to the 32 organs will be summed and the PODA database will be updated. A backup of the database will also be created in another hard drive for data safety in case of accidental storage damage. A 3D dose distribution is available for review and evaluation of organ dose deposition (Figure 9.10b). Based on organ doses deposited in past treatment fractions, linear regression analysis is performed to predict the cumulative organ doses over the entire treatment. The predicted cumulative organ doses are compared with corresponding dose criteria to determine the risk level of each organ. Risk level is defined according to the estimated probability for predicted dose to exceed dose criteria with <5% being Safe, <50% being Risky, and ≥50% being Dangerous. Then, an Extensible Markup Language (XML, supported by mainstream office software such as Microsoft Excel, OpenOffice Calc, and Apple iWork Numbers) format spreadsheet file is automatically created to report the organ dose. Each individual patient's data is shown in an independent sheet (Figure 9.10a, c). Event-specific organ doses are listed in the table. Organ risk is shown in the last column of the table using color blocks as an intuitive view with green representing Safe, orange Risky, and red Dangerous. In particular, organ doses accumulated on a certain day are linked to the specific data folder in the DICOM library for data consistency and reproducibility and also linked to an organ dose histogram for easy evaluation.

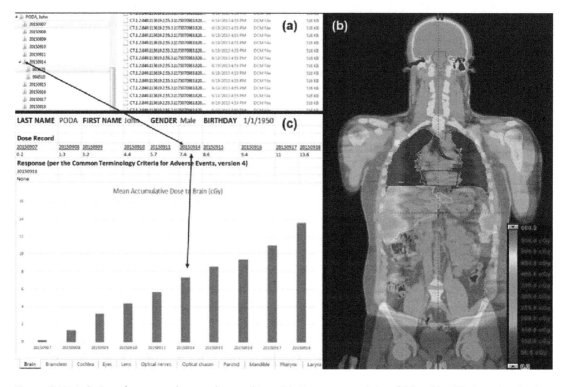

Figure 9.10 A design of a personal organ dose archive with (a) a structured data folder, (b) a 3D dose distribution in patient anatomy, and (c) a personal organ dose data record and report.

9.3.7 EARLY WARNINGS AND INTERVENTION

When the risk level of one or more organs of a patient under tracking is escalated into Dangerous, an early warning will be issued to inform the attending physician. An independent round of calculations will be performed to double check the 3D dose distribution and cumulative organ doses. Once the early warning signal is confirmed to be valid, further actions will be taken to prevent permanent damage to the normal tissues. Possible intervention includes reduction in the use of image-guidance, modification of treatment plan, and so on. Cumulative 3D dose distribution generated by the PODA can be used as the input to an inverse treatment-planning module to initialize the baseline dose. PODA-predicted organ risk level can be used to adjust dose–volume constraints and the weighting factors of OAR. Specifically, stricter constraints and higher weighting factors will be employed for organs whose predicted doses are more likely to exceed the dose criteria.

9.4 NORMAL TISSUE TOLERANCES FOR DIFFERENT TREATMENT SCHEMA

In an attempt to address a clinical need for accurate knowledge of the dose–time–volume relationships for normal tissues in 3D treatment planning, Emami et al. published a highly popular paper based on literature review and their clinical experience up to 1991 [9]. Recently, the QUANTEC group has summarized the currently available 3D dose–volume–outcome data that update and refine early normal tissue tolerance guidelines [10]. Although highly relevant to modern radiotherapy practices, QUANTEC reviews only provide guideline for 18 organs following conventional fractionation. As indicated in the paper by Emami on the tolerance of normal tissues to therapeutic radiation, numerous factors affect radiation-induced complications of normal tissues in any given clinical situation [44]. For example, conventional fractionation usually delivers a dose of 1.8 or 2.0 Gy per day over a course of several weeks, whereas hypofractionated radiation therapy like SBRT delivers a large dose per day over a few days or a week. A difference in the daily dose delivery will lead to quite a different normal tissue dose–volume response, which has been demonstrated in a number of radiobiological experiments and in the clinic. Normal tissue structures can be divided into serial organs and parallel organs. A serial organ, like the spinal cord or brainstem, will fail even if a small volume of it is damaged. A parallel organ, like the lung or liver, has inherent redundancy and will be damaged only if all the functional subunits are damaged [45]. Consequently, specialized dose criteria should be applied to specific clinical situation to maximize the therapeutic ratio of radiation therapy.

In order to assess the clinical effectiveness of the PODA in a variety of clinical scenarios, we build a simplified normal tissue tolerance table to test our hypothesis. Table 9.2 lists the normal tissue tolerances for 28 clinically relevant organs in conventional radiotherapy, IMRT, and SBRT treatments. These tolerance data are compiled from published literature, including QUANTEC group reviews [10,14,33,46–65], Emami et al.'s publications on the tolerance of normal tissue to therapeutic radiation [9,44], the American Association of Physicists in Medicine (AAPM) report [66], and the RTOG reports [67–69]. Tolerance doses for breast, thyroid, pancreas, and spleen are listed as "to be determined" at the bottom of the table because there are no clinical data reported on these organs. Without tolerance data for these four organs, we cannot currently estimate radiation risk for these organs. However, with the implementation of the PODA, we can track and accumulate doses for these organs in 3D and over time for each individual patient. With a large amount of accurate and comprehensive patient organ dose–volume data—together with normal tissue toxicity data obtained through long-term follow-up research—it is likely that the normal tissue tolerance criteria listed in Table 9.2 and QUANTEC guidelines will need to be revised.

Table 9.2 **Summary of normal tissue tolerances for 26 clinically relevant organs during conventional radiotherapy, IMRT, and SBRT treatment schema with 1 fraction, 3 fractions, and 5 fractions**

STRUCTURE	CONVENTIONAL AND IMRT	SBRT, SINGLE FRACTION	SBRT, THREE FRACTIONS	SBRT, FIVE FRACTIONS
Brain	$D_{max} < 60$ Gy	$D_{max} < 60$ Gy	$D_{max} < 60$ Gy	$D_{max} < 60$ Gy
Brainstem	$D_{max} < 54$ Gy	$D_{max} < 15$ Gy	$D_{max} < 23$ Gy	$D_{max} < 31$ Gy
Cochlea	$D_{mean} < 45$ Gy	$D_{max} < 12$ Gy	$D_{max} < 20$ Gy	$D_{max} < 27.5$ Gy
Eyes	$D_{max} < 20$ Gy	$D_{max} < 20$ Gy	$D_{max} < 20$ Gy	$D_{max} < 20$ Gy
Lens	$D_{max} < 3$ Gy	$D_{max} < 3$ Gy	$D_{max} < 3$ Gy	$D_{max} < 3$ Gy
Optical nerves	$D_{max} < 54$ Gy	$D_{max} < 10$ Gy	$D_{max} < 19.5$ Gy	$D_{max} < 25$ Gy
Optical chiasm	$D_{max} < 54$ Gy	$D_{max} < 10$ Gy	$D_{max} < 19.5$ Gy	$D_{max} < 25$ Gy
Parotid	$D_{mean} < 26$ Gy	$D_{mean} < 6$ Gy	$D_{mean} < 16$ Gy	$D_{mean} < 21$ Gy
Mandible	$D_{max} < 70$ Gy	$D_{max} < 17$ Gy	$D_{max} < 44$ Gy	$D_{max} < 58$ Gy
Pharynx	$D_{mean} < 50$ Gy	$D_{mean} < 12$ Gy	$D_{mean} < 31$ Gy	$D_{mean} < 42$ Gy
Larynx	$D_{max} < 63$ Gy	$D_{max} < 16$ Gy	$D_{max} < 41$ Gy	$D_{max} < 55$ Gy
Lungs	$D_{mean} < 20$ Gy	$V_{7.4Gy} < 1000$ cc	$V_{11.4Gy} < 1000$ cc	$V_{13.5Gy} < 1000$ cc
Heart	$D_{mean} < 26$ Gy	$D_{max} < 22$ Gy, $V_{16Gy} < 15$ cc	$D_{max} < 30$ Gy, $V_{24Gy} < 15$ cc	$D_{max} < 38$ Gy, $V_{32Gy} < 15$ cc
Esophagus	$D_{mean} < 34$ Gy	$D_{max} < 19$ Gy	$D_{max} < 27$ Gy	$D_{max} < 35$ Gy
Trachea	$D_{max} < 62$ Gy	$D_{max} < 22$ Gy	$D_{max} < 30$ Gy	$D_{max} < 38$ Gy
Brachial plexus	$D_{max} < 66$ Gy	$D_{max} < 16$ Gy	$D_{max} < 24$ Gy	$D_{max} < 32$ Gy
Liver	$D_{mean} < 25$ Gy	$V_{9.1Gy} < 700$ cc	$V_{17.1Gy} < 700$ cc	$V_{21Gy} < 700$ cc
Stomach	$D_{max} < 54$ Gy	$D_{max} < 16$ Gy	$D_{max} < 24$ Gy	$D_{max} < 32$ Gy
Kidneys	$D_{mean} < 18$ Gy	$D_{mean} < 8$ Gy	$D_{mean} < 10$ Gy	$D_{mean} < 12$ Gy
Spinal cord	$D_{max} < 45$ Gy	$D_{max} < 14$ Gy	$D_{max} < 22$ Gy	$D_{max} < 30$ Gy
Rectum		$D_{max} < 22$ Gy	$D_{max} < 30$ Gy	$D_{max} < 38$ Gy
Bladder	$D_{mean} < 30$ Gy	$D_{max} < 22$ Gy	$D_{max} < 30$ Gy	$D_{max} < 38$ Gy
Small bowel	$D_{max} < 50$ Gy	$D_{max} < 18$ Gy	$D_{max} < 26$ Gy	$D_{max} < 34$ Gy
Colon	$D_{max} < 54$ Gy	$D_{max} < 22$ Gy	$D_{max} < 30$ Gy	$D_{max} < 38$ Gy
Femoral head	$D_{max} < 52$ Gy	$V_{14Gy} < 10$ cc	$V_{21.9Gy} < 10$ cc	$V_{30Gy} < 10$ cc
Penile bulb	$D_{mean} < 52.5$ Gy	$D_{max} < 34$ Gy	$D_{max} < 42$ Gy	$D_{max} < 50$ Gy
Testes	$D_{max} < 1$ Gy	$D_{max} < 1$ Gy	$D_{max} < 1$ Gy	$D_{max} < 1$ Gy
Ovary	$D_{max} < 2$ Gy	$D_{max} < 2$ Gy	$D_{max} < 2$ Gy	$D_{max} < 2$ Gy
Breast	TBD	TBD	TBD	TBD
Thyroid	TBD	TBD	TBD	TBD
Pancreas	TBD	TBD	TBD	TBD
Spleen	TBD	TBD	TBD	TBD

Note: Tolerance data for breast, thyroid, pancreas, and spleen are not yet known. TBD, to be determined.

9.5 CLINICAL APPLICATIONS OF THE PODA FOR PATIENT SAFETY IN RADIOTHERAPY

9.5.1 ORGAN PROTECTION AND RADIATION TREATMENT INTERVENTION

Established organ dose criteria are strictly followed in the clinic when making a treatment plan in order to control the occurrence rate of severe late effects in normal tissues at an acceptable low level. However, as discussed earlier, the planned organ dose may be quite different from the delivered dose because of several factors—such as scatter in the treatment field, leakage from the accelerator head and MLC, morphological change in patient's anatomy, and imaging dose—which are not well considered by current commercial TPS when simulating 3D dose distributions. Unfortunately, these factors are inclined to add to, rather than deduct from, the planned organ dose. For example, if we consider scatter, leakage, and image-guidance procedures in MC simulation, the calculated 3D dose will definitely be higher than the planned dose. A geometrical change in the patient's anatomy is also unfavorable. Shrinkage of tumor and edema resolution in the course of treatment will bring normal tissue much closer to the treatment target where scatter from the treatment field is more intensive [25]. As a result, doses in the normal tissues are likely to increase. Although the contribution of a single factor may be minor, altogether, they could significantly increase the planned organ doses, making normal tissues vulnerable to radiation. The primary goal of the PODA is to provide a possible solution to this problem. By collecting various data generated during treatment and by simulating radiation events with a super-fast GPU-based MC dose engine, we can track the delivered organ doses after every treatment day. Using the accumulated organ dose data during the past treatment fractions, an early warning mechanism could make a prediction about organ risk level at the end of the entire treatment and issue a warning message if organs with high radiation toxicity risk are identified. In cases where an early warning is issued, physicians will be involved to take actions to avoid permanent damage to the normal tissues before it is too late. We believe that PODA provides an additional safeguard mechanism for patients undergoing radiation therapy.

9.5.2 CANCER RISK MONITORING AND CONTROL

Ionizing radiation may induce two kinds of health effects: deterministic effects and stochastic effects. Tissue toxicity is classified as a deterministic effect because it will happen when radiation dose exceeds a threshold level. Radiation-induced cancer is a stochastic effect because its incidence is proportional to the absorbed radiation dose and there seems to be no threshold (at least most recent epidemiological data from life span study of the atomic bomb survivors and the Chernobyl accident emergency workers do not support the existence of a threshold at doses as low as that of several CT scans) [70,71]. Radiation doses deposited in the normal tissues during the radiotherapy treatment, no matter how small they are, will increase the probability of developing cancer according to current knowledge in radiobiology. Therefore, having an accurate and quantitative knowledge of the cancer burden for the individual patients who are receiving radiotherapy and controlling it to an acceptable level is of great importance for the public health. PODA could play a crucial role in this area from the following two aspects. First, using existing epidemiological data and cancer risk models, PODA could estimate cancer risk for all of the 32 organs under tracking at the end of each treatment day. By doing this, we are able to monitor the cancer risk of the patient undergoing radiation therapy and take interventions to reduce the cancer burden if necessary. Just like the early warning mechanism for tissue toxicity, monitoring cancer risk will improve the safety of patients as well in the long run. Second, if PODA is widely used, we could accumulate large amount of epidemiological data because millions of patients are treated with radiation therapy every year all over the world. In comparison, the atomic bomb survivor research project currently has the largest sample size, which is about 120 thousand. Data accumulated through PODA will provide insights into some long-standing fundamental questions such as low dose effects and the organ-specific cancer risk of ionizing radiation. Radiation biology provides the basic principles for radiation therapy. We envision that the implementation of PODA will in turn boost the research in radiation biology.

9.5.3 INTEGRATION INTO A PERSONAL HEALTH DATA ARCHIVE

Basically, an individual's health stems from his or her genetic variation and environmental influences. The response to medical treatment differs from person to person, making it difficult to find a drug or medical device that is effective against a specific disease for all the patients. Personalized medicine is gaining popularity in recent years. It is a medical procedure that separates patients into different groups, with medical treatment tailored to the individual patient based on personal health information such as clinical health records and genetic informatics [72,73]. The structuralized organ dose data in PODA can be integrated into a personal health data archive. On one hand, personal health data can help the PODA make personalized and precise risk prediction and provide an early warning to further improve safety of patients receiving radiation therapy. One the other hand, organ dose data supplements existing personal health data. With available accurate and comprehensive organ dose data, physicians are able to build their diagnosis or prescriptions on a more solid basis.

9.5.4 NORMAL TISSUE TOLERANCE BENCHMARK AND REVISION

Significant progress has been made in our understanding of normal tissue complication and tolerance to therapeutic radiation since the early work of Rubin and Cassarett, to the landmark Emami paper, and the most recent reviews and reports by QUANTEC, Emami, AAPM, and RTOG. However, as indicated in these most recent publications, many questions are still undetermined [44]. Advanced treatment techniques have made the dose distributions extremely complex and multidimensional. The complexity of factors influencing normal tissue toxicity makes it difficult to have data for every clinical scenario in the practice of radiation oncology. To have a deep and comprehensive understanding of this subject and to provide easy-to-use guidelines for clinicians in the daily practice, much more data is urgently needed. PODA provides an ideal platform for collecting a large amount of dose–volume–effects data in the clinic. The PODA database stores accurate 3D organ doses for each individual patient. Late effects of normal tissue can be obtained in the follow-up research and recorded in the database. Linkage between dose–volume data and late effects in the organs can be established. Accumulation of this kind of data can be used to benchmark existing normal tissue tolerance and will help test models that predict normal tissue toxicity in the clinic.

9.6 CONCLUSION AND OUTLOOK

In this chapter, we introduced the PODA, a big data approach to tracking each patient's organ doses associated with the use of sophisticated treatment technologies and image-guidance procedures in modern radiotherapy. With the deployment of this tool in the routine clinical practice, radiation damage to the normal tissues can be efficiently avoided with early warnings and radiation treatments can be better guided with comprehensive knowledge of organ dose accumulation in 3D. The ultimate benefit of PODA is to improve patient safety and reduce side effects for the millions of cancer patients who undergo radiotherapy every year. Data accumulated in the PODA will also facilitate the investigation of low-dose radiation effects and the dose–volume relationship of normal tissue complications.

Basically, PODA compiles and consolidates all the valuable organ dose data into one single archive, data that is often scattered around at various institutions and various electronic medical record (EMR) databases and is often forgotten over time. The strength of PODA stems from the fact that it is designed to accumulate personal organ dose data along one's lifetime accuracy (MC dose simulation), comprehensively (all ionizing radiation events and all relevant organs), and quantitatively (absolute dose distributions in 3D). The PODA built for each individual patient is a permanent and accurate replica of one's journey through time and space, in a dosimetric way. Its power will show up when data pooling and sharing is needed in the clinic to draw any meaningful conclusions with statistical power and certainty. Data sharing and pooling will become very convenient and highly efficient in the era of big data, largely reducing the errors due to lack of data or data loss and helping the clinicians make informed decisions in the clinic.

Besides a direct impact on cancer patients who are receiving radiotherapy, PODA can also be useful in other fields where ionizing radiation plays a role, for example, CT, PET, and fluoroscopy used in diagnostic imaging. Accurate and comprehensive organ dose tracking would be very necessary in order to protect people from normal tissue damage and to reduce second cancer risks in tens of millions of people receiving CT and other radiological procedures worldwide [74–77]. This is especially critical for children, who are more vulnerable to radiation damage than adults [78,79].

Looking forward, as we move into an era of big data, Cloud computing, and personalized medicine, we are very excited about the potential use of PODA. For example, one possible application is that all personal organ dose information may be stored securely in a Cloud computing database for easy access anytime anywhere. Another viable scenario is that every patient will be able to access and receive a copy of his or her own personal organ dose archive in a portable device such as a USB drive for ultra portability. This kind of personal data can also be stored digitally in a wearable device such as Apple Watch or Fitbit. We believe that development of the PODA is a step toward personalized medicine in the near future.

REFERENCES

1. Timmerman, R. et al., Stereotactic body radiation therapy for inoperable early stage lung cancer. *JAMA*, 2010. **303**(11): 1070–1076.
2. Takeda, A. et al., Dose distribution analysis in stereotactic body radiotherapy using dynamic conformal multiple arc therapy. *International Journal of Radiation Oncology Biology Physics*, 2009. **74**(2): 363–369.
3. Grills, I.S. et al., Potential for reduced toxicity and dose escalation in the treatment of inoperable non-small-cell lung cancer: A comparison of intensity-modulated radiation therapy (IMRT), 3D conformal radiation, and elective nodal irradiation. *International Journal of Radiation Oncology Biology Physics*, 2003. **57**(3): 875–890.
4. Holt, A. et al., Volumetric-modulated arc therapy for stereotactic body radiotherapy of lung tumors: A comparison with intensity-modulated radiotherapy techniques. *International Journal of Radiation Oncology Biology Physics*, 2011. **81**(5): 1560–1567.
5. Dawson, L.A. and M.B. Sharpe, Image-guided radiotherapy: Rationale, benefits, and limitations. *Lancet Oncology*, 2006. **7**(10): 848–858.
6. Viani, G.A., E.J. Stefano, and S.L. Afonso, Higher-than-conventional radiation doses in localized prostate cancer treatment: A meta-analysis of randomized, controlled trials. *International Journal of Radiation Oncology Biology Physics*, 2009. **74**(5): 1405–1418.
7. Zelefsky, M.J. et al., High-dose intensity modulated radiation therapy for prostate cancer: Early toxicity and biochemical outcome in 772 patients. *International Journal of Radiation Oncology Biology Physics*, 2002. **53**(5): 1111–1116.
8. Rubin, P. and G.W. Casarett, Clinical radiation pathology as applied to curative radiotherapy. *Cancer*, 1968. **22**(4): 767.
9. Emami, B. et al., Tolerance of normal tissue to therapeutic irradiation. *International Journal of Radiation Oncology Biology Physics*, 1991. **21**(1): 109–122.
10. Bentzen, S.M. et al., Quantitative analyses of normal tissue effects in the clinic (QUANTEC): An introduction to the scientific issues. *International Journal of Radiation Oncology Biology Physics*, 2010. **76**(3 Suppl): S3–S9.
11. Kry, S.F. et al., The calculated risk of fatal secondary malignancies from intensity-modulated radiation therapy. *International Journal of Radiation Oncology Biology Physics*, 2005. **62**(4): 1195–1203.
12. Palm, A. and K.A. Johansson, A review of the impact of photon and proton external beam radiotherapy treatment modalities on the dose distribution in field and out-of-field; implications for the long-term morbidity of cancer survivors. *Acta Oncologica*, 2007. **46**(4): 462–473.
13. Nelms, B.E. et al., Variations in the contouring of organs at risk: Test case from a patient with oropharyngeal cancer. *International Journal of Radiation Oncology Biology Physics*, 2012. **82**(1): 368–378.
14. Jaffray, D.A. et al., Accurate accumulation of dose for improved understanding of radiation effects in normal tissue. *International Journal of Radiation Oncology Biology Physics*, 2010. **76**(3 Suppl): S135–S139.
15. Deng, J. et al., Testicular doses in image-guided radiotherapy of prostate cancer. *International Journal of Radiation Oncology Biology Physics*, 2012. **82**(1): e39–e47.
16. Deng, J. et al., Kilovoltage imaging doses in the radiotherapy of pediatric cancer patients. *International Journal of Radiation Oncology Biology Physics*, 2012. **82**(5): 1680–1688.
17. Zhang, Y. et al., Personalized assessment of kV cone beam computed tomography doses in image-guided radiotherapy of pediatric cancer patients. *International Journal of Radiation Oncology Biology Physics*, 2012. **83**(5): 1649–1654.
18. Hall, E.J. and C.S. Wuu, Radiation-induced second cancers: the impact of 3D-CRT and IMRT. *International Journal of Radiation Oncology Biology Physics*, 2003. **56**(1): 83–88.
19. Hall, E.J., Intensity-modulated radiation therapy, protons, and the risk of second cancers. *International Journal of Radiation Oncology Biology Physics*, 2006. **65**(1): 1–7.

20. Ruben, J.D. et al., The effect of intensity-modulated radiotherapy on radiation-induced second malignancies. *International Journal of Radiation Oncology Biology Physics*, 2008. **70**(5): 1530–1536.

21. Ng, J. et al., Predicting the risk of secondary lung malignancies associated with whole-breast radiation therapy. *International Journal of Radiation Oncology Biology Physics*, 2012. **83**(4): 1101–1106.

22. Donovan, E.M. et al., Second cancer incidence risk estimates using BEIR VII models for standard and complex external beam radiotherapy for early breast cancer. *Medical Physics*, 2012. **39**(10): 5814–5824.

23. Joosten, A. et al., Evaluation of organ-specific peripheral doses after 2-dimensional, 3-dimensional and hybrid intensity modulated radiation therapy for breast cancer based on Monte Carlo and convolution/superposition algorithms: Implications for secondary cancer risk assessment. *Radiotherapy and Oncology*, 2013. **106**(1): 33–41.

24. Murray, L. et al., Second primary cancers after radiation for prostate cancer: A systematic review of the clinical data and impact of treatment technique. *Radiotherapy and Oncology*, 2014. **110**(2): 213–228.

25. Athar, B.S. et al., Comparison of out-of-field photon doses in 6 MV IMRT and neutron doses in proton therapy for adult and pediatric patients. *Physics in Medicine & Biology*, 2010. **55**(10): 2879–2891.

26. Sharma, D.S. et al., Peripheral dose from uniform dynamic multileaf collimation fields: Implications for sliding window intensity-modulated radiotherapy. *British Journal of Radiology*, 2006. **79**(940): 331–335.

27. O'Daniel, J.C. et al., Parotid gland dose in intensity-modulated radiotherapy for head and neck cancer: is what you plan what you get? *International Journal of Radiation Oncology Biology Physics*, 2007. **69**(4): 1290–1296.

28. George, R. et al., Quantifying the effect of intrafraction motion during breast IMRT planning and dose delivery. *Medical Physics*, 2003. **30**(4): 552–562.

29. Jabbari, N., A.H. Barati, and L. Rahmatnezhad, Multiple-source models for electron beams of a medical linear accelerator using BEAMDP computer code. *Reports of Practical Oncology & Radiotherapy*, 2012. **17**(4): 211–219.

30. Tian, Z. et al., An analytic linear accelerator source model for Monte Carlo dose calculations. I. Model representation and construction. arXiv preprint arXiv:1503.01724, 2015.

31. Jia, X., S. Hissoiny, and S.B. Jiang, GPU-based fast Monte Carlo simulation for radiotherapy dose calculation, in *Monte Carlo Techniques in Radiation Therapy*, F.V. Joao Seco, Editor. 2013, Taylor & Francis Group. pp. 283–294.

32. Hissoiny, S. et al., GPUMCD: A new GPU-oriented Monte Carlo dose calculation platform. *Medical Physics*, 2011. **38**(2): 754–764.

33. Jeraj, R. et al., Imaging for assessment of radiation-induced normal tissue effects. *International Journal of Radiation Oncology Biology Physics*, 2010. **76**(3): S140–S144.

34. Jia, X. et al., GPU-based fast Monte Carlo simulation for radiotherapy dose calculation. *Physics in Medicine & Biology*, 2011. **56**(22): 7017.

35. Jia, X. et al., Development of a GPU-based Monte Carlo dose calculation code for coupled electron–photon transport. *Physics in Medicine & Biology*, 2010. **55**(11): 3077.

36. Jahnke, L. et al., GMC: A GPU implementation of a Monte Carlo dose calculation based on Geant. *Physics in Medicine & Biology*, 2012. **57**(5): 1217.

37. Townson, R.W. et al., GPU-based Monte Carlo radiotherapy dose calculation using phase-space sources. *Physics in Medicine & Biology*, 2013. **58**(12): 4341.

38. Tian, Z. et al., Automatic commissioning of a GPU-based Monte Carlo radiation dose calculation code for photon radiotherapy. *Physics in Medicine & Biology*, 2014. **59**(21): 6467.

39. Beltran, C. et al., Clinical implementation of a proton dose verification system utilizing a GPU accelerated Monte Carlo engine. *International Journal of Particle Therapy*, 2016. **3**(2): 312–319.

40. Jia, X. et al., GPU-based fast Monte Carlo dose calculation for proton therapy. *Physics in Medicine & Biology*, 2012. **57**(23): 7783.

41. Poon, E. et al., Patient-specific Monte Carlo dose calculations for high-dose-rate endorectal brachytherapy with shielded intracavitary applicator. *International Journal of Radiation Oncology• Biology• Physics*, 2008. **72**(4): 1259–1266.

42. Tian, Z. et al., Monte Carlo dose calculations for high-dose–rate brachytherapy using GPU-accelerated processing. *Brachytherapy*, 2016. **15**(3): 387–398.

43. Jia, X. et al., Fast Monte Carlo simulation for patient-specific CT/CBCT imaging dose calculation. *Physics in Medicine & Biology*, 2012. **57**(3): 577.

44. Emami, B., Tolerance of normal tissue to therapeutic radiation. *Reports of Radiotherapy and Oncology*, 2013. **1**(1).

45. Kallman, P., A. Agren, and A. Brahme, Tumour and normal tissue responses to fractionated non-uniform dose delivery. *International Journal of Radiation Biology*, 1992. **62**(2): 249–262.

46. Marks, L.B., R.K. Ten Haken, and M.K. Martel, Guest editor's introduction to Quantec: A users guide. *International Journal of Radiation Oncology Biology Physics*, 2010. **76**(3): S1–S2.

47. Viswanathan, A.N. et al., Radiation dose-volume effects of the urinary bladder. *International Journal of Radiation Oncology Biology Physics*, 2010. **76**(3): S116–S122.

48. Lawrence, Y.R. et al., Radiation dose-volume effects in the brain. *International Journal of Radiation Oncology Biology Physics*, 2010. **76**(3): S20–S27.

49. Mayo, C. et al., Radiation dose-volume effects of optic nerves and chiasm. *International Journal of Radiation Oncology Biology Physics*, 2010. **76**(3): S28–S35.

50. Mayo, C., E. Yorke, and T.E. Merchant, Radiation associated brainstem injury. *International Journal of Radiation Oncology Biology Physics*, 2010. **76**(3): S36–S41.

51. Kirkpatrick, J.P., A.J. van der Kogel, and T.E. Schultheiss, Radiation dose-volume effects in the spinal cord. *International Journal of Radiation Oncology Biology Physics*, 2010. **76**(3): S42–S49.

52. Bhandare, N. et al., Radiation therapy and hearing loss. *International Journal of Radiation Oncology Biology Physics*, 2010. **76**(3): S50–S57.

53. Deasy, J.O. et al., Radiotherapy dose-volume effects on salivary gland function. *International Journal of Radiation Oncology Biology Physics*, 2010. **76**(3): S58–S63.

54. Rancati, T. et al., Radiation dose-volume effects in the larynx and pharynx. *International Journal of Radiation Oncology Biology Physics*, 2010. **76**(3): S64–S69.

55. Marks, L.B. et al., Radiation dose-volume effects in the lung. *International Journal of Radiation Oncology Biology Physics*, 2010. **76**(3): S70–S76.

56. Gagliardi, G. et al., Radiation dose-volume effects in the heart. *International Journal of Radiation Oncology Biology Physics*, 2010. **76**(3): S77–S85.

57. Werner-Wasik, M. et al., Radiation dose-volume effects in the esophagus. *International Journal of Radiation Oncology Biology Physics*, 2010. **76**(3): S86–S93.

58. Pan, C.C. et al., Radiation-associated liver injury. *International Journal of Radiation Oncology Biology Physics*, 2010. **76**(3): S94–S100.

59. Kavanagh, B.D. et al., Radiation dose-volume effects in the stomach and small bowel. *International Journal of Radiation Oncology Biology Physics*, 2010. **76**(3): S101–S107.

60. Dawson, L.A. et al., Radiation-associated kidney injury. *International Journal of Radiation Oncology Biology Physics*, 2010. **76**(3): S108–S115.

61. Michalski, J.M. et al., Radiation dose-volume effects in radiation-induced rectal injury. *International Journal of Radiation Oncology Biology Physics*, 2010. **76**(3): S123–S129.

62. Roach, M. et al., Radiation dose-volume effects and the penile bulb. *International Journal of Radiation Oncology Biology Physics*, 2010. **76**(3): S130–S134.

63. Bentzen, S.M. et al., Biomarkers and surrogate endpoints for normal-tissue effects of radiation therapy: The importance of dose-volume effects. *International Journal of Radiation Oncology Biology Physics*, 2010. **76**(3): S145–S150.

64. Deasy, J.O. et al., Improving normal tissue complication probability models: The need to adopt a "Data-Pooling" culture. *International Journal of Radiation Oncology Biology Physics*, 2010. **76**(3): S151–S154.

65. Jackson, A. et al., The lessons of Quantec: Recommendations for reporting and gathering data on dose-volume dependencies of treatment outcome. *International Journal of Radiation Oncology Biology Physics*, 2010. **76**(3): S155–S160.

66. Benedict, S.H. et al., Stereotactic body radiation therapy: The report of AAPM Task Group 101. *Medical Physics*, 2010. **37**(8): 4078–4101.

67. Michalski, J.M. et al., Preliminary toxicity analysis of 3-dimensional conformal radiation therapy versus intensity modulated radiation therapy on the high-dose arm of the Radiation Therapy Oncology Group 0126 prostate cancer trial. *International Journal of Radiation Oncology Biology Physics*, 2013. **87**(5): 932–938.

68. Lee, N. et al., Intensity-modulated radiation therapy with or without chemotherapy for nasopharyngeal carcinoma: Radiation Therapy Oncology Group phase II trial 0225. *Journal of Clinical Oncology*, 2009. **27**(22): 3684–3690.

69. Lilenbaum, R., R. Komaki, and M.K. Martel, Radiation Therapy Oncology Group (RTOG) Protocol 0623: A phase II trial of combined modality therapy with growth factor support for patients with limited stage small cell lung cancer.

70. Ozasa, K. et al., Studies of the mortality of atomic bomb survivors, Report 14, 1950–2003: An overview of cancer and noncancer diseases. *Radiation Research*, 2012. **177**(3): 229–243.

71. Kashcheev, V.V. et al., Incidence and mortality of solid cancer among emergency workers of the Chernobyl accident: Assessment of radiation risks for the follow-up period of 1992–2009. *Radiation and Environmental Biophysics*, 2015. **54**(1): 13–23.

72. Hamburg, M.A. and F.S. Collins, The path to personalized medicine. *New England Journal of Medicine*, 2010. **363**(4): 301–304.

73. Collins, F.S. and H. Varmus, A new initiative on precision medicine. *New England Journal of Medicine*, 2015. **372**(9): 793–795.

74. Brenner, D.J. and E.J. Hall, Computed tomography—An increasing source of radiation exposure. *New England Journal of Medicine*, 2007. **357**(22): 2277–2284.

75. Hall, E.J. and D.J. Brenner, Cancer risks from diagnostic radiology: The impact of new epidemiological data. *British Journal of Radiology*, 2012. **85**(1020): e1316–e1317.

76. Hall, E.J. and D.J. Brenner, Cancer risks from diagnostic radiology. *British Journal of Radiology*, 2008. **81**(965): 362–378.

77. Pearce, M.S. et al., Radiation exposure from CT scans in childhood and subsequent risk of leukaemia and brain tumours: A retrospective cohort study. *Lancet*, 2012. **380**(9840): 499–505.

78. Halperin, E.C. et al., *Pediatric Radiation Oncology*. 2012: Philadelphia, PA: Lippincott Williams & Wilkins.

79. Goske, M.J. et al., The 'Image Gently' campaign: Increasing CT radiation dose awareness through a national education and awareness program. *Pediatric Radiology*, 2008. **38**(3): 265–269.

Big data and comparative effectiveness research in radiation oncology

Sunil W. Dutta, Daniel M. Trifiletti, and Timothy N. Showalter

Contents

10.1 Introduction   145
10.2 Big Data and CER Integration in Health Care   147
10.3 Applying Big Data and CER to Improve Outcomes after Radiation Therapy   148
10.4 Big Data and Patient-Reported Outcomes   149
10.5 Future Directions   150
10.6 Conclusion   150
References   151

10.1 INTRODUCTION

The microscope was invented four centuries ago to see things otherwise invisible to the human eye. In a similar way, Erik Brynjolfsson, an economist at Massachusetts Institute of Technology's Sloan School of Management, believes data measurement is the modern equivalent of the microscope (Lohr, 2012), providing granular perspectives and insights that are not otherwise discoverable. Silicon Valley companies such as Google and Facebook are constantly monitoring user behavior to improve the experience of its users, while also delivering targeted advertisements. The medical field should be adopting data measurement and analysis, like we did with the microscope, to improve patient care and treatment optimization through a robust discovery and validation process.

Big data can be defined generally as information of such quantity that it cannot be analyzed directly by humans. To leverage the potential value of big data for discovery, advanced technologies are being developed to help decode the massive amounts of information. The field of radiation oncology has all the big data elements fit for analysis that can improve patient care, but data access and analysis have not been implemented adequately to leverage data for improvements in care. During the course of a patient's oncology treatment, which can be several weeks in duration, image-based treatment planning and quality assurance are needed to ensure safe radiation delivery daily. As a result, massive amounts of information are created and collected for each patient before, during, and even after treatment is completed. When this information is collected using a robust infrastructure, and if the data are shared, there is potential for improvements on a large population scale that can then ideally be scaled down to a personalized medicine level (Figure 10.1).

Big data research differs from traditional research in that it is not always hypothesis driven (Krumholz, 2014). In traditional research efforts, a hypothesis is usually specified prior to the design and conduct of an experiment, with the analysis and methods focused from the outset on the investigator's hypothesis. Big data research, on the other hand, attempts to identify new questions and to discover new associations among factors by using data analytic techniques to identify trends or relationships within the available data. In this sense, the paradigm of big data research can accelerate discovery by approaching analysis without prespecified hypotheses or assumptions. Big data can be used to find new questions

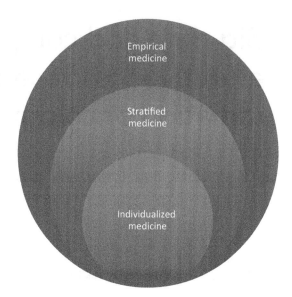

Figure 10.1 The combination of big data and comparative effectiveness research offers potential for discovery across the spectrum from the population level through the individual patient. Big data cohort sizes will be increasingly important at the nucleus of the circle of evidence (individualized medicine).

without the full, preplanned, and resource-intensive efforts required in traditional research methodologies such as a randomized controlled trial. Traditionally, the importance of randomized controlled trials has been especially emphasized in oncology. Although big data methods are not expected to replace randomized controlled trials in oncology, big data approaches may help identify new hypotheses or refine existing hypotheses for testing in clinical trials. Furthermore, big data may provide a framework for extending the potential value of pooled data from clinical trials, which can be recycled in subsequent big data analyses to accelerate discovery.

A robust data infrastructure is needed to realize the potential of big data in health care, and leveraging the electronic medical record (EMR) will be essential for success. The utilization of EMRs increased among office-based physicians from 18% in 2001 to 78% in 2013 (Hsiao and Hing 2014), but there are substantial obstacles in the way of effectively harnessing EMRs for big data research. The EMR contains heterogeneous information, including structured data (e.g., laboratory values, DICOM data sets) and unstructured data (e.g., clinical impression in free text). The heterogeneity presents obstacles for research but also highlights the rich, multifaceted information available in EMR, and it will be important to develop big data analysis techniques that allow for interpretation of this data in ways that cannot be imaged. While each EMR has its own limitations, the entered information contributes significantly to the material that comprises health care big data (Berger and Doban, 2014). EMRs will continue to evolve over time and will require investment in software and hardware devices and engineers to keep up.

Comparative effectiveness research (CER) is an investigative framework that may substantively overlap, complement, and synergize with both big data and traditional research methods. The central aim of CER in medicine is to compare interventions in a practical way to determine what is best on a population and patient level. The Institute of Medicine (IOM) defines CER as "the generation and synthesis of evidence that compares the benefit and harms of alternative methods to prevent, diagnose, treat, and monitor a clinical condition or to improve the delivery of care. The purpose of CER is to assist consumers, clinicians, purchasers, and policy makers to make informed decisions that will improve health care at both the individual and population levels" (Sox, 2010). CER is a new term; however, many popular research topics over the past several decades in health care fit within the framework of CER, including but not limited to traditional tenets of evidence-based medicine and cost-effectiveness analysis. Randomized controlled trials are also consistent with CER methodologies, with an important distinction regarding a focus on effectiveness over efficacy in a CER paradigm. Traditional randomized controlled trials, which are currently the gold standard in evidence-based medicine, often use the *efficacy* of treatments as the primary measure of intervention. CER differentiates itself by measuring the *effectiveness* of interventions, including the benefit and harm to individual patients in real-life scenarios, outside of a controlled environment (Lyman and Levine, 2012).

10.2 BIG DATA AND CER INTEGRATION IN HEALTH CARE

The concept of a learning health system stresses the importance of ongoing reevaluation of health care and offers an idealized platform for applying big data and CER methods. A learning health system model involves the capability to implement the latest evidence efficiently to improve outcomes, focusing on selecting the right health care plan for the individual patient (Berger and Doban, 2014). The IOM recommends a health care infrastructure that will facilitate improved health care outcomes while reducing health care costs (Committee on the Learning Health Care System in America, 2013). In clinical oncology, and in radiation oncology in particular, pooling data from diverse sources and rapid analysis should accelerate innovation in the field (Skripcak et al., 2014). However, data interoperability, privacy concerns, data ownership, publishing rights, and data storage are challenges that need to be addressed (Skripcak et al., 2014).

Analytic approaches have already been developed in other fields that benefit from big data, and these methods are promising for the acceleration of discovery for health care within a learning health system. Big data research methods such as machine learning or data mining can identify causal relationships present in the data generated in EMR and available for analysis within a learning health system framework. Integrating big data analytics with CER methods should have a synergistic effect in that CER methodologies and tenets can be applied to test big data–generated hypotheses in a time-efficient, potentially real-time fashion. The potential hesitation to accept the results of big data analytics—which often start without an initial defined hypothesis and instead may seek to identify novel associations—can be overcome by sequentially validating big data findings using CER methods to directly and rigorously compare interventions by applying the same rigorous criteria we use in designing clinical research studies. An example would be to select patients from a database who would meet proposed randomized trial selection criteria and to test a hypothesis on the data using an observational cohort study design. If the findings are valuable to the medical field, a pragmatic trial can then be performed to test the hypothesis in a controlled setting (Berger and Doban, 2014). In addition, big data and CER methods could be leveraged in the reverse fashion as well to investigate the effectiveness of interventions evaluated in randomized controlled trials in subgroups or situations that differ from the conditions of the randomized controlled trial. In this fashion, big data and CER methods might help inform inferences regarding the external validity of the conclusions of randomized clinical trials of treatment efficacy to treatment effectiveness among real-world patients, including those underrepresented or excluded from cancer trials.

In order to benefit from a learning health system, more work is needed toward EMRs that permit large-scale inquiry and data sharing. The same challenges apply for big data and CER more broadly in oncology, where limited data resources pose substantial limitations on what level of impact can be made by investigators who aim to provide clinically relative evidence and to improve patient outcomes. Jagsi and colleagues reviewed the strengths and weaknesses of the available data sources in oncology (Jagsi et al., 2014). This includes the Surveillance, Epidemiology, and End Results (SEER) program, the National Cancer Database (NCDB), and registries from Centers of Excellence (e.g., National Comprehensive Cancer Network) (Jagsi et al., 2014). Although the large databases have allowed for many radiation oncology–relevant questions to be studied and reported, the depth and richness of details of radiation oncology specific registries are limited. Although costly to develop, radiation oncology–specific registries with complete technical and clinical data could provide invaluable insights regarding the optimization of treatment strategies, identification of patients who would benefit most from treatments, and reduction of the harms of radiation therapy.

The priorities of personalized medicine, which aims to take advantage of the rapid advances in genomics and molecular biology, have been well documented (Council of Advisors on Science and Technology, 2008). In a spatial sense, radiation therapy plans are already personalized in that dose calculations and technical details are patient-specific and customized to patient anatomy and tumor-related factors, similar to surgery. However, there are substantial evidence gaps that limit our ability to define and provide the optimal dose to each patient. For example, recent evidence suggests patients with human papillomavirus (HPV)–positive oropharyngeal cancer can safely receive less than the historically standard radiation dose (i.e., 70 Gy), with high progression-free survival and an improved toxicity profile (Chen et al., 2017). Currently, our field is limited by an inability to add to the evidence basis rapidly for such a clinical scenario in a way that can refine standard of care; the ability to use big data and CER methods to quickly generate evidence on such a specific topic is promising. As our understanding and identification of relevant markers from immunohistochemistry, human genomics, and molecular biology improves, further trials will be warranted to optimize radiation doses.

Personalized medicine does not always aim to create drugs or medical devices unique to a patient, which is sometimes referred to as individual medicine (Trusheim et al., 2007). Instead, it attempts to stratify groups of patients who will benefit from a treatment and sparing those who will not from the side effects and costs from the given treatment. By providing key insights into the effectiveness of radiation therapy in subgroups of patients such as those with a specific constellation of medical comorbidities, big data and CER methodologies may help better stratify patients in a way that leads to better understanding of the balance between the benefits and harms of radiation for a specific patient subgroup and which patients can benefit from advanced radiation techniques.

In the near future, it is likely that many treatment decisions involving radiation therapy will consider either tumor genomics, as a marker of tumor aggressiveness or tumor sensitivity to radiation response, or even patient host genome, as a marker of a patient's normal tissue intrinsic radiosensitivity and susceptibility to radiation-induced adverse effects. Vicini et al. described the progress made in sequencing of DNA and RNA characterization; however, challenges persist with interpreting the data in a way that translates to clinical application (Vicini et al., 2016). Issues include variation in genome sequencing techniques by group. Furthermore, available reference genome databases do not yet have the ancestral information needed to determine statistical significance. For example, variant gene frequencies are not the same in different populations, and nucleotide substitutions in a variable gene differ between populations (Kalow, 1997). Therefore, despite the massive data collected thus far in pharmacogenomics, data measurement is the limiting factor in improving stratification of patients and approaching individualized medicine. To benefit fully from genomics-based information, it will be necessary to combine genomics information with clinical factors (e.g., comorbidities, age) and technical factors. This is a domain where CER and big data methods may be invaluable tools for discovery.

10.3 APPLYING BIG DATA AND CER TO IMPROVE OUTCOMES AFTER RADIATION THERAPY

CER in oncology is placed as a high priority per the IOM. As an example, prostate cancer is considered in the top quartile for priority CER topics. This will require the development of decision-making models based upon both disease factors and patient comorbidities. Xiong et al., in a network meta-analysis attempt to compare randomized trials, found substantial uncertainty regarding the efficacy and safety of different interventions in the treatment of prostate cancer (Xiong et al., 2014). Published data from the ProtecT trial—a prospective randomized trial comparing patients with early stage prostate cancer and a median follow-up of 10 years who underwent radical prostatectomy, radiotherapy, or active surveillance—found no difference in overall or prostate cancer–specific survival between the three groups (Hamdy et al., 2016). Therefore, care must be taken to properly select patients for the appropriate therapy to avoid unnecessary treatment-related morbidity. Patient-reported outcomes from the ProtecT trial showed differences in the timing, severity, and recovery of urinary, bowel, and sexual function between the three treatment groups (Donovan et al., 2016). Decision-making models will need to be continually updated as randomized evidence contributes new information comparing treatment options. For example, the use of high-intensity focused ultrasound (HIFU) is gaining increasing interest as an alternative for the treatment of localized prostate cancer. However, there are no prospective trials comparing its use to surgery, radiotherapy, and active surveillance despite around 65,000 patients with prostate cancer having been treated with HIFU to date (Chaussy and Thuroff, 2017).

Endometrial cancer, like prostate cancer, also has a favorable prognosis when diagnosed at an early stage. On the other hand, there is heterogeneity among histologic subtypes with variable patterns of failure and recurrence among them, particularly in higher-stage disease. Most randomized trials group the various histologic subtypes together (Randall et al., 1995, 2006; Keys et al., 2004; Maggi et al., 2006; Susumu et al., 2008; Nout et al., 2010; Creutzberg et al., 2011). To utilize the available databases, Booth et al. evaluated patients from the NCDB with Stage III endometrial cancer, and found that patients treated with adjuvant-combined chemotherapy and radiotherapy had improved survival compared with patients treated with adjuvant monotherapy (Boothe et al., 2016). Unfortunately, important prognostic factors were not available for analysis, limiting the robustness of the data. Although such analyses are informative, their results should ultimately be used to help direct prospective trials to further define the best treatment options for individual patients (Jhingran and Klopp, 2016). In this case, a randomized phase III trial (GOG 258) was performed comparing chemotherapy with tumor volume–directed irradiation versus chemotherapy alone for advanced endometrial carcinoma. Contrary to the NCDB analysis, early results showed that although combined chemoradiation reduced the rate of local recurrence compared with chemotherapy alone, chemoradiation did not increase regression-free survival (Matei et al., 2017). Although the full results of GOG 258 are not yet published, results so far show that, while thought provoking, there are limitations to database analyses, and standards of care should continue to be developed from randomized evidence when available.

As technology advances, radiotherapy has become more targeted to the individual patient with respect to defining target volumes and sparing normal tissues. In contrast, trial protocols and national guidelines typically prescribe a uniform radiation dose pending several clinical factors and do not take into account tumor biological factors affecting radiosensitivity. Precision medicine trials incorporating genomic alterations into treatment allocation are on the rise; however, most affect chemotherapy delivery, not radiotherapy (Roper et al., 2015). There are genes known to affect radiosensitivity based on their expression, warranting further study (Torres-Roca et al., 2005). Eschrich et al. created a gene expression model using 10 genes to predict pathological response to chemoradiation for rectal and esophageal cancer (Eschrich et al., 2009). By evaluating genes that play a role in radiation signaling pathways, the model was able to significantly separate responders from nonresponders to chemoradiation (Eschrich et al., 2009). Similarly, in the setting of post-operative radiation for prostate cancer, Zhao et al. used a 24-gene Post-Operative Radiation Therapy Outcomes Score (PORTOS) and found that patients with a "high" PORTOS score were less likely to develop metastasis if they received radiotherapy (Zhao et al., 2016). Last, a genomic-adjusted radiation dose (GARD), created from a gene-expression–based radiosensitivity index and the linear quadratic model, was evaluated by Scott et al. using five clinical cohorts, and they found wide variations within populations and tumor types (Scott et al., 2017). The authors found that high GARD scores, and therefore higher therapeutic effect from radiotherapy, predicted improved clinical outcomes for patients. If genomic data from retrospective and prospective trials are used to update and strengthen genomic-based models, there is opportunity to identify those who would benefit most from radiation, as well as potentially adjust radiation dose to the individual level.

10.4 BIG DATA AND PATIENT-REPORTED OUTCOMES

The IOM identified patient-centered care as one of the six aims for high-quality health care in 2001. In the release, patient-centered care was defined as care that is "respectful of and responsive to individual patient preferences, needs, and values, and ensuring that patient values guide all clinical decisions" (Institute of Medicine, 2001). To achieve these aims, patient-reported outcomes (PROs) must be taken into consideration when selecting treatment options for individual patients. Patient registries, particularly cancer registries, are useful in evaluating survival outcomes; however, they do not typically include PROs. There is a need for leveraging big data analytics to evaluate PROs, particularly now that EMRs and patient registries are more widely utilized.

Bandarage et al. reported outcomes from a cohort study by the Victorian Prostate Cancer Registry (Patabendi Bandarage et al., 2016). Within this prostate cancer registry, data are systematically collected by staff members, including a complete summary of patient history, diagnosis, treatment, and, most uniquely, quality of life outcomes reported by patients with prostate cancer. By including PROs in a registry, there is potential to uncover unrecognized, but clinically important, variation in outcomes that relate most to patients. In the Victorian Prostate Cancer registry, authors found a significantly higher rate of "big bother" symptoms following radiotherapy at one regional cancer service compared with the rest of Victoria. These results can be used to further investigate why there are differences in patient outcomes by treatment facility, which may otherwise be ignored.

As information becomes increasingly complex and as we move toward personalized cancer medicine, using big data to evaluate PROs will be important in assessing the successful implementation of genomics-driven medicine. Arora and colleagues articulated a vision for a patient-centered approach to genomic medicine, with genomic data guiding health care alongside the patient's individual comorbidities and preferences and priorities (Arora et al., 2015). They envision a shift away from using a one-size-fits-all approach for patients and instead recognize that there are outliers that require a modified approach, made possible with genomic data. In the same way, the insights provided by big data for radiation therapy must be considered and implemented in a patient-centered way. PROs provide the context for patient-centered application of big data and CER in radiation oncology.

In order to leverage big data for improving PROs, more structured data from patients must be stored and available for analysis, such as the case for the Victorian Prostate Cancer registry. Similar data ideally should be available from all or the majority of patients undergoing cancer treatment. In order to accomplish this, several steps are needed, as outlined by the NIH and other organizations (NIH Advisory Committee to the Director from its Data and Informatics Working Group [DIWG], 2012). Specifically for PROs, trained staff members are needed to document and interpret patient-reported quality-of-life surveys, whether they occur directly during interactions with medical staff or via patient questionnaires. Although this information is usually available in the setting of a clinical trial, more efforts should be made to document PROs for all cancer patients in an accessible format for researchers to analyze.

10.5 FUTURE DIRECTIONS

Although the work described shows a positive trend toward taking advantage of big data to improve patient outcomes, additional effort and resources are needed to improve how we store and process big data. A June 2012 report to the NIH Advisory Committee to the DIWG recommended that the NIH invest in technology to allow researchers to more easily find, access, analyze, and curate research data (NIH Advisory Committee to the DIWG, 2012). To accomplish this, the committee described a serious need for the increased funding of information technology efforts. Specifically, the four recommendation include that the NIH (1) facilitate the broad use of biomedical digital assets by making them findable, accessible, interoperable, and reusable (FAIR); (2) conduct research and develop the methods, software, and tools needed to analyze biomedical big data; (3) enhance training in the development and use of methods and tools necessary for biomedical big data science; and (4) support a data ecosystem that accelerates discovery as part of a digital enterprise (NIH Advisory Committee to the DIWG, 2012). To meet these challenges, the Big Data to Knowledge (BD2K) initiative was created by the NIH. Although the NIH is only a part of the big data ecosystem, their leadership role is crucial in conveying messages, funding, developing metrics for success, and creating infrastructure for long-term solutions within the scientific community (Margolis et al., 2014). An example of a big data system is North Carolina's state-funded Integrated Cancer Information and Surveillance System (ICISS) (Meyer et al., 2014). The primary goal of ICISS is to use big data to facilitate high-impact cancer-focused research. Data are collected and processed from multiple levels, from person-level data to area-level data, and stored in secure data analysis platforms, and a central research tracking system is used to monitor requests for access. Future researchers can use a system like the ICISS as a blueprint for implementing a big data system that will allow for consistent, easily accessible information to analyze.

10.6 CONCLUSION

It is clear that CER and big data provide a framework of complementary methods that can be leveraged for discovery and validation in large data sets. Radiation oncology is a specialty that is uniquely posed to benefit from this synergy based on the fact that the diseases treated with radiotherapy pose many unique challenges and vast amounts of data have been stored in EMRs over recent decades. These clinical notes, DICOM-RT data sets, follow-up images, and laboratory values have been stored and can be utilized to see tremendous gains in cancer care. Moreover, these big data systems and infrastructure for large-scale data collection and interpretation will let investigators make important discoveries with improved external validity and without the time delays characteristic of clinical trials. However, in order to realize this vision, large-scale collaborative data efforts will be necessary.

REFERENCES

Arora NK, Hesse BW, Clauser SB (2015) Walking in the shoes of patients, not just in their genes: A patient-centered approach to genomic medicine. *Patient* 8:239–245.

Berger ML, Doban V (2014) Big data, advanced analytics and the future of comparative effectiveness research. *J Comp Eff Res* 3:167–176.

Boothe D, Orton A, Odei B, Stoddard G, Suneja G, Poppe MM, Werner TL, Gaffney DK (2016) Chemoradiation versus chemotherapy or radiation alone in stage III endometrial cancer: Patterns of care and impact on overall survival. *Gynecol Oncol* 141:421–427.

Chaussy CG, Thuroff S (2017) High-intensity focused ultrasound for the treatment of prostate cancer: A review. *J Endourol* 31:S30–S37.

Chen AM et al. (2017) Reduced-dose radiotherapy for human papillomavirus-associated squamous-cell carcinoma of the oropharynx: A single-arm, phase 2 study. *Lancet Oncol* 18:803–811.

Committee on the Learning Health Care System in America (2013) *Best Care at Lower Cost: The Path to Continuously Learning Health Care in America*. Washington, DC: National Academies Press.

Council of Advisors on Science and Technology (2008) President's Council of Advisors on Science and Technology. Priorities for Personalized Medicine.

Creutzberg CL et al. (2011) Fifteen-year radiotherapy outcomes of the randomized PORTEC-1 trial for endometrial carcinoma. *Int J Radiat Oncol Biol Phys* 81:e631–e638.

Donovan JL et al. (2016) Patient-reported outcomes after monitoring, surgery, or radiotherapy for prostate cancer. *N Engl J Med* 375:1425–1437.

Eschrich SA et al. (2009) A gene expression model of intrinsic tumor radiosensitivity: Prediction of response and prognosis after chemoradiation. *Int J Radiat Oncol Biol Phys* 75:489–496.

Hamdy FC et al. (2016) 10-year outcomes after monitoring, surgery, or radiotherapy for localized prostate cancer. *N Engl J Med* 375:1415–1424.

Hsiao CH, Hing E (2014) Use and Characteristics of Electronic Health Record Systems Among Office-based Physician Practices: United States, 2001–2013. NCHS Data Brief, 143:1–8.

Institute of Medicine (2001) *Crossing the Quality Chasm: A New Health System for the 21st Century*. Washington, DC: National Academies Press.

Jagsi R, Bekelman JE, Chen A, Chen RC, Hoffman K, Shih YC, Smith BD, Yu JB (2014) Considerations for observational research using large data sets in radiation oncology. *Int J Radiat Oncol Biol Phys* 90:11–24.

Jhingran A, Klopp A (2016) Big data to the rescue: Is there a benefit to combined-modality adjuvant therapy in endometrial cancer? *Gynecol Oncol* 141:403–404.

Kalow W (1997) Pharmacogenetics in biological perspective. *Pharmacol Rev* 49:369–379.

Keys HM, Roberts JA, Brunetto VL, Zaino RJ, Spirtos NM, Bloss JD, Pearlman A, Maiman MA, Bell JG, Gynecologic Oncology Group (2004) A phase III trial of surgery with or without adjunctive external pelvic radiation therapy in intermediate risk endometrial adenocarcinoma: A Gynecologic Oncology Group study. *Gynecol Oncol* 92:744–751.

Krumholz HM (2014) Big data and new knowledge in medicine: The thinking, training, and tools needed for a learning health system. *Health Aff (Millwood)* 33:1163–1170.

Lohr S (2012) The age of big data. *The New York Times*.

Lyman GH, Levine M (2012) Comparative effectiveness research in oncology: An overview. *J Clin Oncol* 30:4181–4184.

Maggi R, Lissoni A, Spina F, Melpignano M, Zola P, Favalli G, Colombo A, Fossati R (2006) Adjuvant chemotherapy vs radiotherapy in high-risk endometrial carcinoma: Results of a randomised trial. *Br J Cancer* 95:266–271.

Margolis R, Derr L, Dunn M, Huerta M, Larkin J, Sheehan J, Guyer M, Green ED (2014) The National Institutes of Health's Big Data to Knowledge (BD2K) initiative: Capitalizing on biomedical big data. *J Am Med Inform Assoc* 21:957–958.

Matei D et al. (2017) A randomized phase III trial of cisplatin and tumor volume directed irradiation followed by carboplatin and paclitaxel vs. carboplatin and paclitaxel for optimally debulked, advanced endometrial carcinoma. *J Clin Oncol* 35.

Meyer AM, Olshan AF, Green L, Meyer A, Wheeler SB, Basch E, Carpenter WR (2014) Big data for population-based cancer research: The integrated cancer information and surveillance system. *N C Med J* 75:265–269.

NIH Advisory Committee to the Director from its Data and Informatics Working Group (DIWG) (2012) Draft Report to The Advisory Committee to the Director.

Nout RA et al. (2010) Vaginal brachytherapy versus pelvic external beam radiotherapy for patients with endometrial cancer of high-intermediate risk (PORTEC-2): An open-label, non-inferiority, randomised trial. *Lancet* 375:816–823.

Patabendi Bandarage VR, Billah B, Millar JL, Evans S (2016) Prospective evaluation of patient-reported quality of life outcomes after external beam radiation treatment for prostate cancer in Victoria: A cohort study by the Victorian Prostate Cancer Registry. *J Med Imaging Radiat Oncol* 60:420–427.

Randall ME, Filiaci VL, Muss H, Spirtos NM, Mannel RS, Fowler J, Thigpen JT, Benda JA, Gynecologic Oncology Group S (2006) Randomized phase III trial of whole-abdominal irradiation versus doxorubicin and cisplatin chemotherapy in advanced endometrial carcinoma: A Gynecologic Oncology Group Study. *J Clin Oncol* 24:36–44.

Randall ME, Spirtos NM, Dvoretsky P (1995) Whole abdominal radiotherapy versus combination chemotherapy with doxorubicin and cisplatin in advanced endometrial carcinoma (phase III): Gynecologic Oncology Group Study No. 122. *J Natl Cancer Inst Monogr* 13–15.

Roper N, Stensland KD, Hendricks R, Galsky MD (2015) The landscape of precision cancer medicine clinical trials in the United States. *Cancer Treat Rev* 41:385–390.

Scott JG et al. (2017) A genome-based model for adjusting radiotherapy dose (GARD): A retrospective, cohort-based study. *Lancet Oncol* 18:202–211.

Skripcak T et al. (2014) Creating a data exchange strategy for radiotherapy research: Towards federated databases and anonymised public datasets. *Radiother Oncol* 113:303–309.

Sox HC (2010) Defining comparative effectiveness research: The importance of getting it right. *Med Care* 48:S7–S8.

Susumu N, Sagae S, Udagawa Y, Niwa K, Kuramoto H, Satoh S, Kudo R, Japanese Gynecologic Oncology Group (2008) Randomized phase III trial of pelvic radiotherapy versus cisplatin-based combined chemotherapy in patients with intermediate- and high-risk endometrial cancer: A Japanese Gynecologic Oncology Group study. *Gynecol Oncol* 108:226–233.

Torres-Roca JF et al. (2005) Prediction of radiation sensitivity using a gene expression classifier. *Cancer Res* 65:7169–7176.

Trusheim MR, Berndt ER, Douglas FL (2007) Stratified medicine: Strategic and economic implications of combining drugs and clinical biomarkers. *Nat Rev Drug Discov* 6:287–293.

Vicini P et al. (2016) Precision medicine in the age of big data: The present and future role of large-scale unbiased sequencing in drug discovery and development. *Clin Pharmacol Ther* 99:198–207.

Xiong T, Turner RM, Wei Y, Neal DE, Lyratzopoulos G, Higgins JP (2014) Comparative efficacy and safety of treatments for localised prostate cancer: An application of network meta-analysis. *BMJ Open* 4:e004285.

Zhao SG et al. (2016) Development and validation of a 24-gene predictor of response to postoperative radiotherapy in prostate cancer: A matched, retrospective analysis. *Lancet Oncol* 17:1612–1620.

11 Cancer registry and big data exchange

Zhenwei Shi, Leonard Wee, and Andre Dekker

Contents

11.1	Introduction	154
11.2	Cancer Registry	155
	11.2.1 Different Types of Cancer Registry	155
	11.2.1.1 Hospital-Based CRs	156
	11.2.1.2 Population-Based CRs	156
	11.2.2 Examples of Cancer Registries	156
	11.2.2.1 Surveillance, Epidemiology, and End Results	156
	11.2.2.2 National Radiation Oncology Registry	156
	11.2.2.3 Italian Association of Cancer Registries (AIRTUM)	157
	11.2.2.4 Dutch Surgical Colorectal Audit	157
	11.2.3 Data Sources of CRs	157
	11.2.4 Receiving and Reporting Information in CRs	161
	11.2.4.1 Receiving Information	161
	11.2.4.2 Reporting Information	164
	11.2.5 Assessment of Data Quality within CRs	164
11.3	From CR to Big Data in Radiation Oncology	164
	11.3.1 Barriers of CRs	164
	11.3.2 Big Data in Radiation Oncology	166
11.4	Big Data Exchange in Radiation Oncology	168
	11.4.1 Big Data Collection	168
	11.4.2 Standard and Framework for Data Exchange	169
	11.4.3 Data Pooling Architectures	169
	11.4.3.1 Centralized Architecture	169
	11.4.3.2 Decentralized Architecture	170
	11.4.3.3 Hybrid Architecture	170
	11.4.4 Data Interoperability	171
	11.4.4.1 Semantic Web	171
	11.4.4.2 Ontology	172
	11.4.5 Data Exchange	173
	11.4.5.1 Send Data out	173
	11.4.5.2 Send Questions in	174
	11.4.5.3 Centralized versus Distributed Learning Architecture	174

11.5   Barriers of Big Data Exchange among Multicenters   177
    11.5.1   Administrative Barrier   177
    11.5.2   Ethical Barrier   177
    11.5.3   Political Barrier   177
    11.5.4   Technical Barrier   177
11.6   Conclusion   178
References   178

11.1 INTRODUCTION

A vast stream of data is generated by the routine operations of modern cancer diagnosis and oncologic treatments. Usually, the data is stored electronically but it tends to be scattered across different disciplines (e.g., radiation oncology, radiology, medical oncology, surgery), different data storage platforms (e.g., electronic medical record [EMR], Picture Archiving Communication System [PACS]) and in a wide variety of formats (e.g., Digital Imaging and Communication in Medicine [DICOM], ASCII, PDF) (Deng 2014). In addition, data in cancer care has the particular properties of large volume and complex dependencies between data elements, which is creating growing difficulties for conventional methods of data handling. By handling, we refer to collection, storage, update, and exchange. Although the variety and volume of big data continues to grow exponentially within the field of oncology (Chen et al. 2014), it has not been easy to exploit this rich vein of data to improve patient safety and health outcomes (McNutt et al. 2016).

Data, as a whole, in the clinical routine is one of the most valuable, but most underutilized, assets within radiotherapy and oncology studies (Roelofs et al. 2013). Rapid Learning Health Care (RLHC) (Lambin et al. 2013a) envisions virtuous cycles of rapid knowledge generation and knowledge utilization within health care, by combining routine clinical practice with prospective research studies using a big data exchange paradigm. The resulting interconnected web of data repositories across departments and institutions becomes a global resource to mine for new knowledge, generate hypotheses for novel treatments, and test the effectiveness of interventions in real-world settings. RLHC also leads to a wide range of distributed software applications that can exploit FAIR (findable, accessible, interoperable, and reusable) data repositories (Wilkinson et al. 2016) to create predictive outcome models for clinical decision support (Lambin et al. 2013b) and to discover predictive digital signatures in biomedical images (i.e., radiomics) (Aerts et al. 2014).

To be effective in clinical decision support, predictive models must be able to estimate the probability of a given outcome over a range of clinical situations. Therefore, an approach that integrates the data of many patients over different treatment settings is essential. Predictive outcome models have the potential to improve quality of life, identify patients at high (or low) risk, and to prolong the survival of patients with cancer (Dehing-Oberije et al. 2009, 2010, 2011; Oberije et al. 2014). Some predictive outcome models for various cancer sites can be found at http://www.predictcancer.org. These prediction models support the practice of personalized radiotherapy that is tailored to the individual risk profile of the patient. If multiple treatment possibilities exist where clinical evidence is equivocal, patient and physician preferences for certain outcomes may play a role. The collaborative consultation process between a patient and their treatment physician is known as shared decision making (Elwyn et al. 2012; Stacey et al. 2014). Reliable model-based predictions of potential treatment responses are a prerequisite for information trade-offs between the risks of harm and strength of treatment in the shared decision-making paradigm. In keeping with RLHC, observational data on quality-of-life, patient-reported outcomes, and decision regret should put back into the web of clinical knowledge, so that continually updated models are better able to inform physician's and patients' decisions.

We have prefaced this chapter with a consideration of the wide-ranging opportunities for clinical innovation and improved outcomes that could be possible with comprehensively voluminous, multimodality, and multi-institutional data. Central to this ambition is the integration of all the data that presently remains underutilized in closed, unconnected private repositories. The remainder of this chapter is organized as follows: first, we explore the paradigm of data centralization in the form of a cancer registry (CR). This, whether large or small, represents the current orthodoxy for collection, storage, analysis, and distribution of oncology data. Second, as the multiplicity, volume, and dimensionality of cancer data continues to grow rapidly, we shall argue that a new paradigm for data exchange is urgently needed. We thus describe practical and feasible big data architectures that can be applied to overcome the challenges of poorly connected data repositories in radiation oncology. Third, the remaining barriers impeding universal rollout of big data exchange architecture in radiation oncology field will be examined, as will the unique characteristics of certain data architectures that may be more naturally adapted to overcoming these barriers. The primary aim of this chapter is to support the understanding of readers about big data exchange architectures that facilitate rapid learning in clinical practice and scientific research in the field of radiation oncology.

11.2 CANCER REGISTRY

A CR refers to architectures that are capable of systematically capturing, storing, analyzing, and reporting data on patients with cancer (Santos 1999). Data on cancer occurrence and properties is registered in CRs. A cancer registrar is a person with over-arching responsibility to capture complete, accurate, and timely data on cancer patients. Participation in cancer registries is typically, but not solely, mandated by local legislations.

Data within a CR typically consists of demographic information, medical history, diagnostic investigations, outcomes of therapy, and some follow-up details of patients. CRs may have a variety of functionalities. For example, data within a CR can be used to

    a. Assess treatment outcomes (e.g., survival and toxicity)
    b. Assess disease survivorship aspects such as quality of life
    c. Evaluate efficacy and/or economic impacts of diseases and their interventions
    d. Provide follow-up investigation and guidance
    e. Allocate health resources at regional or country level
    f. Compare quality of treatment and outcomes between care providers
    g. Report on cancer incidence.

11.2.1 DIFFERENT TYPES OF CANCER REGISTRY

A CR may have different specific purposes and functionalities as described above, which primarily depend on local demands and requirements. Two primary types of CR are commonly used worldwide: hospital-based and population-based CRs. Table 11.1 shows a brief description of these two types of CR.

Table 11.1 **Brief description of each type of CR**

TYPES	HOSPITAL-BASED CR	POPULATION-BASED CR
Purposes	• Improvement of cancer care • Administration of cancer information • Clinical research • Training and education	• Cancer prevention and early detection • Determine cancer incidence and trends • Academic research • Assessment of population cancer outcomes
Details	• Maintains data on all cancer patients diagnosed and/or treated at a particular hospital • Provides medical audit-style assessment of outcomes within a particular hospital • Support institutional registries with common standard protocols and integrated data	• Records all new cases in a well defined population (e.g., geographic area) with an emphasis on epidemiology and public healthcare • Informs cancer agencies and organizations of cancer statistics in specific populations • Informs cancer researchers about an unbiased group of cases that can be selected for studies

11.2.1.1 Hospital-based CRs

Hospital-based CRs are established to record the information of cancer patients collected within a specific care setting, such as a major hospital or cancer clinic that aims to offer readily accessible, complete, accurate, and timely information of patients with cancer (Young 1991). This may include, for example, data about diagnosis, treatments, and outcomes. Hospital-based CRs generate reports on the number of cancers observed within a particular hospital per year by site, sex, and age. These reports are very useful for clinical research by comparing the frequency of certain types of cancer within a single hospital to the total number of cancer cases (Santos 1999). Furthermore, it can lead to several potential applications for epidemiological research by (1) providing information on approaches of diagnosis, stage distribution, outcomes to treatment, and overall survival at the hospital level and identifying potential drawbacks of treatments; and (2) predicting future needs for services, equipment, and human resources within a particular cancer center.

Generally, the endpoints of a hospital-based CR are geared toward quality, management, and caseload planning goals. However, many hospital-based CRs are required to submit data to a centralized disease-specific CR as well. For this purpose, the hospital-based CR usually has to collect data elements that are useful for the central CR but may be not immediately useful for the hospital. Therefore, data elements within in a hospital-based CR often cover a wider range than data within a population-based CR (Young 1991).

11.2.1.2 Population-based CRs

Population-based CRs are concerned with collecting data on all new patients with cancer arising in a well-defined community (Santos 1999; Sadjadi et al. 2003; Parkin 2006; Coleman et al. 2011). The primary purpose of population-based CRs is to report cancer incidence and produce analytic findings about cancer in a defined population. In addition, population-based CRs are of benefit in assessment and control of cancer care. Thus, the focus of population-based CRs is on epidemiology research and public health care.

Furthermore, population-based CRs can monitor the occurrence of cancer and prevalence of diseases; thus, they are of importance in planning and evaluating region-based or population-based programs on cancer control through

1. Standardizing treatment priorities and predicting resources needed in the future
2. Examining the effectiveness and appropriateness of screening programs in the community
3. Comparing health care providers in terms of practice and quality
4. Evaluating cancer care population outcomes through survival statistics.

Because it is not always possible to strictly define a catchment population, a hospital-based CR is not necessarily able to provide assessment and statistics on the cancer occurrence in the defined population (Young 1991), which is a major difference between hospital- and population-based CRs.

11.2.2 EXAMPLES OF CANCER REGISTRIES

There are many running CRs in the world. In this section, we will describe a small collection of them, which are from three countries: the United States, Italy, and the Netherlands.

11.2.2.1 Surveillance, epidemiology, and end results

The Surveillance, Epidemiology, and End Results (SEER) program is a source of information on cancer incidence and survival in the United States. The data is published on its Web site (https://seer.cancer.gov/about/overview.html). SEER currently captures and publishes cancer data on incidence and survival from the cancer data sources that cover around 28% of the American population.

The data captured by the SEER Program comprises demographics information, main tumor site, tumor morphology and cancer stage, treatment at first course, and follow-up details. The SEER program is the only program that collects the data on cancer stage at diagnosis and on survival data of cancer patients in the United States. The data is updated annually and published in reports as a public service. The U.S. Census Bureau provides periodic data on the population for the SEER program to use to compute the cancer ratio. Many practitioners and members of the public have used the data reported by the SEER program.

11.2.2.2 National radiation oncology registry

The National Radiation Oncology Registry (NROR; http://www.roinstitute.org/What-We-Do/NROR/Index.aspx) is a collaborative initiative of the Radiation Oncology Institute (ROI) and the American

Society of Radiation Oncology (ASTRO) through guidance and data support from other major stakeholders in oncology. The NROR captures related data on cancer patients' treatment delivery and outcomes; the data are used to improve cancer care. The overarching purposes of the NROR are to (1) compare cancer patients who have similar cancer state or profiles, (2) identify suitable treatment and possible drawbacks in cancer care, and (3) build a CR for health care in a defined population.

The national registry comprises standardized aggregated data on therapies for specific types of cancers. An analysis of the outcomes achieved would yield invaluable benchmarking measures, help define best practices, evaluate the comparative effectiveness of treatments, and identify gaps in quality. However, due to funding limitations, the NROR is no longer active.

11.2.2.3 Italian Association of Cancer Registries (AIRTUM)

The Italian Association of Cancer Registries (AIRTUM), established in Florence, Italy, in 1997, aims to coordinate multiple cancer registries in Italy. For more details, see http://www.registri-tumori.it/cms/en. The linkage created by the association supports research, editorial output, and methodological development of the various member registries. The association is connected to equivalent bodies in other countries at European and global levels. Statistics about the distribution of cancer in the areas covered by the member registries covers

a. Incidence—the number of new cancer cases per year
b. Prevalence—the number of Italians who have a particular cancer
c. Mortality—the various different causes of death for Italians registered in the CR
d. Trends—whether the number of cancer cases has been increasing or decreasing with respect to preceding years
e. Survival—how long Italians survive after treatments for cancer
f. Comparison of registries—whether the impacts of cancer are uniform across Italy
g. International comparisons—how the situation in Italy compare with the rest of the world

11.2.2.4 Dutch surgical colorectal audit

The Dutch Surgical Colorectal Audit (DSCA) (Van Leersum et al. 2013) was established by the Association of Surgeons of the Netherlands (ASN) in 2009 and aims at surveillance, assessment, and improvement of colorectal cancer treatment. For more details, see https://www.dica.nl/dlca.

All Dutch hospitals that perform bowel cancer and rectal cancer surgery, participate in the Web-based quality registration. At present, more than 60,000 treatments have been registered. This registration allows quality benchmarking, where hospitals compare the quality of their cancer care with those of others. The comparisons are statistically corrected for differences in level of care (local case-mix) and random sampling, which renders the analysis meaningful and "fair." The system has been able to propose possible improvements to individual colorectal surgeons, while the national professional association facilitates and monitors these improvements.

One of the main contributions of DSCA is that it leads to effective multicenter surgical collaboration. Because the ASN has an important role in audits, all colorectal surgeons in the Netherlands have participated in the collaboration.

11.2.3 DATA SOURCES OF CRs

Data sources of CRs are external data resources available to the registry that are used for the collection and verification of cancer-related information. According to the relation between the data elements and the goals of CRs, there are two types of data sources: primary and secondary data sources (Gliklich et al. 2014). Primary data sources refer to the data collected for the immediate goals of the CR. The data collection from primary data sources can promote elements of data quality such as completeness, validity, and reliability. This data collection is implemented via a standard protocol. The protocol is intended to enforce the same procedures and data format used in all CRs and patients, which ultimately benefits data analysis, tracking, and integration. Due to the auditability of collected data, the entered data can typically be traced back to an individual patient. Finally, the quality of primary data sources is usually better than secondary data sources because of automatic quality control procedures or follow-up checks made by data managers.

The initial purposes of secondary data sources are not for cancer registration (e.g., data generated in routine medical practice and insurance claim forms). The data in secondary data sources is usually stored electronically and can be accessed through appropriate permissions (Gliklich et al. 2014).

To ensure that few cancer cases are missed and that the quality of data (i.e., dimensionality, completeness, accuracy, timeliness) remains high, CRs usually collect the data of patients with cancer from multiple sources. For instance, data sources of a population-based CR often refer to cancer centers, general practitioners, screening programs, coroners' recording systems, health insurance companies, and other CRs. However, as a disadvantage, the use of multiple sources of information raises concerns about receiving multiple notifications of the same cancer patient. To avoid this problem, the data of the same patient existing within multiple data sources should be linkable, thereby eliminating duplicate registry.

It might be generally presumed that it is simple to maintain a population-based CR when sub-registries (e.g., hospital-based CRs) can openly transfer identifiable data such that the population-based CR only needs to integrate the incoming data. However, this is not always possible in real-world scenarios. In practice, the population-based CR still needs to collect overlapping data from numerous data sources. The reasons are twofold. First, patients with an eligible condition might never attend a contributing hospital, therefore the population-based CR needs to use multiple sources to prevent eligible cases being missed. Second, patients may attend more than one hospital over different parts of their treatment pathway. The use of multiple sources is then of benefit of identifying duplicate registrations or missing registrations of the same patient. Although it is not always possible to collect all data from all data sources in practice, the aim is still to use as many cancer data sources as possible. As described in (Gliklich et al. 2014), advantages and disadvantages of key sources are presented in Table 11.2.

Table 11.2 **Advantages and disadvantages of key data sources**

DATA SOURCE	ADVANTAGES	DISADVANTAGES
Patient report	• Unique perspective based on patient experience of disease and treatment. • Information on treatments not necessarily prescribed by clinicians. • Obtaining information about intended compliance. • Useful when timing of follow-up may not be concordant with timing of clinical encounter. • Can capture patient and/or caregiver outcomes.	• Literacy, language, physician access or other barriers that may result in under-enrolment of some sub-populations. • Validated data collection equipment may need to be established. • Patients may refuse to participate in follow-up study. • Limited confidence on clinical information and utilization information.
Clinician report	• More specific information than available from coded data or medical record. • Tends to be more objective and focused on impacts on care delivery.	• Clinicians are highly conscious of administrative to burden. • Potential inconsistencies in capture of patient signs, symptoms, use of non-prescribed therapy.
Medical chart abstraction	• Information on routine medical care, with more clinical context than coded claims. • Potential for comprehensive view of patient medical and clinical history.	• The underlying information is not always collected in a systematic way. For example, a diagnosis of bacterial pneumonia by one physician may be based on a physical exam and patient report of symptoms, while another physician may record the diagnosis only in the presence of a confirmed laboratory test.

(Continued)

Table 11.2 (*Continued*) **Advantages and disadvantages of key data sources**

DATA SOURCE	ADVANTAGES	DISADVANTAGES
	• Use of abstraction and strict coding standards (including handling missing data) increases the quality and interpretation of data abstracted.	• It is difficult to interpret missing data. For example, absence of a specific symptom in the visit record would not indicate whether the symptom was truly absent or that the physician did not actively inquire about this specific symptom or set of symptoms. • Data abstraction is more (technical) resource intensive. • Complete medical and clinical history may not be available (e.g., new patient to clinic).
Electronic health records (EHRs)	• Information on routine medical care and practice, with more clinical context than coded claims. • Potential for comprehensive view of patient medical and clinical history. • Effective access to medical and clinical data. • Use of data transfer and coding standards (including handling of missing data) will increase the quality of data abstracted.	• Underlying information from clinicians is not collected using uniform decision rules. (See example under "Medical chart abstraction.") • Consistency of data quality and breadth of data collected varies across sites. • Difficult to handle information uploaded as image files into the EHRs (e.g., scanned clinician reports) vs. direct entry into data fields. • Historical data capture may require manual chart abstraction prior to implementation date of medical records system. • Complete medical and clinical history may not be available (e.g., new patient to clinic). • EHR systems vary widely. If data come from multiple systems, the registry should plan to work with each system individually to understand the requirements of the transfer.
Institutional databases	• Diagnostic and treatment information (e.g., pharmacy, laboratory, blood bank, radiology). • Resource utilization data (e.g., days in hospital). • May incorporate cost data (e.g., billed and/or paid amounts from insurance claims submissions).	• Important to be knowledgeable about coding systems used in entering data into the original systems. • Institutional or organizational databases vary widely. The registry should plan to work with each system individually to understand the requirements of the transfer.

(Continued)

Big data in radiation oncology

Table 11.2 (*Continued*) **Advantages and disadvantages of key data sources**

DATA SOURCE	ADVANTAGES	DISADVANTAGES
Administrative databases	• Useful for tracking health care resource utilization and cost-related information. • Range of data includes anything that is reimbursed by health insurance, generally including visits to physicians and allied health providers, most prescription drugs, many devices, hospitalization(s), if a lab test was performed, and in some cases, actual lab test results for selected tests (e.g., blood test results for cholesterol, diabetes). • In some cases, demographic information (e.g., gender, date of birth from billing files) can be uploaded. • Potential for efficient capture of large populations.	• Represents clinical cost drivers vs. complete clinical diagnostic and treatment information. • Important to be knowledgeable about the process and standards used in claims submission. For example, only primary diagnosis may be coded and secondary diagnoses not captured. In other situations, value-laden claims may not be used (e.g., an event may be coded as a "nonspecific gynecologic infection" rather than a "sexually transmitted disease"). • Important to be knowledgeable about data handling and coding systems used when incorporating the claims data into the administrative systems. • Can be difficult to gain the cooperation of partner groups, particularly in regard to receiving the submissions in a timely manner.
Death indexes	• Completeness—death reporting is mandated by law in many countries, such as the United States. • Dependable alternative source for mortality tracking (e.g., if a patient was lost to follow-up). • National Death Index (NDI) — centralized database of death records from State vital statistics offices; database updated annually. • NDI causes of death relatively reliable (93–96%) compared with State death certificates. • Social Security Administration's (SSA) Death Master File—database of deaths reported to SSA; database updated weekly.	• Time delay—indexes depend on information from other data sources (e.g., State vital statistics offices), with delays of 12 to 18 months or longer (NDI). It is important to understand the frequency of updates of specific indexes that may be used. • Absence of information in death indexes does not necessarily indicate "alive" status at a given point in time. • Most data sources are country specific and thus do not include deaths that occurred outside of the country. • As of November 2011, Death Master File no longer includes protected State records.

(*Continued*)

Big data in radiation oncology

Table 11.2 (*Continued*) **Advantages and disadvantages of key data sources**

DATA SOURCE	ADVANTAGES	DISADVANTAGES
Existing registries	• Can be merged with another data source to answer additional questions not considered in the original registry protocol or plan. • May include specific data not generally collected in routine medical practice. • Can provide historical comparison data. • Reduces data collection burden for sites, thereby encouraging participation	• Important to understand the existing registry protocol or plan to evaluate data collected for element definitions, timing, and format, as it may not be possible to merge data unless many of these aspects are similar. • Creates a reliance on the other registry. Other registry may end. • Other registry may change data elements (which highlights the need for regular communication). • Some sites may not participate in both. Must rely on the data quality of the other registry.

11.2.4 RECEIVING AND REPORTING INFORMATION IN CRs

Because a CR is an organization for systematically processing the data of patients with cancer, it is very important to determine (1) the data items that a CR will receive from various data sources, and (2) the data items that a CR will report to other organizations. Received information by CRs refers to the information (e.g., demographic information, medical history, diagnostic discoveries, therapeutic information, follow-up details) collected by CRs from multiple sources or reported by certain institutions. On the other hand, reported information by CRs is simply concerned with summary analyses generated by the CRs, which can be deemed the functional output of the CRs. Because CRs can be different from each other in terms of aims, functions, local needs, and requirements, receiving and reporting information might also be different. The details of receiving and reporting information of CR are as follows.

11.2.4.1 Receiving information

The data elements collected by a CR are directly related to its aims and functionalities. The primary purpose of a hospital-based CR is administration of patients with cancer in a particular site. On the other hand, the main goal of a population-based CR is to produce statistics about cancer occurrence in a defined population. Thus, choosing data elements for a CR requires considering many aspects such as data reliability, the necessity in analyzing treatment responses, and even the cost of data collection (Gliklich et al. 2014).

11.2.4.1.1 Basic data elements

Although we have to determine the purposes and functionalities of a CR before specifying the data elements, some basic common elements exist in most CRs. The basic data elements that have been described in (Gliklich et al. 2014) are listed in Table 11.3, with the caveat that many directly identifying data elements will often be coded or partitioned in a secure part of the registry for privacy purposes.

11.2.4.1.2 Optional data items

Adding additional collection elements will increase the complexity and cost of registration (MacLennan 1991). Therefore, when designing registration forms and before performing registration, one should first

Table 11.3 **Examples of possible basic data elements**

Registrar information	• Registrar contact information • Contact information (e.g., address, telephone and email) of another individual who can be reached for follow-up
Patient information	• Patient identifiers (e.g., name, age, date of birth, place of birth, Social Security number) • Permission/consent • Source of enrollment (e.g., provider, institution, phone number, address, contact information) • Enrollment criteria • Sociodemographic characteristics, including race, gender, and age or date of birth • Education and/or economic status, insurance, etc. • Place of birth • Location of residence at enrollment • Source of information • Country, State, city, county, ZIP Code of residence.

consider whether a CR really needs certain data elements and whether it can sustain the cost associated with collecting these data element. Apart from the basic data elements, other data elements may be needed based on the specific design and purpose of a registry. Table 11.4 shows a collection of possible data elements that are described in (Gliklich et al. 2014).

Data element selection is primarily dependent on the goals of a CR, the approaches used for data collection, and the resources available. Many cancer CRs have failed because they attempted to capture too many data elements. The focus of the CR should be on the quality of data rather than the quantity. As we described in Section 2, those successful and productive CRs only collected a limited amount of information for each patient.

Table 11.4 **Examples of optional data elements**

PRE-ENROLMENT HISTORY	
Medical history	• Morbidities/conditions • Onset/duration • Severity • Treatment history • Medications • Adherence • Health care resource utilization • Diagnostic tests and results • Procedures and outcomes • Emergency room visits, hospitalizations (including length of stay), long-term care, or stays in skilled nursing facilities • Genetic information • Comorbidities
Environmental exposures	• Places of residence • Hazardous occupations? • Exposure to occupational hazards?
Patient characteristics	• Development (pediatric/adolescent) • Functional status (including ability to perform tasks related to daily living), quality of life, symptoms

(Continued)

Table 11.4 (*Continued*) **Examples of optional data elements**

PRE-ENROLMENT HISTORY	
	• Health behaviors (alcohol, tobacco use, physical activity, diet) • Social history • Marital status • Family history • Work history • Employment, industry, job category • Social support networks • Economic status, income, living situation • Sexual history • Foreign travel, citizenship • Legal characteristics (e.g., incarceration, legal status) • Reproductive history • Health literacy • Social environment (e.g., community services) • Enrollment in clinical trials (if patients enrolled in clinical trials are eligible for the registry)
Provider/system characteristics	• Geographical coverage • Access barriers • Quality improvement programs • Disease management, case management • Compliance programs • Information technology use (e.g., computerized physician order entry, e-prescribing, electronic medical records)
Follow-up/ Outcomes	• Safety: adverse events (see Chapter 12) • Quality measurement/improvement: key selected measures at appropriate intervals • Effectiveness and value: intermediate and endpoint outcomes; health case resource use and hospitalizations, diagnostic tests and results. Particularly important are outcomes meaningful to patients, including survival, symptoms, function, and patient-reported outcomes, such as health-related quality-of-life measures. • Natural history: progression of disease severity; use of health care services; diagnostic tests, procedures, and results; quality of life; mortality; cause/date of death • Economic status • Social functioning
Other potentially important information	• Changes in medical status • Changes in patient characteristics • Changes in provider characteristics • Changes in financial status • Residence • Changes to, additions to, or discontinuation of exposures (medications, environment, behaviors, procedures) • Changes in health insurance coverage • Sources of care (e.g., where hospitalized) • Changes in individual attitudes, behaviors

Big data in radiation oncology

11.2.4.2 Reporting information

The most important purpose of a CR is to perform statistics on cancer occurrence, treatment and outcomes in a particular region or population (Powell 1991). Therefore, collation, examination, and explanation of the captured data are the main reporting tasks of a CR.

Information reported by CRs is often presented by means of cancer incidence reports, practice and treatment outcomes reports, and scientific publications. Results and conclusions are usually documented in reports and subsequently published to users. Generally, the reports contain background information about registration, procedures of registration, population of covering, data quality (e.g., completeness and validity), and results of analysis. The population-based CR should perform basic statistics that are primarily about the distribution of the tumor in the community. The data and findings may be displayed in various types of format such as tabular and graphical forms, by which the readers can draw their own conclusions according to their interests.

11.2.5 ASSESSMENT OF DATA QUALITY WITHIN CRs

CRs have evolved beyond a data provider that reports cancer incidence within a well-defined population (Parkin 2006). By linking sufficient resources, a CR is useful in many aspects of the cancer control domain, such as identification of causes of specific cancer, assessment of screening programs, and improvement of cancer care (Armstrong 1992; Parkin, 2008).

The functionalities of a modern CR and its capacity to perform cancer control activities are highly dependent on the quality of data within the CR. Three dimensions of data quality have been introduced in the earlier publication (Storm 1996): comparability, completeness, and validity. As described in (Bray and Parkin 2009), timeliness is another key indicator of data quality for CRs.

In order to assure data quality, quality control plays as an important role. Theoretically, it is possible that a CR can collect very high-quality data (similar to clinical trials) without extensive quality control processes, but this is seldom the case in real-world scenarios. There is no large-scale database that can be perfect in regard to completeness and validity. Therefore, routine quality control is a necessary step to identify the area needing improvement. The quality control can help with data interpretation and may further indicate a need for procedural changes (Navarro et al. 2010).

11.3 FROM CR TO BIG DATA IN RADIATION ONCOLOGY

11.3.1 BARRIERS OF CRs

We have seen that CRs can act as a valuable oncology data resource at several geographic scales and between multiple collaborating cancer centers. It remains a complex and costly process to maintain a CR, and the constraints on its architecture are obvious, that is, it is difficult to be scaled up to manage big data in radiation oncology. A centralized data repository needs to coordinate uniform collection at many different data sources and manage exchanges between different types of storage formats. This requires every contributing party to first agree on, and then rigidly adhere to, the same data collection instrument and the same internal data structure. The types of statistical analysis is constrained by the limited data fields collected and further restricted by the operational objective(s) of the CR. Although some attempts have been made by certain CRs to make data available for research, the elements generally have to be extracted as specialized queries by a data manager at the CR.

A large amount of human resources (e.g., registrars and data managers) and infrastructure (e.g. servers, user interfaces) is required to build and maintain a CR, adding significantly to its financial cost. Furthermore, rigidly structured collaborations among many departments are needed to support a CR. For example, a population-based CR requires first that local hospitals, cancer centers, and other institutions extract and upload (automatically or entering by hand) very specific data collection forms pertaining to the condition of interest. Then the collected data has to be processed and audited internally in the CR. Finally, with a significant time latency, results from a rigidly prescribed statistical analysis are reported to higher authorities and made available to the public.

Crucially, one of the most significant barriers to universal adoption of CRs pertains to how big data can be flexibly and securely shared among a large network of cancer institutes or research programs to answer a broad range of clinically relevant questions. First, the data is neither readily findable nor discoverable; no general query process is available to physicians and researchers to ask if data pertaining to their clinical question reside in the CRs.

Second, data in CRs are generally inaccessible to physicians and researchers due to concerns over patient privacy, data sharing, or intellectual property rights. Where an entity outside the CR might be permitted to request some data, internal resources of the CR are required to program a data extraction query specific to each request received. For instance, researchers in a U.S. cancer center may be interested in the overall survival outcomes of lung cancer patients in a reasonably similar cohort in the Netherlands. Assuming the investigators in the United States even know what data fields to ask for, they would not be permitted to structure any external query that might address questions such as the following: (1) Is the case-mix of Dutch and U.S. populations approximately comparable? (2) What types of oncological and surgical interventions do Dutch lung cancer patients receive? (3) Do the follow-up protocols in both populations overlap to a sufficient degree to even contemplate a comparative study? In Dutch law, it is not allowed to share the data even for the purpose of clinical academic research. Patient privacy laws prevent effective data sharing, even in a highly geographically localized setting (e.g., two neighboring cancer hospitals in the Netherlands).

Third, even if the administrative, legal, and political barriers that prevent data sharing can be overcome, the technical challenge remains of reading data across incompatible computer systems that reside in many different formats. The exchange of "data dictionaries" between any two centers may permit a level of syntactic interoperability, but the parties also need to know the exact structure of how the data elements are actually stored in memory in order to construct a query. Exchange protocols such as Health Level-7 (HL7) and Fast Healthcare Interoperability Resources (FHIR) attempt to offer a degree of interoperability. However, each data transaction must be specifically customized to each collaboration site, rather than creating a single universal query that can work across all collaboration sites.

Therefore, one may not only consider a CR as a particular organization or entity (such as a hospital or a governance authority), but rather the CR must be considered as an architectural archetype for progressively more centralized data collection, storage, management, analysis, and dissemination. At each layer of centralization, the variety and contextual richness of the data is incrementally lost, such that only the broadest and least-detailed summary statistics can be reported at the uppermost level (for example, a national cancer registry). The reduction in dimensionality and contextual depth is compensated by an increasing universality in population coverage.

Such a trade-off in CR architecture may be visualized as in Figure 11.1. At the first level, individual institutions (e.g., clinics and hospitals) are the point of generation of data that lies closest to the patient. Data collection is generally immediate and highly multidimensional (e.g., clinician notes, nursing

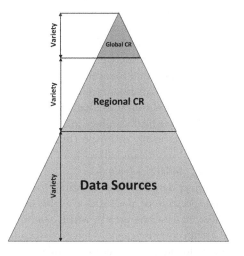

Figure 11.1 Illustration of Cancer Registry (CR) in terms of hierarchies of data collection. Data sources (clinics, hospitals) hold records of highest granularity (individuals) and greatest variety. At the level of a Regional CR, data from multiple sources are aggregated, but this typically reduces the granularity and variety of data elements. At the next level of aggregation, such as Global CRs, more universal coverage of the cancer cancers may be achieved across multiple regions and/or countries, but generally with greatly reduced dimensionality of data.

observations, diagnostic imaging scans, treatment information, follow-up observations). The data is typically summarized and related to the parameters that may address governance, quality, and cost information. One level further, say at the level of a geographic region, each population-based CR may collect the data from every hospital in the region, but only for the conditions of specific interest to itself. The regional population-based CRs covers many more cancer cases and aggregates certain types of data elements, but this also results in increasing the fragmentation of clinical data as well as even further loss of dimensionality. Finally, population-based CRs typically summarize their data further to a global CR. Following the same process as the previous step, the global CR enables more universal coverage of cancer cases but with reduced data elements. The height of each layer represents the variety of the data (e.g., richness and dimensionalities). The bottom "layer" is thickest, indicating that it consists of the most varieties of data elements. In contrast, the upper "layer" is thinnest, indicating that it includes fewer dimensionalities of the data.

Such data collection architecture can result in three serious issues:

1. The central CR cannot collect all the data, part of which are very important for modern cancer treatments and control (e.g., images data), from those data sources.
2. It usually spends much expense and takes a long time to report data to central CRs and collect data from data sources.
3. There is no complete network between different data sources, which seriously impedes linking, integrating, and exchanging data.

The volume of data in oncology is large and increasing rapidly, and it is already well beyond the processing capability of humans. Some companies (e.g., Google) are able to show that handling this volume of data is no longer a technical hurdle. A major problem is that most of the data in oncology is still unstructured, that means the clinical knowledge implicit in the data cannot be explicitly mined by the use of machines. Without major disruption to the way data is stored and the manner in which data is used to generate clinical knowledge, essential insights for personalized medicine and truly participative decision support systems (that involve predictive modeling of treatment effects) will remain beyond our reach.

11.3.2 BIG DATA IN RADIATION ONCOLOGY

With the advances in information technology, computing power, and digital medical equipment, the amount and types of data elements generated from just a single patient during routine cancer treatment are increasing in terms of observational, biological, genetic, imaging, and omics data. However, the data is usually collected at various points of care and then stored in varying formats inside separate databases. As discussed in the preceding sections, this presents significant challenges when trying to aggregate, analyze, and disseminate this data through traditional methods such as CRs. To employ these data to improve cancer treatment and health care of patients, an architecture specifically designed for big data is needed.

Big data research refers to the collection and analysis of a large volume of data elements and interrelationships that are difficult to discover through traditional methods. Big data approaches have been used in many areas of medicine, including comparison of innovative techniques (Chen 2014) and treatment modalities (Aneja and Yu 2014) in the field of radiation oncology.

A large volume of data has been generated within the radiation oncology field—mainly due to the frequent use of innovative techniques (e.g., medical imaging) in diagnostic and therapeutic procedures—during the routine practice of modern radiotherapy treatment (Lustberg et al. 2017). The data within radiation oncology displays all of the most significant hallmarks of big data, that is

1. The use of data-intensive imaging modalities (*volume*),
2. Imaging generates a large amount of data per time interval (*velocity*),
3. There is an increasingly diverse spectrum of modalities available (*variety*), and
4. Objectivity and quality of collected data vary greatly, which can influence accurate analysis (*veracity*).

As illustrated in Figure 11.2, a CR architecture is generally able to cover almost 100% of cancer cases (*volume*), but only about 3% of potentially prognostic factors are collected, along with a moderate data loss rate of approximately 20%. A different data-generating paradigm in oncology (i.e., clinical trials), typically enrolls around 3% all eligible cancer cases (Murthy et al. 2004; Movsas et al. 2007; Grand and O'Brien; 2012). However, due to stringent quality assurance and strict protocolization, nearly all of the factor of interest are recorded (*variety*) for only a relatively low (perhaps around, 5%) missing data rate.

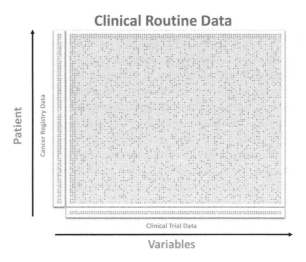

Figure 11.2 Schematic illustration of data fragmentation. Clinical trials generally record numerous variables per patient with high completeness, but only cover a small segment of the population. Conversely, a cancer registry tends to record a few key variables for a broad spectrum of the population. However, the schematic shows that these approaches miss an immense pool of data generated in routine clinical work, which however tends to be poorly structured and suffers from missing values.

What we conceive of as "big data" in oncology is represented by the entire graphic in Figure 11.2, where a vast volume and high throughput of data is continuously being generated by routine operations of modern cancer diagnosis and treatments. It is clear that neither CRs nor clinical studies adequately cover the full range of variety and volume of the clinical information available. However, as indicated in the figure, the consistency of data capture in the real world is generally low, thereby resulting in a high rate of missing data (perhaps around 80%). Further, unlike the rigid protocols required by clinical trials and registry submissions, interobserver biases and divergent interpretations are additional data quality issues. With the increasing use of automation and electronic data capture technology in oncology clinics, we expect that the potential bias and variation in the data (i.e., *veracity*) will continue to improve.

There is strong motivation within the oncology field to exploit hitherto underutilized real-world data. Improvements in treatment outcomes guided by big data utilization, in combination with registry studies and clinical trials, are widely seen as the most effective avenue of delivering value-based health care (Larson 2013; Murdoch and Detsky 2013; Khoury and Ioannidis 2014; Schneeweiss 2014). Big data therefore lies at the core of a personalized approach to medicine that leads to increased value of treatment for a given financial outlay. The question of how to use real-world big data to improve cancer control and reduce treatment-related toxicity has gained increasing traction over time. Closely related questions of interest concern improving data collection coverage in real clinical settings, imputation of missing data, and validation of outcome predictions in a wide variety of clinical settings.

As one of its primary objectives, big data research is expected to find the multiple clinical biological, and treatment variables that are related to treatment outcomes (e.g., overall survival, toxicity) (Rosenstein et al. 2016). This benefits the creation of better predictive models that promote the advances of personalized therapies for each individual patient (e.g., delivering more aggressive therapies where needed and less aggressive treatments when appropriate). In order to develop reliable and robust prognostic outcome models, data sharing and exchange is required between multiple institutions. The reasons are threefold:

1. Obviously, each institution has a limited capability of data collection. The amount of new patients who are diagnosed or treated in a clinic may range from hundreds to thousands per year. Thus, it is reasonable to estimate that the amount of cancer cases stored in-house for an individual clinic's past one or two decades within a clinic should be 20 to 200 thousands. Modern analytics methods (e.g., machine learning) are poised to satisfy the promise of identifying and guiding the response to variables influencing treatment outcomes of patients. However, the value of these analytics methods are highly

dependent on the volume of data used for learning. As an illustrative example for size of training data, the hotly debated artificial intelligence (AI) computer program in 2016, AlphaGo (designed by Alphbet Incorporated's Google DeepMind in London) was initially trained to mimic human gameplay from a historical games database containing approximately 30 million moves (Metz 2016). The data volume within an individual institution is often not sufficient to build such reliable and robust predictive outcome models through modern machine learning algorithms. Thus, data integration from multiple centers is necessary to develop realistic, understandable, and robust predictive outcome models.

2. Data collected within a particular hospital usually refer to local patients with some specific types of cancer. For example, the incidences of esophageal cancer vary widely among countries, with approximately half of all cases occurring in China. For other countries, it is difficult to collect sufficient data elements of esophageal cancer data for both volume and dimensionalities due to the very small number of patients with esophageal cancer. To generate prognostic outcome models, data exchange between two or more institutions seems a feasible and efficient approach.

3. As the predictive models are trained through local cohorts, it may be robust to predict the treatment outcome (e.g., overall survival) of unseen cancer cases within a local population. However, the predictive performance may be poor when applied to other populations, which impedes the usability and extension of the predictive model. Therefore, external validation is always necessary to measure the performance of a predictive model.

Because of the reasons explained above, data exchange between multiple clinics is a necessary procedure in radiation oncology field. Hence, there is a need to develop robust data exchange architectures instead of traditional methods to handle big data within the radiation oncology field. The future data architectures must be able to (1) scale to process ever-increasing amounts of data (*volume*); (2) have the throughput capacity to deal with high rates of data generation, in particular from imaging modalities (*velocity*); (3) process many different types of data into a form that is amenable to machine-based analysis (*variety*); and (4) intercept issues of data quality (e.g., bias, nonreproducibility and abnormality) (*veracity*).

11.4 BIG DATA EXCHANGE IN RADIATION ONCOLOGY

11.4.1 BIG DATA COLLECTION

Within radiation oncology, multiple types of data are routinely generated in the clinic from a variety of sources, which is the basis for big data research and provides an opportunity to improve cancer care.

One important source of big data is the patient demographics and clinical baseline factors obtained at the very beginning of the radiation oncology process. This information includes information about family history and personal health status. Crucially, this form of big data also consists of clinical observations and a baseline for treatment-related outcomes (especially comorbidities before treatment) by which the effectiveness of cancer interventions are evaluated.

Furthermore, an increasingly important source of big data, especially in regard to volume and variety, is the data stream produced by radiological and diagnostic imaging modalities. In radiation oncology, this routine data generation involves CT, PET, and MRI. The volume of image-based data is increasing rapidly due to the use of daily verification imaging with the patient lying in the intended treated position (e.g., for cone beam CT). It is no surprise that the largest (by volume) repositories of data in radiation oncology are PACS.

Radiotherapy treatment planning is a highly sophisticated and computationally intensive process that generates big data in the form of organ delineations, beam geometry, radiation energy, collimation settings, and spatial dose distribution. This data generally resides within the radiotherapy Treatment Planning System (TPS). With the growing trend toward adaptive radiotherapy responding to real-time imaging at the point of treatment delivery, this volume of data is also set to grow rapidly.

Last, a rapidly developing data source is the result of digital pathology and high-throughput specimen analysis from medical laboratories. This includes genomics, proteomics, metabolomics, histologics, and hematologics.

Table 11.5 provides an overview of many of the possible radiotherapy research data types.

Table 11.5 **Radiotherapy research data types**

DATA TYPE	DATA EXAMPLES
Baseline clinical data	Demographics information, TNM-stage, date of diagnosis, histopathology
Diagnostic imaging data	Diagnostic CT, MR and PET imaging
Radiotherapy treatment planning data	Delineation sets, planning-CT, dose matrix, beam set-up, prescribed dose and fractions
Radiotherapy treatment delivery data	Cone beam CTs, orthogonal EPID imaging, delivered fractions
Non-radiotherapy treatment data	Surgery, chemotherapy
Outcome data	Survival, local control, distant failure, toxicity, quality of life
Follow-up imaging data	Follow-up CT, MR and PET imaging
Biological data	Sample storage, shipping, tracing and lab results
Additional study conduct data	Study design, protocol, eligibility criteria

Source: Skripcak, T. et al., *Radiother. Oncol.*, 113, 303–309, 2014.

11.4.2 STANDARD AND FRAMEWORK FOR DATA EXCHANGE

Efforts have been made to standardize data exchange between medical information archival systems with the DICOM standard and HL7 interoperability standard.

DICOM (Mildenberger et al. 2002) refers to a standard in medical imaging, which is supported by all imaging systems in the medical field and used widely. DICOM has two identities: a type of file format and a network communication protocol. First, medical image systems generate DICOM files containing patient information (e.g., name, identifier, sex, date of birth) and then acquires information of medical image systems and corresponding settings. The images are stored in DICOM files. Second, the DICOM protocol can be employed to exchange data (e.g., image or patient information) between different systems that are connected to the network within the hospital. DICOM has shown its ability to improve data exchange in medicine field. Radiotherapy data are commonly exchanged using a subset of DICOM often referred to as DICOM-RT.

HL7 (Dolin et al. 2001, 2006) refers to a widely accepted standard-setting organization that provides standards to define the protocol, language, and data type used for information communication among different systems. The most used version of HL7 is version 2 with which only a limited and not semantically rich data can be exchanged. HL7 version 3 had a much wider scope but is generally considered a failed standard due to its complexity and limited uptake. HL7 FHIR is the most recent standard and is receiving a lot of positive attention from the community and has resulted in real-world implementations by medical vendors.

11.4.3 DATA POOLING ARCHITECTURES

As described above, data integration is a necessary procedure of modern studies in the radiation oncology field. A data pooling architecture is used for data processing, storage, management, and exchange within an individual institution or between multiple institutions. In this section, we will describe three data pooling architectures (i.e., centralized, decentralized, and hybrid architecture), of which the infrastructures are shown in Figure 11.3.

11.4.3.1 Centralized architecture

A centralized architecture has complete physical control over the data pooled in a centralized repository. There is no direct real- or near real-time connection among participating institutions and operations (e.g., push/pull transactions and auditing occur in a central server). Although the architecture of the centralized model is simple, it can raise several issues including privacy and anonymization, duplication of data, mapping the local data to the central data model (usually resulting in manual data entry/copying), and intellectual property (IP) rights (Skripcak et al. 2014). The most significant advantage of the centralized model is that the data is stored in a centralized repository, which makes data management and access easy.

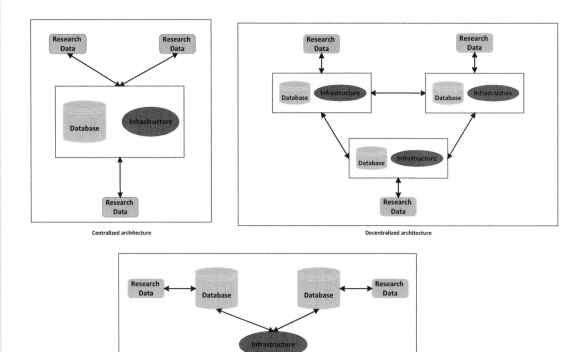

Figure 11.3 Illustration of three types of data pooling architectures. The centralized architecture stores data in a single centralized repository with one homogenized data structure and common infrastructure imposed on all data contributors. A decentralized architecture permits multiple data structures and more than one infrastructure, so collaboration has to be based on agreed standards for data exchange. A hybrid (or federated) architecture tries to combine elements from the previous two; heterogeneous data structures co-exist independently of each other, but a common infrastructure provides interoperability and connectivity between the databases.

The CR architecture described in Section 2 can be considered as an instance of the centralized model. A centralized CR collects data of patients with cancer from different sub-data-sources (e.g., hospitals, cancer centers, other institutions) through regular standards. Afterward, the centralized CR reports the analysis based on these data to the public and local government. However, an institution is not usually allowed to access the data within another institution due to data privacy of patients or local laws. It means there is usually no direct connection between these data sources.

11.4.3.2 Decentralized architecture

A decentralized architecture enables data exchange to occur among multiple institutions without any mediators, which is project-based via direct communication (Skripcak et al. 2014). However, the infrastructure required to enable data exchange must first be established at each site and comply with a standard exchange protocol. Shared data may be persistent (i.e., stored after exchange) or volatile (i.e., nothing is stored after exchange).

11.4.3.3 Hybrid architecture

A hybrid architecture attempts to combine the strengths of centralized and decentralized architectures. Data is transferred through direct communication across multiple sites. The difference between the hybrid and decentralized architectures is that the information on infrastructure, data representation format, controlled terminologies, and other required metadata are stored in a central server, making the maintenance and modification of data exchange settings easier. Another advantage of a hybrid architecture is that big data generated in a local context is conceptually centralized but stored locally (i.e., inside the hospital or clinic). This hides the local complexities that differ for every clinic, below the level of multicenter sharing. This can happen using the agreed-upon terminology. We prefer decentralized and hybrid architectures for data exchange across multiple centers based on the present IT systems within hospitals (Skripcak et al. 2014).

11.4.4 DATA INTEROPERABILITY

Data interoperability has an important role in data exchange among multicenters: It is concerned with the capacity of a system to read and understand data transferred from another system. To implement complex and comprehensive data analyses, the data sources are required to be made fully interoperable across multiple information technology (IT) systems. Data interoperability consists of two main subprinciples: (1) to enable data exchange between multiple institutions, all institutions need to have syntactic interoperability in reference to establishing uniform data formats and exchange protocols. In other words, data representation for writing and reading information should be identical among all institutions, (2) syntactic and semantic interoperability should be in place, as described by Valentini et al. (2012). The aim of data semantic interoperability is to make data **consistently** understandable by machines (Skripcak et al. 2014).

In a real-world scenario, it is difficult to achieve data interoperability because of privacy of patient data, local policy, and even technical issues. To enable data interoperability among different IT platforms, certain technologies (e.g., Semantic Web and ontology) have been applied to big data exchange in the radiation oncology field.

11.4.4.1 Semantic web

The Semantic Web (also known as "Linked Data") is an extension of the Web via many standards by the World Wide Web Consortium (W3C). The standards boost the development of data formats and communication protocols on the Web. Among the various data formats in the Semantic Web, Resource Description Framework (RDF) is the most fundamental format and is commonly used. The rationale behind the RDF data model is that any arbitrary statement about resources within the web can be represented by a simple triple (i.e., subject, predicate, and object). Any levels of complexity in the descriptions of resources are possible using multiple lines of triples. The subject and object here can be considered as two resources. The predicate is the property of the subject and represents the relation between the subject and object. For example, a patient's survival age, biological sex, and type of carcinoma can be described in the RDF format. Figure 11.4 shows the virtual representation of this ontology.

Storing data in the Semantic Web-based triple store is of great advantage for data access, usage, and exchange when compared with a relational database (e.g., SQL). A situation that arises often is data exchange between two hospitals (A and B), which use different relational databases (i.e., the name of each column may be different as well as the internal linkages within the respective databases). Given that each database likely comprises many thousands of rows, the problem of integrating these two relational tables is a serious problem and requires both parties to have in-depth knowledge of the other's relational data structure. Querying a nonexistent field in the other party's database may cause the query to crash. Therefore, one cannot add, delete, or otherwise move records around without informing each other. The need for such knowledge necessarily precludes the ability to preserve privacy and data confidentiality. In contrast, it is trivially simple to integrate data that are stored in triple stores. Because everything is stored as statements with only three pieces of information per statement, a single additional line with a reference from one database into a unique resource identifier in the other is typically all that is needed. In triple stores, there is no constraint that a data structure must be known in advance for a query to work (the query returns null instead of crashing the system), and therefore lines can be added, removed, or ordered in any fashion without affecting the query.

Consider, as an example, that the data on a patient is stored in a relational table called "Patient_1" with several columns and that all the radiomics features of the same patient are stored in another separate table called "Radiomics_Patient_1." Linking the two tables is often not an easy task, which means that we have to add new columns of each radiomics feature in "Patient_1" table and enter the value. However, if these the data of patient and his/her radiomics features are stored in separate RDF triple stores, data integration is trivial. To link the two triple stores, we just need one single sentence to define the triple: "Patient_1 (subject) has Radiomics Feature (predicate) Radiomics_Patient_1 (object)" via unique resource identifiers

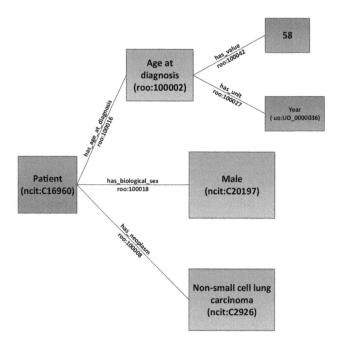

Figure 11.4 Graphical representation of the hypothetical statement "A 58-year old male patient with confirmed non-small cell lung cancer" in the Resource Description Framework (RDF) format. All resources (i.e data elements) are represented in the form of "subject-predicate-object" triplets. Subjects and objects are typically members of a class of entities which are identified by unique codes, such as "Patient = ncit:C16960," "Male = ncit:C20197," and "non small cell lung cancer = ncit:C2926." The relationships between class entities are defined using predicates such as "has age at diagnosis," "has biological sex," and "has neoplasm," each also having their own unique codes. Values or measurements such as "58" are literal quantities, but may themselves possess relationships defining the type of literal (integer) and its relevant units (years).

(URI) that represent patient_1 and his/her radiomics features. Now, the two databases are linked in the entirety of their data elements. For this triple, the class entities are defined in the Radiation Oncology Ontology (ROO) and the Radiomics Ontology (RO).

Based on current developments in technology, the Semantic Web is a more feasible and flexible choice for data representation in the radiation oncology field. As the data is represented through ontologies, the Semantic Web enables seamless linkage of data that are stored in different data platforms. Another benefit of the Semantic Web is as that searching and accessing data can be done through web technologies that are known to be extremely scalable. Recently HL7 FHIR has also been specified in the RDF format.

11.4.4.2 Ontology

An ontology refers to a terminology dictionary that defines the commonly used entities and relationships between entities in a particular domain. It includes machine-interpretable definitions of basic concepts in the domain and relations among them. There are several ontologies available in radiation oncology field such as the ROO (https://www.cancerdata.org/roo), the National Cancer Institute Thesaurus (NCIT) ontology (Sioutos et al. 2007), and the International Classification of Diseases (ICD) ontology. When linking an entity to an ontology, the unique code in the corresponding ontology will be used to replace the value stored within the local database. The main purposes of using ontologies are as follows:

1. One of the primary purposes of using ontology is to share the common meaning of information among humans or machines (Musen 1992; Gruber 1993). For instance, the literal representation of a patient's biological sex may be "female" or "male." When linking to the NCIT ontology, codes C16576 and C20197 are used to represent "female" and "male," respectively. One of the advantages is that a machine only needs to recognize the ontology codes that represent the common concept (e.g., biological sex) and ignore the literal representations stored within local databases (e.g., "f/m", "female/male" and "0/1").

2. Knowledge defined in one domain can be reused in another domain through the use of ontologies. For instance, assume different domains all needed to represent the concept of treatment of oncology including radiotherapy, chemotherapy, chemo radiotherapy, and surgery. If some group built an ontology on treatment of oncology, this knowledge can be simply reused in the others' domains. In addition, it is common to reuse a general ontology (e.g., NCIT ontology).

3. The differentiation of the domain knowledge from the operational knowledge is another important purpose. As an example for radiotherapy, we can define as domain knowledge that a planning target volume is based on margins around the clinical target volume while still allowing local operational knowledge from a trial or institution to describe if and how the margins were applied.

4. If a declarative explanation of the term is available, it is possible to analyze domain knowledge. For example, a radiation ontology specifying that radiation on the chest may result in nonbacterial radiation pneumonitis might be used in another context to prescribe steroids instead of antibiotics in these patients.

11.4.5 DATA EXCHANGE

As mentioned above, different types of data have been routinely generated in a clinic from a variety of sources, which can be of benefit to cancer care. In order to ultimately reach FAIR data (Wilkinson et al. 2016), data exchange among different data sources is a necessary and important step. Therefore, architectures of data exchange should be built based on the rules such as efficiency, safety, and veracity. For the purpose of data exchange, two manners (i.e., manual and automatic data exchange) are used for both internal and external data exchange. This operation can be considered as "send data out," which will be described in Section 4.5.1. However, "send data out" is not suitable or efficient for exchanging big data between multicenters in one or more countries because of reasons such as IP rights, local policy, and patient data privacy. Instead of sending data out, it is better to keep data inside hospitals and "send questions in." This method is known as "distributed learning," of which the details will be described in Section 4.5.2. Finally, the comparison between centralized and distributed multicenter architectures for big data exchange will be described in Section 4.5.3.

11.4.5.1 Send data out

11.4.5.1.1 Internal exchange

In many situations, text data of patients with cancer are generated in a hospital, such as (1) personal information forms on demographic information and medical histories filled by patients, (2) diagnosis and treatment notes created by doctors, and (3) follow-up details. These data can be in free text or structured questionnaires, which are entered in departmental (e.g., Oncology Information System) and/or hospital systems (e.g., electronic health record [EHR] system) for storage, management, and analysis. The data are exchanged, usually through the EHR platform, from one department to another within a hospital.

In addition, some types of data generated in the process of radiotherapy can be transferred to computers automatically, including imaging scans (e.g., CT, PET, MRI), tumor delineations, and treatment plans. One of the common properties of these data is that they are generated by electronic equipment (e.g., scanners and computers). The medical imaging data is usually transferred using the DICOM protocol to a central imaging archive (e.g., PACS). Another type of data that is often digitally exchanged is laboratory results.

11.4.5.1.2 External exchange

For data exchange among multiple centers, one commonly used approach is based on multicenter clinical trials. Data are often transferred via mail, fax, and email between two or more sites or to a central location. Although email is popular and convenient for receiving and sending information, it may result in three serious issues when it involves medical data exchange: (1) missing data and data security, (2) there are no standards for data exchange via email, and (3) transfer efficiency will be very low for a large volume of data (e.g., image data).

Another approach of external data exchange is through a Web-based application. For example, Openclinica (https://www.openclinica.com) has been developed for clinical data aggregation. Indeed, the use of Web-based products for data exchange have some benefits such as reliability and flexibility, although their use still can result in privacy-related issues.

Unlike in clinical research, data exchange for health care is not well developed and regular mail and faxes are still often the main manners of information exchange. Hence, there is a need for a more efficient and powerful approach for big data exchange among multiple centers.

11.4.5.2 Send questions in

As described above, data sharing between multiple centers often raises privacy-related issues. A better manner is to keep data in hospitals and "send questions in" (Lindell and Pinkas 2000; Wiessler 2013; Damiani et al. 2015) rather than sharing data. The primary goal of data sharing is to mine knowledge from others' data. If there is an approach that can answer the research questions without allowing data outside of the hospitals, sending data out is unnecessary. The distributed solution allows advanced data analysis (e.g., knowledge sharing). Mathematical models are trained on local databases and shared with other hospitals. Because models contain only the "answer" to the question while the research data are kept within the hospital, using a mathematical model avoids privacy-related issues. Only some aggregate parameters are transferred between multicenters to reach the global convergence (consensus) of the mathematical model. This approach is known as "distributed learning" (Lambin et al. 2013a).

In addition, this distributed learning can be implemented on a Web-based learning environment (e.g., Varian Learning Portal). The learning platform can be considered as the master that merges knowledge models learned from different participating sites and continuously updates the model when more data are available (Skripcak et al. 2014).

11.4.5.3 Centralized versus distributed learning architecture

Modern medical research has to process an increasing number of data generated from many fields such as medical imaging, genomic, and proteomics. However, the reality is that an individual hospital only has data on a limited number of patients, which may be not sufficient to medical research. From the experience of machine learning in other fields, we know one needs a sufficient number of events to build a reliable predictive model for cancer treatments. In general, the more data collected from different sources, the more robust a predictive model is. Thus, cooperation between two or more hospitals is needed to collect more data regarding patients with cancer. The architectures of centralized and distributed learning among multiple centers has been described by van Soest et al. (2015).

Figure 11.5 shows the general overview for the centralized multicenter architecture. This approach allows participating sites to build the institutional architectures based on local policies. In addition to the entry points of all institutions, there are two key components within a centralized learning architecture: (1) a central machine learning server is the place where learning occurs; and (2) a central collection point is responsible to perform the horizontal accumulation of data between all sites. As an example to explain the learning process, first, participating Site A sends an algorithm to the central machine learning server. Second, the central server implements calculation of this algorithm on the centralized data repository. Finally, the results are sent back to Site A after the calculation is completed.

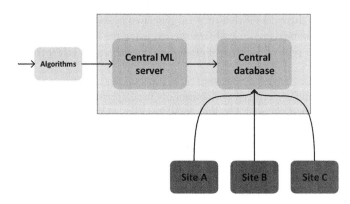

Figure 11.5 Schematic of a Centralized Multicentre Learning Infrastructure. The centralized learning approach forces each site to contribute structured data into a central accumulation point, whereupon algorithms can operate on the data via a machine-learning (ML) server co-located with the data to obtain the global result based on all the data.

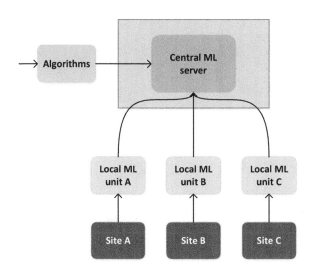

Figure 11.6 Schematic of a Distributed Multicentre Learning Infrastructure. In the distributed learning approach, a central machine-learning (ML) server splits the intended algorithm to each of the sites. Each copy of the algorithm operates only on the site-specific data via its own local ML unit. Results of computation from each site are returned independently and asynchronously to the Central ML server, which then has to process multiple site results into a single global result. Where necessary, the Central ML server may iterate through the site computation steps many times.

On the other hand, the distributed learning architecture between multiple centers is different in terms of the places where computations happen (Meldolesi et al. 2014; Damiani et al. 2015; van Soest et al. 2015). Figure 11.6 displays the general overview of distributed learning architecture. We can see that the local unit has been added and the central federation point has been removed. In this architecture, the responsibility of the central machine learning server is only coordination. First, a site submits an algorithm or query to the central machine learning server. The algorithm or query is then split into several small sub-tasks. Second, the small sub-tasks are packed and sent to local machine learning units within each site. They will query the local data that is stored in RDF triple stores and the sub-algorithms are implemented in the local sites. The local application learns a model from local data. Third, the central machine learning server will merge all the results that have been computed on the local machine learning units of all sites (van Soest et al. 2015). Finally, if the preset criteria are met, a final model is generated. If not, the central machine learning server will send the models to all sites for re-learning until the preset convergence criteria have been reached.

The significant difference between distributed and centralized machine learning architecture is the transfer of data versus the transfer of model weights. When performing centralized learning, data leaves the hospital and is sent to the machine learning system. In contrast, data is kept within the hospital when we perform distributed learning. In this setting, the volume of data that is needed to be transferred is decreased compared with the centralized learning architecture; however, the transfer efficiency per task is increased. Boyd et al. (2011) and Wu et al. (2012) have given a complete explanation of how distributed machine learning algorithms work in their publications.

As proof that the concepts covered in Section 11.4 are indeed practical, a real-world machine learning project on distributed databases, known as Computer Assisted Theragnostics (CAT), has been proposed as illustrated schematically in Figure 11.7. First, data (e.g., image, genomics) are collected from a variety of data sources within each site and stored in local databases. Second, these data within each site are converted into the RDF data format and stored in the Semantic Web–based triple stores. Third, as shown in Figure 11.7, researchers in (for example) Rome can send research questions to (for example) Oxford via a global learning server. Then, learning happens on the local learning server in Oxford. After finishing the local learning, the results are sent back to the global learning server. Finally, researchers in Rome can query the global server for

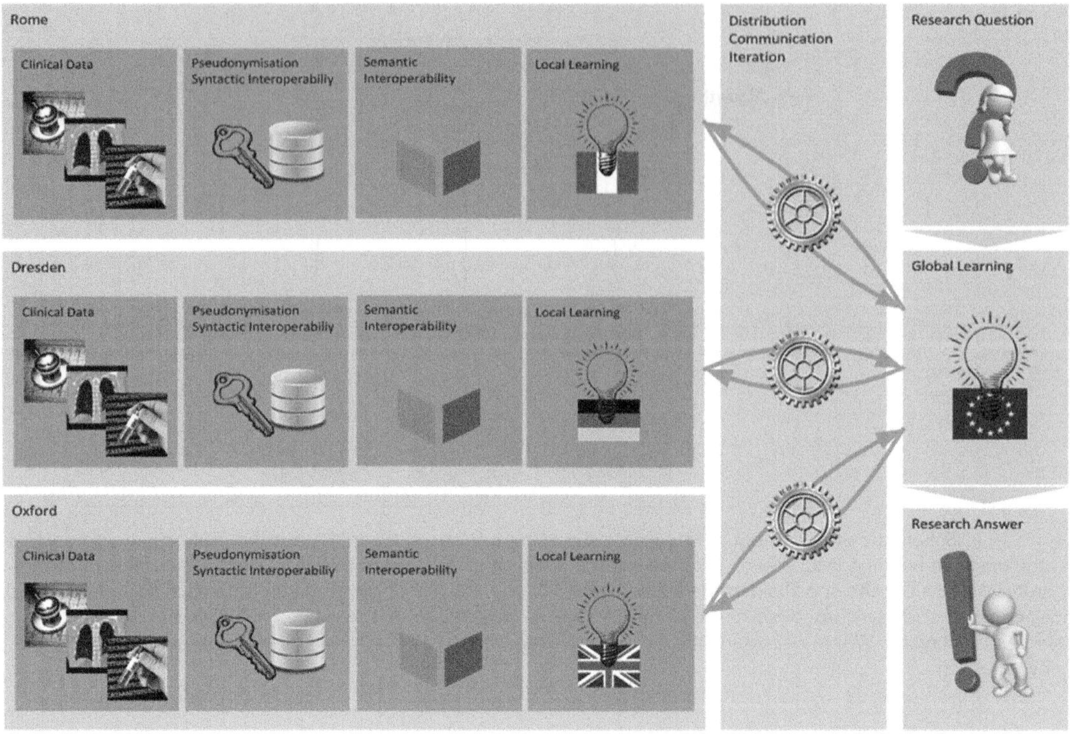

Figure 11.7 The diagram of a real-world machine learning project on decentralized databases, known as Computer Assisted Theragnostics (CAT). The schematic shows three hypothetical sites separated by language and geography, as well as site-specific data structures and local institutional infrastructures. The CAT workflow at each site always involves the same general steps of (i) removing personal identifiable information and mapping the local data into a tentative data structure, (ii) providing semantic interoperability using ontologies and describing the local data in Resource Description Framework (RDF), and finally (iii) connecting the site RDF data to a distributed multicentre learning infrastructure such as the CAT system.

the answers to the research questions they proposed. This pipeline enable the communication between two or more institutions that have participated in the CAT project. Data exchange via the CAT architecture leads to a few benefits as:

1. Because the information shared in the CAT pipeline is the result (i.e., answers of research questions), not the real data, data exchange and knowledge mining occur without leaving data outside the local hospitals, avoiding data privacy-related issues.
2. All the data items stored in different databases within each hospital are standardized by the same ontology, leading to linked data.
3. In the example above, researchers in Rome only needed to send the algorithms or query (known as a "research question"), which is a small package (around hundreds of kilobytes), compared with sending the real data (a large volume).

A few projects (e.g., I2B2 [https://www.i2b2.org], EURegiOnal Computer Assisted Theragnostics [EuroCAT] [Deist et al. 2017], VAlidation of high TEchnology [VATE] [Meldolesi et al. 2014]) have demonstrated the feasibility of distributed multicenter machine learning. In addition, if we implement the procedures of distributed learning correctly, the results can be the same with as the centralized learning approach (Wiessler, 2013). Furthermore, it is possible to improve the robustness of prognostic models through validating on external data sets (Dekker et al. 2013), which has been demonstrated based on the results of the EuroCAT project (Deist et al. 2017).

11.5 BARRIERS OF BIG DATA EXCHANGE AMONG MULTICENTERS

This section will describe the barriers of big data exchange among multiple institutions in the radiation oncology community. The exchange barriers involves four primary aspects: administrative, ethical, political, and technical barriers. The details are as follows.

11.5.1 ADMINISTRATIVE BARRIER

Two main administrative barriers impede data exchange among multiple institutions: data completeness and bias. First, it is usually not possible to collect every data element of an individual patient and not all data elements are applicable, resulting in many missing data for the patient. This means there are no data to exchange. Second, system settings vary widely across sites. The bias in routine operation, protocols, and equipment settings can all result in differences between data across different sites. Data bias highly impedes data integration among multiple sites.

11.5.2 ETHICAL BARRIER

The ethical issues refer to data privacy and the reuse of research data. There may be a large difference in privacy explanation, application of confidentiality, and legislation between counties and even between ethical committees (Skripcak et al. 2014).

For both centralized and distributed multicenter infrastructures, privacy preservation is a major topic that must be considered. If correctly implemented as the applications described in preceding section, the distributed solution is generally more secure than the centralized solution because only a few parameters are transferred among multiple centers rather than the real data, although we cannot draw the conclusion that the issues on privacy preservation have been overcome via the distributed solution. Actually, there are no standard methods to solve privacy-related issues currently. Various stakeholders will always have to find a balance between the value of information and anonymity of participating patients.

11.5.3 POLITICAL BARRIER

In some scenarios, people do not want to share their data to others because of the issues related to the culture and local policy. Thus, there is a need for more high-quality published research articles that completely prove the benefits of data exchange (e.g., efficiency, robustness, security) to try to persuade data holders to participate in the collaborative research community and subsequently share their data.

11.5.4 TECHNICAL BARRIER

Even if these administrative, ethical, and political issues are solved, the technical barriers such as interoperability between clinical departments and lack of a uniform standard of data collection may still impede data exchange among multiple institutions. First, obtaining internal clinical IT systems interoperability is important for the generation of local anonymized data sets, which enables a universal access to integrate research data. These data are managed by an institutional data warehouse and can be findable through the corresponding semantic models (ontologies). However, it is often difficult to reach this interoperability in real-world scenarios because of the differences in their support of internationally standardized protocols, formats, and semantics. Although these problems can be solved, they often require an investment in resources that is not available in the operational setting.

Second, because the radiotherapy terminologies dictionary and ontologies are still under development, it is difficult to ensure that an element has a unique term and definition (Skripcak et al. 2014). As an example described above, different hospitals may use various representations to describe the biological sex of a patient such as "female and male," "f and m," or "0 and 1." When performing computations on data from the two different origins, two incompatible representation will be encountered. The best way to overcome this issue is to link clinical variables to ontologies that can provide the standard definitions of these variables. The biological sex of a patient "female" and "male" are represented by NCIT codes C16576 and

C20197, respectively. Thus, the meaning of a clinical variable is only related to its ontological code rather than the literal representation.

11.6 CONCLUSION

In this chapter, we have seen that how different types of CR work as an organization for cancer data collection, management, storage, analysis, and exchange. **However, the architecture of a CR is not easily scalable to exploit some properties of big data in radiation oncology, most notably Volume, Velocity and Variety**. In addition, it remains a complex and costly process to maintain a CR, which needs a large amount of human resources (e.g., registrars and managers) and infrastructure (servers and user interfaces).

One of the most important challenges to the universal adoption of CRs is how big data can be flexible and securely shared among a large network of cancer institutes or research programs to answer a broad range of clinically relevant questions. Thus, there is a need to build robust data exchange architectures to handle big data within the radiation oncology field instead of using traditional methods (e.g., CR). "Send data out" is the commonly used approach of data exchange via mail, fax, email, or Web-based applications although it may result in privacy-related data issues. The best manner to avoid privacy-related issues, while exploiting multicenter data, is to avoid sending data outside of the hospitals. This can be achieved through applying the method "send questions in," which is also known as distributed learning. The distributed approach only transfers a subprocess machine learning algorithm to a specific hospital and sends the results back to the sender rather than transferring real data. It means that data/knowledge exchange occurs without allowing data outside of the hospital. Many collaborative projects (e.g., I2B2, EuroCAT, VATE) have demonstrated the feasibility of this distributed learning architecture in handling big data in radiation oncology field. If the procedures are implemented correctly, it can produce the same results as a centralized learning architecture.

REFERENCES

Aerts HJWL, Velazquez ER, Leijenaar RTH, Parmar C, Grossmann P, Cavalho S, Bussink J et al. (2014) Decoding tumour phenotype by noninvasive imaging using a quantitative radiomics approach. *Nat Commun* 5:4006.

Aneja S, Yu JB (2014) Comparative effectiveness research in radiation oncology: Stereotactic radiosurgery, hypofractionation, and brachytherapy. *Semin Radiat Oncol* 24:35–42.

Armstrong BK (1992) The role of the cancer registry in cancer control. *Cancer Causes Control* 3:569–579.

Boyd S, Parikh N, Chu E, Peleato B, Eckstein J (2011) Distributed optimization and statistical learning via the alternating direction method of multipliers. *Foundations and Trends® in Machine Learning* 3:1–122.

Bray F, Parkin DM (2009) Evaluation of data quality in the cancer registry: Principles and methods. Part I: Comparability, validity and timeliness. *Eur J Cancer* 45:747–755.

Chen AB (2014) Comparative effectiveness research in radiation oncology: Assessing technology. *Semin Radiat Oncol* 24:25–34.

Chen M, Mao SW, Liu YH (2014) Big data: A survey. *Mobile Netw Appl* 19:171–209.

Coleman MP, Forman D, Bryant H, Butler J, Rachet B, Maringe C, Nur U et al. (2011) Cancer survival in Australia, Canada, Denmark, Norway, Sweden, and the UK, 1995–2007 (the International Cancer Benchmarking Partnership): An analysis of population-based cancer registry data. *Lancet* 377:127–138.

Damiani A, Vallati M, Gatta R, Dinapoli N, Jochems A, Deist T, Van Soest J, Dekker A, Valentini V (2015) Distributed learning to protect privacy in multi-centric clinical studies. In: *Conference on Artificial Intelligence in Medicine in Europe*, pp. 65–75: Springer.

Dehing-Oberije C, Aerts H, Yu S, De Ruysscher D, Menheere P, Hilvo M, van der Weide H, Rao B, Lambin P (2011) Development and Validation of a Prognostic Model Using Blood Biomarker Information for Prediction of Survival of Non–Small-Cell Lung Cancer Patients Treated With Combined Chemotherapy and Radiation or Radiotherapy Alone (NCT00181519, NCT00573040, and NCT00572325). *International Journal of Radiation Oncology* Biology* Physics* 81:360–368.

Dehing-Oberije C, De Ruysscher D, Petit S, Van Meerbeeck J, Vandecasteele K, De Neve W, Dingemans AM et al. (2010) Development, external validation and clinical usefulness of a practical prediction model for radiation-induced dysphagia in lung cancer patients. *Radiother Oncol* 97:455–461.

Dehing-Oberije C, Yu S, De Ruysscher D, Meersschout S, Van Beek K, Lievens Y, Van Meerbeeck J et al. (2009) Development and external validation of prognostic model for 2-year survival of non-small-cell lung cancer patients treated with chemoradiotherapy. *Int J Radiat Oncol Biol Phys* 74:355–362.

Deist TM, Jochems A, van Soest J, Nalbantov G, Oberije C, Walsh S, Eble M, Bulens P, Coucke P, Dries W (2017) Infrastructure and distributed learning methodology for privacy-preserving multi-centric rapid learning health care: EuroCAT. *Clin Transl Radiat Oncol* 4:24–31.

Dekker A, Nalbantov G, Oberije C, Wiessler W, Eble M, Dries W, Janvary L, Bulens P, Krishnapuram B, Lambin P (2013) PD-0496: Multi-centric learning with a federated IT infrastructure: application to 2-year lung-cancer survival prediction. *Radiother Oncol* 106:S193–S194.

Deng J (2014) Big data in radiation oncology: Challenges and opportunities. *Cancer Sci Res Open Access* 1:1–2.

Dolin RH, Alschuler L, Beebe C, Biron PV, Boyer SL, Essin D, Kimber E, Lincoln T, Mattison JE (2001) The HL7 clinical document architecture. *J Am Med Inform Assoc* 8:552–569.

Dolin RH, Alschuler L, Boyer S, Beebe C, Behlen FM, Biron PV, Shabo Shvo A (2006) HL7 clinical document architecture, release 2. *J Am Med Inform Assoc* 13:30–39.

Elwyn G, Frosch D, Thomson R, Joseph-Williams N, Lloyd A, Kinnersley P, Cording E, Tomson D, Dodd C, Rollnick S, Edwards A, Barry M (2012) Shared decision making: A model for clinical practice. *J Gen Intern Med* 27:1361–1367.

Gliklich RE, Dreyer NA, Leavy MB (2014) Data sources for registries. In: *Registries for Evaluating Patient Outcomes: A User's Guide* (Gliklich, R. E. et al., eds), pp. 127–144. Government Printing Office, Rockville, MD.

Grand MM, O'Brien PC (2012) Obstacles to participation in randomised cancer clinical trials: A systematic review of the literature. *J Med Imaging Radiat Oncol* 56:31–39.

Gruber TR (1993) A translation approach to portable ontology specifications. *Knowl Acquis* 5:199–220.

Khoury MJ, Ioannidis JP (2014) Medicine. Big data meets public health. *Science* 346:1054–1055.

Lambin P, Roelofs E, Reymen B, Velazquez ER, Buijsen J, Zegers CM, Carvalho S et al. (2013a) Rapid learning health care in oncology—An approach towards decision support systems enabling customised radiotherapy. *Radiother Oncol* 109:159–164.

Lambin P, van Stiphout RGPM, Starmans MHW, Rios-Velazquez E, Nalbantov G, Aerts HJWL, Roelofs E et al. (2013b) Predicting outcomes in radiation oncology-multifactorial decision support systems. *Nat Rev Clin Oncol* 10:27–40.

Larson EB (2013) Building trust in the power of "big data" research to serve the public good. *JAMA* 309:2443–2444.

Lindell Y, Pinkas B (2000) Privacy preserving data mining. In: *Annual International Cryptology Conference*, pp. 36–54. Springer.

Lustberg T, van Soest J, Jochems A, Deist T, van Wijk Y, Walsh S, Lambin P, Dekker A (2017) Big data in radiation therapy: Challenges and opportunities. *Br J Radiol* 90:20160689.

MacLennan R (1991) Items of patient information which may be collected by registries. In: *Cancer Registration: Principles and Methods* vol. 95 (Jensen, O. M., ed), pp. 43–63: IARC, Lyon, France.

McNutt TR, Moore KL, Quon H (2016) Needs and challenges for big data in radiation oncology. *Int J Radiat Oncol Biol Phys* 95:909–915.

Meldolesi E, van Soest J, Alitto AR, Autorino R, Dinapoli N, Dekker A, Gambacorta MA, Gatta R, Tagliaferri L, Damiani A (2014) VATE: VAlidation of high TEchnology based on large database analysis by learning machine. Colorectal Canc 3(5):435–450.

Metz C (2016) In major AI breakthrough, Google system secretly beats top player at the ancient game of go.

Mildenberger P, Eichelberg M, Martin E (2002) Introduction to the DICOM standard. *Eur Radiol* 12:920–927.

Movsas B, Moughan J, Owen J, Coia LR, Zelefsky MJ, Hanks G, Wilson JF (2007) Who enrolls onto clinical oncology trials? A radiation patterns of care study analysis. *Int J Radiat Oncol* 68:1145–1150.

Murdoch TB, Detsky AS (2013) The inevitable application of big data to health care. *JAMA* 309:1351–1352.

Murthy VH, Krumholz HM, Gross CP (2004) Participation in cancer clinical trials: Race-, sex-, and age-based disparities. *JAMA* 291:2720–2726.

Musen MA (1992) Dimensions of knowledge sharing and reuse. *Comput Biomed Res* 25:435–467.

Navarro C, Martos C, Ardanaz E, Galceran J, Izarzugaza I, Peris-Bonet R, Martinez C, Spanish Cancer Registries Working Group (2010) Population-based cancer registries in Spain and their role in cancer control. *Ann Oncol* 21(Suppl 3):iii3–iii13.

Oberije C, Nalbantov G, Dekker A, Boersma L, Borger J, Reymen B, van Baardwijk A et al. (2014) A prospective study comparing the predictions of doctors versus models for treatment outcome of lung cancer patients: A step toward individualized care and shared decision making. *Radiother Oncol* 112:37–43.

Parkin DM (2006) The evolution of the population-based cancer registry. *Nat Rev Cancer* 6:603–612.

Parkin DM (2008) The role of cancer registries in cancer control. *Int J Clin Oncol* 13:102–111.

Powell J (1991) Data sources and reporting. In: *Cancer Registration: Principles and Methods*, vol. 95 (Jensen, O. M., ed), pp. 29–42. IARC, Lyon, France.

Roelofs E, Persoon L, Nijsten S, Wiessler W, Dekker A, Lambin P (2013) Benefits of a clinical data warehouse with data mining tools to collect data for a radiotherapy trial. *Radiother Oncol* 108:174–179.

Rosenstein BS, Capala J, Efstathiou JA, Hammerbacher J, Kerns SL, Kong FM, Ostrer H et al. (2016) How will big data improve clinical and basic research in radiation therapy? *Int J Radiat Oncol* 95:895–904.

Sadjadi A, Malekzadeh R, Derakhshan MH, Sepehr A, Nouraie M, Sotoudeh M, Yazdanbod A et al. (2003) Cancer occurrence in Ardabil: Results of a population-based cancer registry from Iran. *Int J Cancer* 107:113–118.

Santos SI (1999) The role of cancer registries. In: *Cancer Epidemiology, Principles and Methods* (Santos, S. I., ed), pp. 385–403. IARC, Lyon, France.

Schneeweiss S (2014) Learning from big health care data. *N Engl J Med* 370:2161–2163.

Sioutos N, de Coronado S, Haber MW, Hartel FW, Shaiu WL, Wright LW (2007) NCI Thesaurus: A semantic model integrating cancer-related clinical and molecular information. *J Biomed Inform* 40:30–43.

Skripcak T, Belka C, Bosch W, Brink C, Brunner T, Budach V, Buttner D et al. (2014) Creating a data exchange strategy for radiotherapy research: Towards federated databases and anonymised public datasets. *Radiother Oncol* 113:303–309.

Stacey D, Legare F, Col NF, Bennett CL, Barry MJ, Eden KB, Holmes-Rovner M et al. (2014) Decision aids for people facing health treatment or screening decisions. *Cochrane Db Syst Rev* 1.

Storm HH (1996) Cancer registries in epidemiologic research. *Cancer Causes Control* 7:299–301.

Valentini V, Schmoll H-J, van de Velde CJ (2012) *Multidisciplinary Management of Rectal Cancer: Questions and Answers*. Springer Science & Business Media Cham, Switzerland.

Van Leersum NJ, Snijders HS, Henneman D, Kolfschoten NE, Gooiker GA, ten Berge MG, Eddes EH et al. (2013) The Dutch surgical colorectal audit. *Eur J Surg Oncol* 39:1063–1070.

van Soest JP, Dekker AL, Roelofs E, Nalbantov G (2015) Application of machine learning for multicenter learning. In: *Machine Learning in Radiation Oncology*, pp. 71–97. Springer, Cham, Switzerland.

Wiessler W (2013) PO-0886: Privacy-preserving, multi-centric machine learning across hospitals and countries: Does it work? *Radiother Oncol* 106:343.

Wilkinson MD, Dumontier M, Aalbersberg IJ, Appleton G, Axton M, Baak A, Blomberg N et al. (2016) The FAIR Guiding Principles for scientific data management and stewardship. *Sci Data* 3:160018.

Wu Y, Jiang X, Kim J, Ohno-Machado L (2012) Grid Binary LOgistic REgression (GLORE): Building shared models without sharing data. *J Am Med Inform Assn* 19:758–764.

Young JL (1991) The hospital-based cancer registry. In: *Cancer Registration: Principles and Methods*, vol. 95 (Jensen, O. M., ed), pp. 177–184: IARC, Lyon, France.

12 Clinical and cultural challenges of big data in radiation oncology

Brandon Dyer, Shyam Rao, Yi Rong, Chris Sherman, Mildred Cho, Cort Buchholz, and Stanley Benedict

Contents

12.1 Overview of the Clinical and Cultural Challenges of Big Data in Radiation Oncology 182
12.2 Challenge of Capturing and Assessing Big Data Throughout the Radiation Oncology Clinical Workflow 183
    12.2.1 Physician Referral Challenges and Electronic Health Record 183
    12.2.2 Time Demands of Electronic Health Record Data Entry 183
    12.2.3 Discussion of Specific Challenges for Capturing All the Data within the Radiation Oncology Process: Simulation, Planning, Quality Assurance, Treatment, Follow-Up 183
12.3 Challenges in Data Integration with Outside Departments: Introducing Pathologic Data Analytics and Watson Oncology 184
    12.3.1 Watson Oncology 185
    12.3.2 Watson Oncology for Radiation Oncology 186
12.4 Industry and Academic Centers' Solutions to Big Data Mining and Decision Support 186
12.5 Challenges of Patient Privacy and Data Security—Balancing Access and Control 189
    12.5.1 Overview of HIPAA and Its Meaningful Use 189
    12.5.2 Innovations in PHI Acquisition: A Discussion on Best Practices for Capturing PHI in a Clinic Using Handhelds and Wearables and Wireless Location Tracking 189
    12.5.3 Patient Data Storage and Accessibility: Best Practice for On-Premises, Off-Premises, and Cloud Storage 190
    12.5.4 PHI Reporting and Visualization 192
12.6 The Challenge of Information and Health Literacy in the Era of Big Data: Patient Involvement in Decision Making and Feedback 194
    12.6.1 Patient Health Literacy 194
12.7 The Ethics of Predictive Analytics for Precision Medicine in Radiation Oncology 195
    12.7.1 Ethical Challenges Arising from Learning Health Care Systems and Big Data 195
    12.7.2 Informed Consent 195
    12.7.3 Privacy 196
    12.7.4 Justice and Fairness in the Distribution of Benefits and Harms 196
    12.7.5 Dual Use of Predictive Analytics for Precision Medicine 196
12.8 Upcoming Big Data Challenges: Low Hanging Fruit and Apples Just a Bit Out of Reach 197
References 197

12.1 OVERVIEW OF THE CLINICAL AND CULTURAL CHALLENGES OF BIG DATA IN RADIATION ONCOLOGY

From a high level, patient well-being, the healthy-aging process, and medical care is a spectrum of health supported by four medical pillars, including (1) staying healthy (generally being young); (2) identifying health problems early; (3) detecting, diagnosing, managing, and monitoring any diseases; and (4) identifying a proper and appropriate methodology for treating diseases. Across these pillars of health, as each patient proceeds on their journey, they generate a wealth of new data inside the health care system(s). This includes a wide array of radiologic imaging, pathology reports, cytogenetics, tissue banking, flow cytometry, physician documentation, multidisciplinary tumor board discussions, nursing and therapist notes on patient progression and status, and specialist input (Figure 12.1). During this process, massive amounts of data are collected and stored in a digital format largely because the demonstration of meaningful use has been tied to medical financial reimbursement under the Patient Protection and Affordable Care Act (ACA) and because the Centers for Medicare and Medicaid Services (CMS) has moved 90% of the fee-for-service requirements into quality metrics. Unfortunately, data collection is heterogeneous in terms of both input and output formatting (and access), rendering the data impossible to effectively manage, analyze, or data mine or use to generate clinical decisions from due to a lack of user interface, data volume, and required computational power (software and hardware). Programs to effectively manage this data within the radiation oncology setting are being developed, and they will require a system that provides adequate capacities for sorting through high data storage, with acceptance of data variety, as well as high speed for data processing and analyzing. The need for these high-level programs is why the radiation oncology field has introduced big data analytics, with the hopes of it improving clinical outcomes and optimizing the patient–physician experience and overall effectiveness.

With the increasing health care focus on value-based medicine through demonstrable metrics (e.g., patient-reported outcomes and satisfaction, clinical medical outcomes) and seeking to offer personalized medical treatment, data-driven decision making is vital. Fortunately, in the past several years, many novel technologies have emerged from industry and academic centers to solve the computational barriers that big data analytics present, thus enabling greater breadth of clinical applications in medicine. Several institutions have taken the lead in this technology innovation to bring big data strategies into the radiation

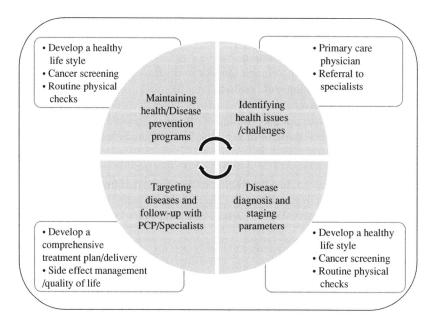

Figure 12.1 The four domains of well-managed patient care: maintaining/prolonging health, detecting health problems, diagnosing and evaluating disease, and treating diseases with continuing clinical follow-up.

oncology clinic, but challenges continue in every step of the health care pathway (Cottle et al. 2013). In this chapter, we will explore some of the challenges and hurdles, both clinical and cultural, for big data in radiation oncology. The emphasis for this perspective focuses on issues with (1) data acquisition throughout the patient clinical flow, (2) limitations in current industry and academic center solutions, (3) the crucial need to maintain patient privacy and security, (4) addressing patient information and health literacy, and (5) ethical and social concerns of precision medicine.

12.2 CHALLENGE OF CAPTURING AND ASSESSING BIG DATA THROUGHOUT THE RADIATION ONCOLOGY CLINICAL WORKFLOW

"Big data is like teenage sex: everyone talks about it, nobody really knows how to do it, everyone thinks everyone else is doing it, so everyone claims they are doing it …" (Ariely 2013).

12.2.1 PHYSICIAN REFERRAL CHALLENGES AND ELECTRONIC HEALTH RECORD

Quality care for patients with cancer often requires access to specialty providers. Patients who do not receive specialty care in a timely manner are more likely to have poor outcomes and report lower health care satisfaction. Furthermore, impediments to specialty referral may delay treatment in patients with cancer, a disease in which timeliness can be a critical determinant of survival and quality of life.

A recent study by Kwon et al. (2015) demonstrated that greater availability of medical records—a marker for access to information technology—was associated with lower reported incidence of referral barriers. Physicians who reported high availability of medical records may practice with an electronic health record (EHR) and/or electronic referral system, both of which potentially facilitate care coordination. Given the growing interest in improving the coordination of primary and specialty care—especially through accountable care organizations (ACOs) and patient-centered medical homes (PCMH)—data suggest that implementation of information technology may serve as a strategy for reducing barriers to specialist referral for cancer patients.

12.2.2 TIME DEMANDS OF ELECTRONIC HEALTH RECORD DATA ENTRY

The drive to shift the nation's health care system away from fee-for-service and toward rewarding quality and outcomes was spurred on by the ACA and CMS tying reimbursement to demonstrable medical benefit. Although sound in theory, this poses significant challenges for stand-alone medical clinics, rural medical practitioners, and independent physicians. The burden of the ACA/CMS changes primarily arises from the requirement to collect patient data so that the provider can monitor outcomes, according to Reed Tinsley, a Houston-based practice consultant. "There is always more money behind knowing the clinical outcomes and data," he says. "A lot of doctors are saving payers money and not getting a piece of the pie" (Ritchie et al. 2014).

Health information technology tools such as EHRs have the potential to significantly improve care delivery and patient outcomes. However, physicians who have adopted EHRs continue to struggle to efficiently use these systems because of the difficulty of dividing their time between the patient and the computer. The average physician spends 30%–60% of a patient encounter looking directly at the EHR, with the majority of that time spent typing in an office layout that does not allow the patient to remain engaged through screen sharing, according to 2013 research from the Journal of General Practice (Sinsky et al. 2016; Bendix et al. 2017).

12.2.3 DISCUSSION OF SPECIFIC CHALLENGES FOR CAPTURING ALL THE DATA WITHIN THE RADIATION ONCOLOGY PROCESS: SIMULATION, PLANNING, QUALITY ASSURANCE, TREATMENT, FOLLOW-UP

Radiation oncology simulation, planning, treatment delivery, and follow-up processes have become sophisticated to the point that the entire workflow is captured, stored, and available for analysis. The hope is that with this type of data acquisition future patients will benefit from an iterative process improvement where everything from the referral through the completion of treatment has been streamlined, optimized, value-assessed, and cost-reduced for maximum efficiency and clinical outcome, with treatment still revolving

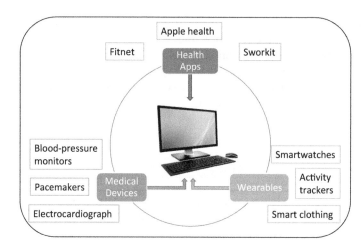

Figure 12.2 A the potential data acquisition points for a single patient: medical devices, wearables, and health applications on mobile devices.

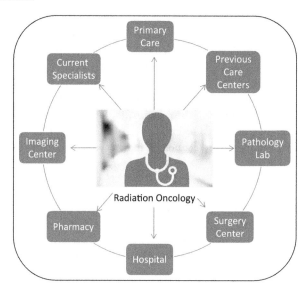

Figure 12.3 Access to health records and patient information comes from a wide and diverse set of health care institutions, including hospital, pharmacy, specialists, primary care, and laboratory.

around the patient. Figures 12.2 and 12.3 demonstrate the breadth and depth of a typical radiation therapy patient's data. This includes a broad spectrum of potential data sources from the patient—wearables, downloads from their home medical devices, and data stored from health applications on their mobile phones—all of which may prove valuable in the future for analysis. The challenge of bringing together the diverse institutions that store their own data across platforms of multiple institutions that may be currently or in the past involved in the care of a patient is presented in the Figure 12.3.

12.3 CHALLENGES IN DATA INTEGRATION WITH OUTSIDE DEPARTMENTS: INTRODUCING PATHOLOGIC DATA ANALYTICS AND WATSON ONCOLOGY

The increasing volume of data from a multitude of sources offers opportunities for improving health care. This is where "the data" begins. At the single-patient level, it is the generation of "small data." With increased data collection, tabulation and intelligent storage, and the combination of cohort/population data this leads to "big data." With big data intact, it is then possible to retrieve information through data mining, feature extraction, and processing. Through further modeling and algorithm generation, it is possible to generate knowledge, and subsequently, a radiation oncology application through the application of hypotheses and machine learning (Regge et al. 2017). Historically, pathology has been a transactional enterprise: The lab receives an order, collects the specimen, performs the test, and reports the results. However, pathology can go beyond that traditional role and provide the benefit of furthering analysis of the data in the context of other information about the patient or about the population. To this end, many large cancer centers now have tissue biorepositories and tissue banking, allowing for retrospective analysis and providing decision support for future research endeavors. This additional service will lead to greater clinical and economic value.

Predictive analytics is another way that pathology can generate new knowledge and improve radiation oncology outcomes. Other industries have applied machine learning and visual graphics to gain new insights. For example, Memorial Sloan Kettering Cancer Center (MSKCC) and others are teaming up with IBM Watson Health to sift through mountains of data to help physicians come to the right diagnosis more quickly and to provide the most likely effective interventions at the right time. In Figure 12.4, we show that big data and medical informatics requires the confluence and contributions of different fields to successfully apply information technology in medicine (Kagadis et al. 2008).

Specifically, informatics tools are necessary to support the quality and outcome of radiation oncology practices at the point of delivery. To achieve this goal, radiation oncology professionals must develop, and communicate with professionals in informatics to develop tools that facilitate treatment decision making in concert with the patient, and promote medical practice at the cutting edge of technologies and patient outcomes. By integrating databases across disciplines, it is possible to elevate both science and the quality of patient care (Chetty et al. 2015).

12.3.1 WATSON ONCOLOGY

Watson Oncology is a cognitive computing system designed to support the broader oncology community of physicians as they consider treatment options with their patients. MSKCC clinicians and analysts have

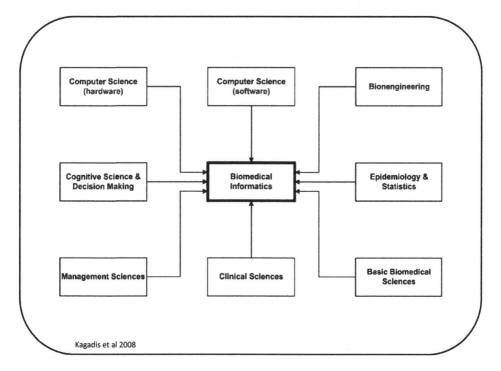

Kagadis et al 2008

Figure 12.4 How big data and medical informatics require the confluence and cohesive work from different fields to successfully apply information technology in medicine. (From Kagadis, G. C. et al., *Med. Phys.* 35, 119–127, 2008.)

Big data in radiation oncology

partnered with IBM to train Watson Oncology to interpret cancer patients' clinical information and identify individualized, evidence-based treatment options that leverage specialists' decades of experience and research. Watson Oncology was developed to summarize key medical attributes of a patient and provide information to oncologists to help them deliver treatment options based on iterative feedback and machine learning from MSKCC oncologists.

Watson Oncology ranks the treatment options, linking the derived treatment decisions to peer-reviewed studies that have been curated by MSKCC. Watson Oncology also provides a large corpus of medical literature for a physician to consider, drawing on more than 300 medical journals, more than 200 textbooks, and nearly 15 million pages of text to provide insights about different treatment options.

As Watson Oncology's "teacher," the faculty at MSKCC aimed to create a powerful resource that will help inform treatment decisions for those who may not have access to a specialty center like MSKCC. With Watson Oncology, they believe they can decrease the amount of time it takes for the latest research and evidence to influence clinical practice across the broader oncology community, help physicians synthesize available information, and improve patient care (Monegain 2016; IBM 2018).

12.3.2 WATSON ONCOLOGY FOR RADIATION ONCOLOGY

Some 50,000 oncology research papers are published each year, according to PubMed, and by 2020 medical information is projected to double every 73 days—outpacing the ability of a human to keep up with the proliferation of medical knowledge (Densen 2011). In addressing this issue, Swedish radiation oncology information technology vendor Elekta is collaborating with artificial intelligence (AI) kingpin IBM Watson Health to offer Watson Oncology as part of Elekta's cancer care systems. Elekta will market Watson Oncology as an AI-based clinical decision support system paired within Elekta's digital cancer care systems, including its Oncology Information System.

12.4 INDUSTRY AND ACADEMIC CENTERS' SOLUTIONS TO BIG DATA MINING AND DECISION SUPPORT

The integration of electronic medical records (EMRs), a picture archiving and communication system (PACS), radiotherapy, and other clinical systems data into a patient-centric view enable multidisciplinary collaboration. However, at present, the amount of cross-platform integration is somewhat limited compared with other technology developments in place. For example, a robust clinical dashboard displays vital information to provide clinical context for radiologists and to ensure that dictated impressions are as accurate as possible. Such a tool for the oncologist does not exist, and remains an area open for further development and optimization. In the setting of multidisciplinary specialty care, clinicians could gain valuable, metric-driven clinical decision support (CDS) to make informed patient management decisions. The preparation and execution of critical clinical multidisciplinary care processes, such as the Tumor Board, are streamlined and enlivened with these multi-informational platforms (Figure 12.5).

Current imaging workflows do not allow for volumetric analysis of lesions due to the time-intensive nature of performing three-dimensional (3D) segmentation. However, some in industry (HealthMyne) have developed the Quantitative Imaging Decision Support platform and Precise Metrics functionality enabled by a simple gesture launching sophisticated algorithms that execute a 3D segmentation in seconds. When compared with expert radiologists, these new programs have shown improved accuracy and consistency, which can greatly reduce intra- and inter-reader variability in key measurements needed to determine treatment response. The workflow is also made more efficient with image registration and lesion propagation techniques (Figures 12.6 and 12.7).

These applications create a full complement of quantitative metrics—ranging from simple diameter measurements used in therapy-response assessments like response evaluation criteria in solid tumors (RECIST) to a comprehensive radiomic profile—that can influence patient management decisions. Furthermore, the incorporation of even relatively simple metrics such as these—in addition to radiomics—has been shown to have diagnostic, prognostic, and prescriptive capabilities (Yip and Aerts 2016; Gillan et al. 2016). In addition, radiologists can create customized metrics of personal importance at the *Point of Read*, and clinicians

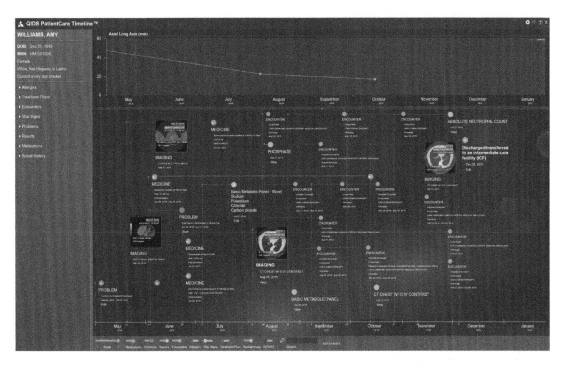

Figure 12.5 The comprehensive big view dashboards for patient; this one includes radiologic images for a fictious patient with updates, metrics, assessments, and a wide array of analytics. (Courtesy of HealthMyne, Madison, WI.)

SEMI AUTOMATIC MEASUREMENT

With a simple gesture a Radiologist identifies a lesion. Algorithms automatically find lesion boundaries and identify the entire volume with greater accuracy and consistency than manual delineation.

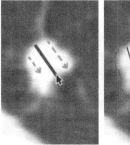
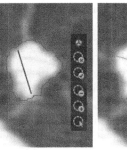

QUANTITATIVE METRIC EXTRACTION

Diameter measurements, evidence-based, volume based metrics and a full radiomic profile (500+ metrics) are automatically extracted and immediately available for treatment decisions.

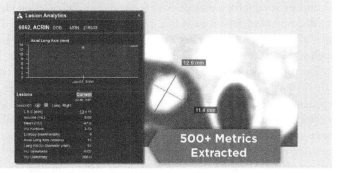

Figure 12.6 Typical radiometric quantitative results that may be made available to the radiation oncologist allow for continuity in the analytics for each ordered study. (Courtesy of HealthMyne, Madison, WI.)

Big data in radiation oncology

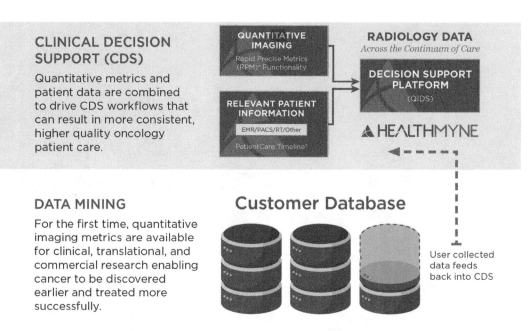

Figure 12.7 Clinical decision-making and data-mining support that allow feedback into the system for machine learning parameters and identifiers of linked and associated data for assessing patient outcomes and treatment effectiveness. (Courtesy of HealthMyne, Madison, WI.)

on the Multidisciplinary Care team can do the same at the *Point of Care*. This fosters digital collaboration across the continuum of care, positively affecting patient management decisions.

Multiple clinical decision support workflows are transformed by combining quantitative imaging metrics with relevant patient information. These workflows include automatic tumor staging, radiotherapy (RT) dose overlays, specialized patient-centric reporting, cancer screening programs, incidental findings management, clinical trials process streamlining, and precision medicine initiatives. These gains advance the Radiation Oncology mission to detect cancer as early as possible to enhance the probability of treating it successfully and meeting the goals of delivering consistent, high-quality care in line with the Quadruple Aim: "adding provider satisfaction to the triple goals of enhancing patient experience, improving population health, and reducing costs" (Bodenheimer and Sinsky 2014).

Currently, imaging metrics are hard to extract for research and discovery because they are hidden and trapped in dictated textual reports. In addition, large volumes of relevant patient data cannot be generated because existing software is cumbersome and slow. Every lesion identified adds more than 500 metrics to the discoverable database providing a wealth of curated radiomic data for clinical, translational, and commercial research. As discoveries are made, this data can instantly be identified in the clinical workflow and applied to today's patient for delivery of personalized care.

The complexity of a radiation oncology digital health record is compounded by the various phases of care that include consultation, treatment, follow-up, quality of life, and survival. The breadth and depth of a patient's digital health record in radiation oncology begins with consultation, which includes the patient's medical and surgical history, imaging, and pathology and other labs, and these may have been provided by an array of public or private institutions. The next phase of the patient record is treatment, and this includes surgical procedures for the current situation, radiotherapy, and systemic therapy, which may cross various platforms within the same institution. After treatment, the patient is seen in follow-up, and here we expect assessments to continue at the treating institution and perhaps at other locales as well. Finally, and this is key, patient quality of life and survival are assessed and this information is hopefully provided for outcomes research and on-going evaluations at all of the contributing institutions. An overview of this personalized care continuum using multiple cross-platform databases of typical patients for a radiotherapy course is presented in Figure 12.8.

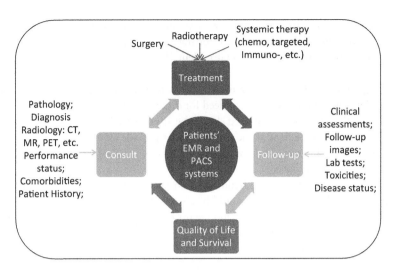

Figure 12.8 Overview of personalized care using a database of prior patients for a typical radiotherapy patient. (Courtesy of T. McNutt, Hopkins inHealth 2015.)

12.5 CHALLENGES OF PATIENT PRIVACY AND DATA SECURITY—BALANCING ACCESS AND CONTROL

Information technology design and security pose a challenge in radiation oncology, as it does in every other part of the cancer center.

12.5.1 OVERVIEW OF HIPAA AND ITS MEANINGFUL USE

The Health Insurance Portability and Accountability Act of 1996 (HIPAA) is U.S. legislation that provides data privacy and security provisions for safeguarding medical information. Whereas the HIPAA Privacy Rule deals with protected health information (PHI) in general, the HIPAA Security Rule (SR) deals with electronic protected health information (ePHI), which is essentially a subset of what the HIPAA Privacy Rule encompasses. According to the HIPAA SR, technical safeguards are "the technology and the policy and procedures for its use that protect electronic protected health information and control access to it."

Although the chances of a clinical practice being audited for HIPAA violations are slim, keeping patient information secure is growing more complicated (Rieders and Monroy 2014). Since 2009, there have been more than 800 patient data breeches and 29 million patient records have been affected by HIPAA violations, according to the 2013 Redspin Breach Report (Dolan 2014).

HIPAA violations may seem like a large-organization problem, but considering that many breaches are a result of employee theft and carelessness, smaller practices are also at risk. One issue practitioners face is that it becomes harder to keep track of electronic communication within the practice when patients and staff often have mobile devices and can be unaware of how easily HIPAA rules can be violated. In addition, for HIPAA and in demonstrating meaningful use, many physicians are not aware that the ability to complete and keep updated a security risk analysis—intended to identify risks to record security—is required and that it is a primary reason why practices fail meaningful-use audits.

12.5.2 INNOVATIONS IN PHI ACQUISITION: A DISCUSSION ON BEST PRACTICES FOR CAPTURING PHI IN A CLINIC USING HANDHELDS AND WEARABLES AND WIRELESS LOCATION TRACKING

Patients are granted both privacy and security rights as it relates to their personal health records and any data associated with their care. This includes, but is not limited to, appointment times and dates, records of family history, prescriptions, and procedures or elements of care. These are important considerations when collecting, storing, and handling patient data.

For the security portion, there are some primary components to consider. The first component of security is anonymization; aggregate data collected for statistical purposes can be anonymized such that no single data point can be traced back to a particular patient. The next component is encryption and secure auditing.

For the sake of digital data collection using modern tools such as handhelds like tablets, wearables, and/or wireless location tracking, there will be a need for discerning between personalized data versus anonymized data. The majority of the data collected should be anonymized for the sake of creating aggregate data models that can be used to see data in the aggregate that provide a more cohesive view than what can be gathered from a single set of personalized data points. Some of the data collected would be personalized and stored on a patient-by-patient basis.

Data points captured with wearables and handhelds can fall into both categories, depending on the data itself. For instance, a wearable heart monitor would be a valuable source of personalized data that would give a health care provider meaningful information regarding a single patient, whereas a timer/pedometer designed to track steps and the duration of and or during an office visit would serve better as a source of anonymous data from which to build an aggregate record.

Handheld devices such as tablets and wireless location tracking devices may be used to collect regular per-visit self-reported check-up information for a patient to inform how that patient's condition may have changed between visits. This data can then be used for comparison with an aggregate map of patient outcomes and self-reporting data of other patients with shared diagnosis, treatment protocol, age, sex, and history.

To improve individual patient experience and outcomes, and overall performance and treatment planning, it is necessary to take a scientific approach to the collection of data in a manner that affords patients the greatest ability to inform their own care while serving the need for a greater collection of aggregate data. Handhelds, wearables, and patient and health care provider tracking within a care facility can be combined to accumulate a wealth of data that, if handled properly, would not impinge upon the rights of the patients to the security and accessibility of their personal health care data as defined in HIPAA.

12.5.3 PATIENT DATA STORAGE AND ACCESSIBILITY: BEST PRACTICE FOR ON-PREMISES, OFF-PREMISES, AND CLOUD STORAGE

Storage of personalized patient data can be managed in one of three general ways, on-premises, off-premises, and Cloud storage. Each manner of storage has its own strengths and weaknesses as it pertains to security and accessibility. In Figure 12.9, we show a table depicting the strengths and weaknesses for each of the approaches to data storage.

Accessibility in terms of patient data speaks to how easily data can be accessed from disparate locations, as well as how easily the data is accessed and presented. In general, the greater the security requirements to access a set of data, the less accessible that data becomes. As an example, securing your online email account with a password reduces its accessibility by a measure of how much effort it takes to remember and successfully enter the password. Having to enter a password is a commonly accepted trade-off for security, but there are also additional nonhostile ways to secure data while preserving relative ease of access. Security and accessibility must be balanced such that the security needs are met without making the system frustrating or unwieldy to use.

Security in general means placing blocks on who can access data, when, and how. For the sake of fulfilling the HIPAA regulatory guidelines, it also means logging and tracking who and how patient data is accessed. This means that any and all interactions with the data by anyone have to be recorded and that record must be able to be made available to a patient on request.

On-premises storage affords the greatest physical security of data, but that can be a limitation for accessibility. An on-premises storage solution would involve having a machine or set of machines that are physically present at the health care facility. Having the machines themselves, and the data they store, be located within a single facility that operates both as the premises that patients visit and the data storage location means that there is very little risk of external intrusion or data corruption from an outside source. It also means that accessing the patient data from an external location, or any other patient care facility,

	On Premises (Intranet)	Off Premises (Co-lo)	Cloud
Top Security and Data Vulnerabilities	> Social Engineering > Data loss > Malicious Insiders > Physical intrusion > Internal man in the middle attack	> Social Engineering > Inadequate user security practices > Insecure external interfaces > System vulnerabilities > Malicious Insiders > Advanced Persistent Threats > Data loss	> Social Engineering > Inadequate personal security practices > Insecure external interfaces > System vulnerabilities > Malicious Insiders > Data loss or corruption
Accessibility advantages	> Speed within a local network is potentially higher and more reliable > Physical access to the machine or storage medium reduces the risk of loss of internal access due to security protocols	> Ability to access data remotely from multiple locations > Ability to aggregate data from multiple sources increasing access to data in general > Ability to gain physical access to hardware if necessary	> Ability to access data remotely from multiple locations > Ability to aggregate data from multiple sources increasing access to data in general > Availability of speed and processing power scaling reduces risks of access being interrupted by high-traffic.
Accessibility disadvantages	> No or limited external or multi-site access. > No or limited scaling to handle high-traffic > No or limited sharing or aggregating of data between service or care locations	> Architecture for scaling likely limited and not as complete as for cloud > Geographically distributed access could suffer slower access > Necessary security protocols can potentially impose a heavy burden on users	> No or very limited physical access to hardware as its generally virtualized > Necessary security protocols can potentially impose a heavy burden on users

Figure 12.9 Strengths and weaknesses of data storage solutions. (Courtesy of C. Buchholz, SingleMind Consulting 2018.)

would be more challenging and potentially reduces the security gains of having the data stored locally. It can also mean a greater risk of internal data integrity and potential loss if a catastrophic event such as a fire or a mechanical or electronic failure were to damage the machines or data stores themselves. However, the greatest risk to an on-premises data storage solution is social engineering (Warwick 2016). An administrator, engineer, or anyone who has basic access to the physical hardware could be co-opted, coerced, or even unwittingly conned into either granting access to an outside source or obtaining and distributing data in ways they should not. Recording and logging access to the data, from a technical perspective, is not really any easier with an on-premises solution unless access to the data is physically and logically locked to a single terminal. In which case, the only way to see the data would be to access that single terminal interface. Best practice for managing on-premises data storage involves restricting who has direct access to data and when. This can be done by limiting physical access to points of entry, as well as securing terminal access through the use of account control and/or key fobs or biometric security protocols. It also calls for a training program for anyone who is granted access for the sake of providing them tools and resources to recognize social engineering attempts, how to avoid falling prey to them, and protocols for reporting them.

Off-premises and Cloud storage share a number of advantages in terms of accessibility, but also incur a greater cost and risk to the data security and integrity. Having the data stored off-premises means that the physical location of the machine or machines that house the patient data either are in a single dedicated space such as a co-location facility or are distributed between several disparate facilities. Having a single dedicated space for the off-premises machines means that there is a single point of failure in terms of both data integrity and security. It also means a similar risk to the data integrity in the event of a natural disaster or catastrophic mechanical or electronic failure, but this can be mitigated with advanced backup systems and security and safety protocols put in place by a top-tier co-location facility. For the sake of security, both single location and distributed location off-premises solutions impose a cost of requiring a higher level of technology in place to secure the data. Encrypting the data in place is, by itself, not enough, because the systems storing the data must allow external access. This can be secured by way of a collection of modern security systems such as hardware and software firewalls, secure token authorization, and secure socket communications. Best practice methods for securing an off-premises data storage solution involve encrypting the data local to each instance, creating regular archives for data integrity, and securing data against loss as well as creating a secure service layer that operates as an interface for external access. The secure service layer could have access to it restricted to specific access points such as particular network and/or even specific machines.

Cloud data storage offers the greatest flexibility in terms of accessibility, as well as the highest protection for data integrity and accessibility in terms of loss or corruption of data in the case of internal failures or catastrophic events. It also incurs the highest cost and risk in terms of external security requirements. With a Cloud storage solution, the data is recorded, encrypted, and shared between a wide array of disparate machines distributed across multiple facilities and networks. This means that the data can be accessed from almost anywhere quickly and efficiently. It also means that, in order to secure it and also allow for the logging of access, there would need to be an advanced set of systems in place to afford that security. There would need to be an advanced system that managed the data storage, there would also need to be an advanced system that handled the access to that data. It would also mean that certain types of aggregate data could be more easily collected from multiple sources and interacted with collectively rather than having to manage a secondary system or set of procedures to allow for combining data sources. The greatest advantage of Cloud-based solutions is that the aggregate data can more easily be processed, thereby supporting the ability of machine learning and advanced systems to leverage the statistical patterns that emerge when a large enough block of data is collected. Best practice methods for securing a Cloud data storage solution are similar to those of an off-premises solution but would include the resources and advanced architecture afforded by utilizing an existing infrastructure.

12.5.4 PHI REPORTING AND VISUALIZATION

All good science is predicated upon the collection, observation, and evaluation of data: testing a hypothesis against a model or a working system to observe the outcomes. The larger the data set, the greater the certainty that an outcome is reproducible. Better yet, the more points of data that intersect, the greater the ability to extrapolate new hypotheses. Pattern recognition and observation of intersections are significantly improved when modern machine learning and data visualization are incorporated into any scientific approach. Modern machine learning strives to tailor multiple algorithms and approaches to best serve the needs of a particular problem. In the realm of machine learning, there is an adage of "no free lunch." This means that no one approach will solve every problem. With the advent of considerably greater access to Cloud computing and with the ability to more easily and cost-effectively scale computing power, data scientists have learned that they can apply multiple machine or deep-learning approaches at a time and contrast and compare results to best solve a particular data problem. Figure 12.10 depicts the high-level relationships between the three primary components of this field of science.

Many disparate data points can be mapped either to an individual patient, or anonymously to patients with shared conditions, history and treatments. Even similar visit patterns and lifestyles can illuminate how treatments are distributed or received. These things, along with patient outcomes and self-reported data can be collected and aggregated across a wide array of patient and care-provider interactions. Once this data is gathered and collated, applying machine learning systems that can observe and identify obscure patterns within massive data sets could result in novel or even revolutionary insight into how to most efficiently treat individual patients or even expose inefficiencies in the operation of a particular clinic. Figure 12.11 depicts a classic case of machine learning in the pursuit of pattern recognition.

Leveraging machine learning systems to build maps between disparate data sets can allow for a greater ability to visualize patterns in a multidimensional fashion. With the democratization of this technology that has come about over the last few years, it is more feasible than ever. Advanced graphical representation of data maps grants an observer better control and can be used to garner a greater understanding of how

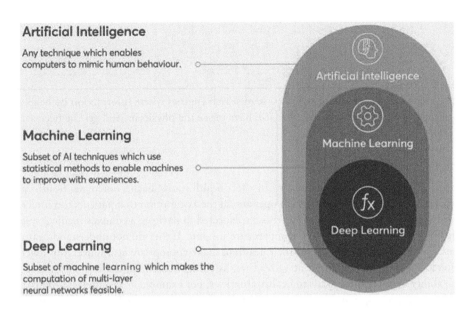

Figure 12.10 High-level relationships between artificial intelligence, machine learning, and deep learning. (Courtesy of C. Buchholz, SingleMind Consulting 2018.)

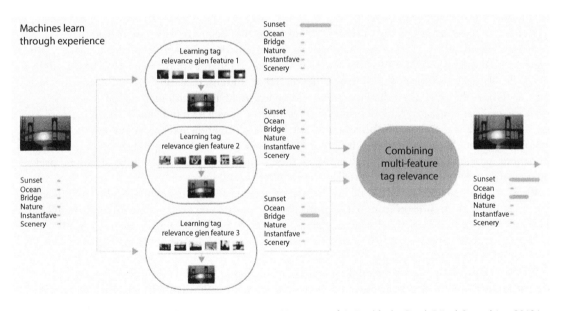

Figure 12.11 Machine learning and pattern recognition. (Courtesy of C. Buchholz, SingleMind Consulting 2018.)

an individual patient may experience treatment (Mehl 2017). Graphical data modeling systems such as Tableau can be utilized to build the graphical representation of an expansive and complex data set (Tableau 2017). Data modeling could also be utilized to evaluate disparate treatment regimens against self-reporting data as well as outcomes in order to identify possible improvements in a given process or even weaknesses in a specific care facility.

Advanced visualization of data can also afford a care provider greater insight into a particular patient's journey through a treatment program. Both within a single facility, and across disparate care facilities or practitioners.

Big data in radiation oncology

12.6 THE CHALLENGE OF INFORMATION AND HEALTH LITERACY IN THE ERA OF BIG DATA: PATIENT INVOLVEMENT IN DECISION MAKING AND FEEDBACK

Patients in the twenty-first century make great use of search engines where information is cheap, yet proper application of the knowledge is expensive (you still have to see the physician, and get the necessary diagnostic tests, referrals, imaging, and so on).

12.6.1 PATIENT HEALTH LITERACY

Limited health literacy is a hidden epidemic. It can affect health status, health outcomes, health care use, and health costs. The entire health care system operates on the assumption that patients can understand complex written and spoken information. Patients are expected to navigate a complex medical system and then manage more and more of their often complex care at home. If they do not understand health information, they cannot take necessary actions for their health or make appropriate health decisions (Weiss 2007).

Health literacy requires a complex group of reading, listening, analytical, and decision-making skills, as well as the ability to apply these skills to health situations. For example, it includes the ability to understand instructions on prescription drug bottles, appointment slips, medical education brochures, doctor's directions and consent forms, and the ability to negotiate complex health care systems.

More recent definitions focus on specific skills needed to navigate the health care system and the importance of clear communication between health care providers and their patients. Both health care providers and patients play important roles in health literacy. The number of different definitions for health literacy demonstrate how the field has evolved.

Health literacy is defined by the Institute of Medicine as the "degree to which individuals have the capacity to obtain, process, and understand basic health information and services needed to make appropriate health decisions" (IOM, 2014) The National Network of Libraries in Medicine (NNLM, a subdivision of the National Institutes of Health) provides several definitions, including: (1) "the degree to which individuals have the capacity to obtain, process, and understand basic health information and services needed to make appropriate health decisions," (2) "the degree to which an individual has the capacity to obtain, communicate, process, and understand basic health information and services in order to make appropriate health decisions," and (3) "the use of a wide range of skills that improve the ability of people to act on information in order to live healthier lives." These skills include reading, writing, listening, speaking, numeracy, and critical analysis, as well as communication and interaction skills (NNLM 2018). The Medical Library Association's (MLA) definition is more detailed and further defines health information literacy as "the set of abilities needed to: recognize a health information need; identify likely information sources and use them to retrieve relevant information; assess the quality of the information and its applicability to a specific situation; and analyze, understand, and use the information to make good health decisions" (MLA 2013).

Provider explanation, and patient understanding, of complex oncologic treatment pathways can be overwhelming despite well-documented informed consent and written consent information (Long 2001). In addition, available patient resources are often beyond what patients are capable of comprehending, with some studies suggesting that less than 1% of the currently available patient education materials meet the American Medical Association and National Institutes of Health recommendations for patient education resources (Prabhu et al. 2016). To further enhance provider communication and patient engagement, methods are needed that enhance patient engagement, provider efficiency, and patient understanding for overall treatment success.

The ACA focused attention on major reforms, including greater use of evidence-based medicine, shared decision making, comparative effectiveness research, and transparency of cost and quality information. Collectively, these efforts are at the heart of evidence-based health care. Yet it is the acceptance of evidence-based health care by the consumer that will greatly determine its success or failure. If consumers reject it, do not understand it, or see it as an inappropriate way to make decisions, then the movement and the efforts may fail (Graham and Brookey 2008).

As with anyone providing services to customers, physicians have always wanted their patients to be satisfied. But government programs such as the Physician Quality Reporting System—which indirectly ties Medicare reimbursements to patient satisfaction scores—as well as the growth of websites devoted to evaluating doctors, the need to keep patients satisfied has taken on new urgency. However, is that need affecting physician medical decision-making?

The answer is "yes," according to many practitioners. "Physicians are judged by their performance grade card, so to protect their livelihood they tend to modify their practice patterns, to be more productive and improve their scores," says David Fleming, MD, MACP, president of the American College of Physicians. "In all too many practices medicine is devolving into a metrics-centered business, rather than a patient-centered profession" (Terry et al. 2014). "The challenge for clinicians is that the goal of patient satisfaction isn't always aligned with the goal of providing high-value care," says Joshua J. Fenton, MD, MPH, associate professor of family and community medicine at the University of California, Davis. "I assume this is true in other specialties, but in primary care, there can be tension between what a patient wants and expects and what the provider believes to be clinically important and evidence-based" (Terry et al. 2014).

12.7 THE ETHICS OF PREDICTIVE ANALYTICS FOR PRECISION MEDICINE IN RADIATION ONCOLOGY

12.7.1 ETHICAL CHALLENGES ARISING FROM LEARNING HEALTH CARE SYSTEMS AND BIG DATA

Many of the current challenges in the clinical translation of precision medicine have arisen because of two significant changes to the context of biomedical research. The first is the shift toward learning health care systems, which facilitates the translation of research findings by providing real-world settings in which to evaluate their external validity. However, the learning health care system also represents a blurring of traditional boundaries between research and clinical care, which raises fundamental questions about the ethical responsibilities of researchers, caregivers, and health care institutions (Kass et al. 2013). The second significant contextual change is the availability and use of big data, which are easily collected, stored, and shared in digital form, and allow linking of data collected in health care settings such as electronic health records (EHRs) with personal data collected outside of the health care context such as from smartphones, fitness trackers, and direct-to-consumer commercial diagnostic tests such as those offered by 23andMe.

Achieving the vision of precision medicine depends critically on these two contextual changes (Collins and Varmus 2015). If patients are to be stratified according to a large number of individual characteristics in order to tailor diagnoses, prognoses, treatments, or preventive measures, large data sets will need to be gathered, combined, and analyzed. Although precision medicine is currently largely a research activity, translation to clinical practice will require data capture and analytic systems to be integrated into the core of health care systems and be designed to learn by continually updating its databases and algorithms.

These paradigmatic shifts to clinical translation thus raise new questions about the appropriate ethical and regulatory frameworks to apply. In the United States, relevant regulation includes the Common Rule (45 CFR 46) or the FDA equivalent, HIPAA, and the Health Information Technology for Economic and Clinical Health (HITECH) Act, among others. However, novel aspects of the learning health care system and big data require researchers to consider specific ethical issues that may not directly be addressed by regulation. These include informed consent, privacy, justice and fairness, and the potential for dual use.

12.7.2 INFORMED CONSENT

The general requirements for informed consent for research conducted in the United States that is subject to federal regulation (generally at institutions that receive federal funding) are described in section 46.116 of the Code of Federal Regulations under Part 46 Protection of Human Subjects. The generally accepted elements of informed consent that are detailed in 46.116 are that (1) the nature of the research and the participant's involvement are fully disclosed, (2) there is adequate comprehension on the part of the potential participant or proxy decision-maker, and (3) the participant's choice to participate is voluntary. All three elements are necessary to support truly informed consent. Yet, translational research in precision medicine can pose challenges to all three elements. If research is conducted in a health care setting, especially if informed consent is obtained by patients' doctors, the very fact that the patient is participating in a research study is potentially obscured, even if the "intervention" is routine clinical care. It may be less obvious to patients in the health care context that their EHR data is being used for purposes other than direct health care or that data collected from nonmedical sources is being combined with EHR data and analyzed for research. It may even be less obvious that there is the potential for the research data to be shared outside of the caregiver group or patient's health system. Patients might assume that risks are low because research is occurring in the context of regular health care. Finally, participants might not realistically have given voluntary consent for a research study conducted at their health care institution or clinic, especially in cases where their entire clinic or hospital is assigned to use particular interventions or procedures in a study using cluster randomization (Kelley et al. 2015; Lee et al. 2016).

12.7.3 PRIVACY

Privacy issues are of course of concern, especially with the common use of digital information (Mittelstadt and Floridi 2016). The research uses of big data, which for precision medicine can include highly granular biological data now regularly deposited in EHRs such as genome sequence, images from CT scans or MRIs, and microscopic images of biopsy tissue, as well as environmental and social data such as geographic locations, employment, or diet now provide highly detailed profiles of individuals such that are uniquely identifying, even if standard identifiers of personal health information such as names, addresses, and telephone, social security, and medical record numbers are stripped. Although the ease of re-identification of personal genome data led to calls for policy changes, the potential for loss of privacy of big data must be balanced by consideration of the benefits of researchers' access to granular data and incorporation of technical solutions such as the federated ecosystems for data sharing (Global Alliance for Genomics and Health 2016) and encrypted identifiers (Hansson et al. 2016). Some have recommended that "collectors of predictive analytics data notify patients that the data gathered on them in the course of regular health care may be used in de-identified form in predictive analytics models" (Cohen et al. 2014).

12.7.4 JUSTICE AND FAIRNESS IN THE DISTRIBUTION OF BENEFITS AND HARMS

Perhaps one of the major challenges to effective and ethical clinical translation of precision medicine is the lack of diversity of research participants (Popejoy and Fullerton 2016), which will potentially lead to inaccurate findings in underrepresented populations such as higher rates of false-positive diagnoses (Manrai et al. 2016) and genomic variants of uncertain significance (Caswell-Jin et al. 2018). Although digitization of health care provides greater access to data by researchers, patient populations who do not have access to health systems with substantial infrastructure for data capture, storage, and analysis and are already on the wrong side of the digital divide will face increased disparities if their data are not included to teach learning health care systems.

Justice and fair distribution of harms, as well as benefits, are especially important ethical principles to consider in the design of predictive analytics (Association for Computing Machinery 2018). The potential for algorithms used in prediction to exacerbate existing social biases, especially with respect to legally protected classes such as race and sex, has been identified as a serious issue for computer scientists to address (Lum and Johndrow 2016). The "black box" nature of predictive algorithms, especially those that employ machine learning and unsupervised methods in particular (Deo 2015), mean that clinicians are not able to provide accountability for algorithm-aided decision making, posing challenges to the exercise of professional responsibility (Price 2015). Computer scientists are developing ways to improve transparency and accountability of machine learning algorithms to address this problem (Bastani et al. 2017).

12.7.5 DUAL USE OF PREDICTIVE ANALYTICS FOR PRECISION MEDICINE

Finally, an often-overlooked issue in the application of predictive analytics to precision medicine is the general problem of "dual use." Whereas the term dual use is typically used to describe malicious uses of technology that was intended for beneficial uses, such as genetic modification of bacteria to improve crop yields being maliciously used to create biological weapons, the concept here applies more broadly. The vision of precision medicine is to provide individuals with diagnosis, treatment, and prevention measures that lead to better health through the identification of effective interventions (Collins and Varmus 2015). However, the same predictive analytic approaches used for precision medicine can be used as the justification for withholding of or denial of care (Bates et al. 2014) and is foreshadowed by the potential use of whole genome sequencing to justify limitation of high-cost and resource-intensive care such as organ transplantation for critically-ill children (Char et al. 2015). Elimination of inappropriate treatment through precision medicine is a worthy and ethical goal if it is in service of improving patient care, even if it results in withholding of treatment that is likely to be ineffective or directs patients to palliative care, for example. The danger is the subversion of precision medicine strategies toward goals such as reimbursement or cost containment by such things as identifying high-cost patients in order to reduce their utilization of services. The patient perspective must be prioritized over administrative or purely economic outcomes in design and validation of predictive analytic models if they are to ethically serve patients (Cohen et al. 2014).

12.8 UPCOMING BIG DATA CHALLENGES: LOW HANGING FRUIT AND APPLES JUST A BIT OUT OF REACH

There are a number of large national funding mechanisms through the National Cancer Institute (NCI) Informatics Technology for Cancer Research (https://itcr.cancer.gov/funding-opportunities) for algorithm development, prototyping and hardening, enhancement and dissemination, and sustainment. In addition, recognition that leverage given to motivated people (physicians and scientists) with a big stick (big data) applied in earnest can move mountains, the American Board of Medical Specialties (AMBS) allowed clinical informatics to become a board-certified medical subspecialty in 2011, and fellowships in clinical informatics are becoming available around the country (e.g., Oregon Health & Science University; Massachusetts General Hospital; Stanford University; University of California, San Francisco).

Health care could realize $300 billion in savings annually by leveraging big data. Part of this is based on reducing waste and delays, but it is also based on delivering appropriate treatment to the right patients (IHTT 2013).

In closing, this chapter has exposed a wide array of challenges of big data in radiation oncology. The complex and evolving issues of diverse patient data platforms, clinical workflows, data mining, decision support, information literacy, and ethics and social concerns of big data technology in radiation oncology have all been touched upon for consideration. In that regard, perhaps Dr. Brian D. Kavanagh, MD, MPH, FASTRO and Chair of the American Society of Radiation Oncology (ASTRO) Board of Directors, put it best in his address of the future challenges of radiation oncology with his unique and inspiring perspective. Dr. Kavanagh stated, "Our field has been inventing and embracing new technology ever since the days of Marie Curie and Wilhelm Röntgen […]The natural next steps in our technological evolution will include enhanced auto-contouring of tumors and normal tissues, improved remote patient safety monitoring, centralized high-speed planning and automated interactive health records that instantaneously generate guideline-based treatment regimens tailored to a patient's TNM stage, comorbidities, tumor genetics and performance status. Think of how much we gain in that future vision. We will improve patient outcomes with standardization. We can provide greater access in remote locations. And we will free ourselves of some of the more tedious chores of our profession so that we can have more time to spend talking with our patients and knowing their stories and supporting them in ways that robots will never be able to do" (Kavanagh 2017).

REFERENCES

Association for Computing Machinery (2018). *Code of Ethics and Professional Conduct: Draft 3*, https://ethics.acm.org/2018-code-draft-3/.

Ariely D (2013). *Big Data is Like Teenage Sex*, https://twitter.com/danariely/status/287952257926971392?lang=en.

Bastani O, Kim C, Bastani H (2017). *Interpretability via Model Extraction*, https://arxiv.org/abs/1706.09773.

Bates DW, Saria S, Ohno-Machado L, Shah A, Escobar G (2014). Big data in health care: Using analytics to identify and manage high-risk and high-cost patients. *Health Aff (Millwood)* 33:1123–1131.

Bendix J, Krivich RS, Martin K, Mazzolini C, Shryock T (2017). *Top 10 Challenges Facing Physicians in 2018*. http://www.medicaleconomics.com/medical-economics-blog/top-10-challenges-facing-physicians-2018.

Bodenheimer T, Sinsky C (2014). From triple to quadruple aim: Care of the patient requires care of the provider. *Ann Fam Med* 12:573–576.

Caswell-Jin JL, Gupta T, Hall E, Petrovchich IM, Mills MA, Kingham KE, Koff R et al. (2018). Racial/ethnic differences in multiple-gene sequencing results for hereditary cancer risk. *Genet Med* 20:234–239.

Char DS, Cho M, Magnus D (2015). Whole genome sequencing in critically ill children. *Lancet Respir Med* 3:264–266.

Chetty IJ, Martel MK, Jaffray DA, Benedict SH, Hahn SM, Berbeco R, Deye J et al. (2015). Technology for innovation in radiation oncology. *Int J Radiat Oncol Biol Phys* 93:485–492.

Cohen IG, Amarasingham R, Shah A, Xie B, Lo B (2014). The legal and ethical concerns that arise from using complex predictive analytics in health care. *Health Aff (Millwood)* 33:1139–1147.

Collins FS, Varmus H (2015). A new initiative on precision medicine. *N Engl J Med* 372:793–795.

Cottle M, Hoover W, Kanwal S, Kohn M, Strome T, Treister N (2013). Transforming Health Care through Big Data Strategies for leveraging big data in the health care industry. Institute for Health Technology Transformation, http://ihealthtrancom/big-data-in-healthcare.

Densen P (2011). Challenges and opportunities facing medical education. *Trans Am Clin Climatol Assoc* 122:48–58.

Deo RC (2015). Machine learning in medicine. *Circulation* 132:1920–1930.

Dolan P (2014). *Be Proactive to Avoid HIPAA Violations, http://www.medicaleconomics.com/health-care-information-technology/be-proactive-avoid-hipaa-violations.*

Gillan C, Giuliani M, Harnett N, Li W, Dawson LA, Gospodarowicz M, Jaffray D (2016). Image guided radiation therapy: Unlocking the future through knowledge translation. *Int J Radiat Oncol Biol Phys* 96:248–250.

Global Alliance for Genomics and Health (2016). Genomics. A federated ecosystem for sharing genomic, clinical data. *Science* 352:1278–1280.

Graham S, Brookey J (2008). Do patients understand? *Perm J* 12:67–69.

Hansson MG, Lochmuller H, Riess O, Schaefer F, Orth M, Rubinstein Y, Molster C et al. (2016). The risk of re-identification versus the need to identify individuals in rare disease research. *Eur J Hum Genet* 24:1553–1558.

IBM (2018). *IBM Watson Health*, https://www.ibm.com/watson/health/.

IHTT (2013). *Transforming Health Care Through Big Data: Strategies for Leveraging Big Data in the Health Care Industry*, http://c4fd63cb482ce6861463-bc6183f1c18e748a49b87a25911a0555.r93.cf2.rackcdn.com/iHT2_BigData_2013.pdf.

IOM National Academies of Science, Engineering, Medicine: Health and Medicine Division (2004). *Health Literacy: A Prescription to End Confusion*, http://www.iom.edu/Reports/2004/Health-Literacy-A-Prescription-to-End-Confusion.aspx.

Kagadis GC, Nagy P, Langer S, Flynn M, Starkschall G (2008). Anniversary paper: Roles of medical physicists and health care applications of informatics. *Med Phys* 35:119–127.

Kass NE, Faden RR, Goodman SN, Pronovost P, Tunis S, Beauchamp TL (2013). The research-treatment distinction: A problematic approach for determining which activities should have ethical oversight. *Hastings Cent Rep Spec* 43:S4–S15.

Kavanagh B (2017). Are We Ready For The Self-Driving Car? How technology will continue to shape the radiation oncology workforce. In, Winter Edition: ASTRO.

Kelley M, James C, Alessi Kraft S, Korngiebel D, Wijangco I, Rosenthal E, Joffe S, Cho MK, Wilfond B, Lee SS (2015). Patient perspectives on the learning health system: The importance of trust and shared decision making. *Am J Bioeth* 15:4–17.

Kwon DH, Tisnado DM, Keating NL, Klabunde CN, Adams JL, Rastegar A, Hornbrook MC, Kahn KL (2015). Physician-reported barriers to referring cancer patients to specialists: Prevalence, factors, and association with career satisfaction. *Cancer* 121:113–122.

Lee SS, Kelley M, Cho MK, Kraft SA, James C, Constantine M, Meyer AN, Diekema D, Capron AM, Wilfond BS, Magnus D (2016). Adrift in the gray zone: IRB perspectives on research in the learning health system. *AJOB Empir Bioeth* 7:125–134.

Long LE (2001). Being informed: Undergoing radiation therapy. *Cancer Nurs* 24:463–468.

Lum K, Johndrow J (2016). *A Statistical Framework for Fair Predictive Algorithms*, https://arxiv.org/abs/1610.08077.

Manrai AK, Funke BH, Rehm HL, Olesen MS, Maron BA, Szolovits P, Margulies DM, Loscalzo J, Kohane IS (2016). Genetic misdiagnoses and the potential for health disparities. *N Engl J Med* 375:655–665.

Mehl V (2017). *Using Data to Improve Patient Care*, https://www.hopkinsmedicine.org/news/articles/using-data-to-improve-patient-care.

Mittelstadt BD, Floridi L (2016). The ethics of big data: Current and foreseeable issues in biomedical contexts. *Sci Eng Ethics* 22:303–341.

MLA Medical Library Association (2013). *What is Health Information Literacy?* https://www.mlanet.org/resources/healthlit/define.html.

Monegain B (2016). *IBM Watson, Quest Diagnostics, Memorial Sloan Kettering Cancer Center, MIT, Harvard Combine Forces for Massive Oncology, Precision Medicine Initiative*, http://www.healthcareitnews.com/news/ibm-watson-quest-diagnostics-memorial-sloan-kettering-cancer-center-mit-harvard-combine-forces.

NNLM (2018). *Health Literacy Definitions*, https://nnlm.gov/initiatives/topics/health-literacy.

Popejoy AB, Fullerton SM (2016). Genomics is failing on diversity. *Nature* 538:161–164.

Prabhu AV, Hansberry DR, Agarwal N, Clump DA, Heron DE (2016). Radiation oncology and online patient education materials: Deviating from NIH and AMA recommendations. *Int J Radiat Oncol Biol Phys* 96:521–528.

Price II W (2015). Black-box medicine. *Harv J Law Technol* 28:(2)419–467.

Regge D, Mazzetti S, Giannini V, Bracco C, Stasi M (2017). Big data in oncologic imaging. *Radiol Med* 122:458–463.

Rieders L, Monroy M (2014). *Creating Your Practice's 'Bring Your Own Mobile Device' Policy*, http://medicaleconomics.modernmedicine.com/medical-economics/news/creating-your-practices-bring-your-own-mobile-device-policy.

Ritchie A, Marbury D, Verdon D, Mazzolini C, Boyles S (2014). *Shifting Reimbursement Models: The Risks and Rewards for Primary Care*, http://medicaleconomics.modernmedicine.com/medical-economics/content/tags/aca/shifting-reimbursement-models-risks-and-rewards-primary-care?page=0%2C1.

Sinsky C, Colligan L, Li L, Prgomet M, Reynolds S, Goeders L, Westbrook J, Tutty M, Blike G (2016). Allocation of physician time in ambulatory practice: A time and motion study in 4 specialties. *Ann Intern Med* 165:753–760.

Tableau (2017). *Advanced Charting—APAC*, https://www.tableau.com/learn/webinars/advanced-charting-apac.

Terry K, Ritchie A, Marbury D, Smith L, Pofeldt E (2014). Top 6 Practice Management Challenges Facing Physicians in 2015, http://medicaleconomics.modernmedicine.com/medical-economics/news/top-6-practice-management-challenges-facing-physicians-2015?page=0%2C3.

Warwick A (2016). Social Engineering is Top Hacking Method, Survey Shows, https://www.computerweekly.com/news/4500272941/Social-engineering-is-top-hacking-method-survey-shows.

Weiss B (2007). Health literacy and patient safety: Help patients understand, 2 Edition. Chicago, IL: American Medical Association Foundation and American Medical Association.

Yip SS, Aerts HJ (2016). Applications and limitations of radiomics. *Phys Med Biol* 61:R150–R166.

Radiogenomics

Barry S. Rosenstein, Gaurav Pandey, Corey W. Speers, Jung Hun Oh, Catharine M.L. West, and Charles S. Mayo

Contents

13.1 Use of Genomics to Guide Treatment Response and Predict Normal Tissue Toxicity 201
13.2 Development of Genome-Based Tests to Predict Tumor Response 202
    13.2.1 Gene Signatures 202
    13.2.2 RadiotypeDx 203
    13.2.3 Decipher 203
    13.2.4 Post-operative Radiation Therapy Outcomes Score 203
    13.2.5 Radiation Sensitivity Index 203
    13.2.6 Genomic-Adjusted Radiation Dose 204
    13.2.7 Hypoxia Signatures 204
13.3 Prediction of Normal Tissue Toxicity 204
    13.3.1 The Radiogenomics Consortium 204
    13.3.2 Big Data Projects in Radiogenomics 205
    13.3.3 Large-Scale Big Data Radiogenomics Projects Currently in Progress 206
13.4 Approaches to Radiogenomics Using Machine Learning 207
    13.4.1 Single-SNP Association Test 208
    13.4.2 Machine Learning-Based Predictive Model 208
    13.4.3 Random Forest Model 208
    13.4.4 Penalized Model 209
    13.4.5 Preconditioning Random Forest Regression 209
    13.4.6 Biological Data Analysis 209
    13.4.7 Predictive Model Optimization Based on Biological Processes 210
13.5 Other Applications of Machine Learning in Biomedical Sciences 210
    13.5.1 Protein Function Prediction 210
    13.5.2 Discovery of Genetic Interactions 211
    13.5.3 Biomarker Discovery and Personalized Medicine 212
13.6 Summary and Future Directions 213
References 213

13.1 USE OF GENOMICS TO GUIDE TREATMENT RESPONSE AND PREDICT NORMAL TISSUE TOXICITY

The concept that genetic/genomic alterations (either DNA sequence or subsequent expression) may function as surrogate biomarkers of disease response or normal tissue toxicity underpins the field of radiogenomics (Rosenstein 2017). Prior to the genomics era, and for over half a century, research aimed at predicting tumor and normal tissue response to radiation was dominated by *in vitro* culture of malignant and normal cells (Fertil and Malaise 1981; West et al. 1991; Burnet et al. 1992; Geara et al. 1993; Steel 1993; Johansen et al. 1994; Begg et al. 1999). Although beyond the scope of this chapter, these experiments provided the foundation upon which the linear quadratic model was derived, but the methods had a limited ability to adequately and reliably model the heterogeneous response of tumors and normal tissue to ionizing radiation. With advances in knowledge enabled in part through the human genome project, investigators identified more sophisticated ways of both describing and predicting radiotherapy responses. This advance, coupled with technological advances that allow for a more complete evaluation of DNA, RNA, protein, and cellular metabolism, has led to the development of "omics"-based approaches for prediction of radiotherapy outcomes (Torres-Roca et al. 2005; Weichselbaum et al. 2008; Eschrich et al. 2009, 2012; Servant et al. 2012; Speers et al. 2015; Yard et al. 2016; Zhao et al. 2016; Scott et al. 2017).

Pioneering work in the area of genomic-based signature development focused on prognostication and response to systemic therapy (van de Vijver et al. 2002; van't Veer et al. 2002; Paik et al. 2004, 2006; Glas et al. 2006; Albain et al. 2010; Knauer et al. 2010; Tang et al. 2011; Dowsett et al. 2013). These early-era investigators used genomic-based approaches to predict response to chemotherapy, hormone therapy, or to determine prognosis if adjuvant therapy was omitted after surgery. Indeed, several functional genomic assays are currently in clinical practice that function either as biomarkers to aid in determining prognosis independent of treatment or as predictive biomarkers that are useful in directing appropriate clinical management (Sparano et al. 2015). Initially these genomic tests were restricted to the expression or mutation of a single gene. Examples of such biomarkers are included as part of the American Society for Clinical Oncology Assays and Predictive Markers resource page (ASCO 2017). More recently, however, as sequencing and high-throughput assaying techniques improved and diminished in cost, these genomic tests have become more sophisticated and indications for their use broadened. In the breast cancer prognostic and predictive biomarker space alone, OncotypeDx, ProSigna, MammaPrint, EndoPredict, Mammostrat, and Breast Cancer Index all represent genomic tests with either prognostic or predictive capability that aid in risk stratification beyond standard clinic-pathologic parameters (van de Vijver et al. 2002; Piening et al. 2009; Albain et al. 2010; Tang et al. 2011; Servant et al. 2012; Dowsett et al. 2013; Sparano et al. 2015; Lalonde et al. 2016). While each test varies in its clinical indication, utility, and genomic makeup, many have been successfully validated as possessing clinical utility and have been adopted, to varying degrees, into clinical practice. Indeed, some of these tests are now part of recommended workup by the national and international "best practice" guidelines of the National Comprehensive Cancer Network (NCCN) and other professional societies (Network 2017).

However, the concept of using genomic-based approaches for predicting systemic therapy response is not unique. Although similar genomic tests for predicting tumor response or normal tissue toxicity after radiation have been slower in development, within the last several years an increasing number of tests with varying clinical indications and utility were reported (Eschrich et al. 2012; Speers et al. 2015; Strom et al. 2015; Yard et al. 2016; Zhao et al. 2016; Scott et al. 2017). Groups have described gene expression–based genomic tests that predict the likelihood of breast cancer patients benefiting from adjuvant radiotherapy. Multiple gene signatures based on RNA expression have been derived that reflect hypoxia and are not only prognostic for tumor outcomes following radiotherapy but also predict benefit from adding hypoxia-modifying treatments (Eustace et al. 2013; Thomson et al. 2014; Irlam-Jones et al. 2016; Toustrup et al. 2016; Yang et al. 2017). These signatures have entered the clinical arena and are being evaluated in clinical trials.

13.2 DEVELOPMENT OF GENOME-BASED TESTS TO PREDICT TUMOR RESPONSE

13.2.1 GENE SIGNATURES

One of the first potentially useful signatures was developed by researchers who analyzed the gene expression of 143 early stage breast cancer samples from patients treated with radiation after surgery (Niméus-Malmström et al. 2008). This group identified an 81-gene signature (these genes being selected as the "most strongly associated" with recurrence) that outperformed clinical and pathologic characteristics alone for predicting local recurrence events. A subsequent signature was also derived from the gene expression profiling of human breast tumors (Weichselbaum et al. 2008). These investigators identified a signature predictive not only of radiation response, but also of chemosensitivity, and found that this was heavily reliant on genes associated with interferon signaling. Despite early enthusiasm, both of these signatures failed external validation in data sets in which they were not trained, and other investigators have reported an inability to derive a radiation response signature using patient tumors for the basis of their signature development (Kreike et al. 2006; Nuyten et al. 2006; Piening et al. 2009). Additional research continues in an effort to develop prognostic signatures in breast cancer that are associated with local recurrence after surgery alone and which are independent of radiation treatment. Though undergoing further refinement, these signatures have not always performed well in external validation (Kreike et al. 2006; Nuyten et al. 2006; Niméus-Malmström et al. 2008; Weichselbaum et al. 2008; Piening et al. 2009; Servant et al. 2012).

13.2.2 RADIOTYPEDX

A different approach was employed by investigators who used breast cancer cell line expression data and radiation response as the basis for the development of a radiation response signature (Speers et al. 2015). The resulting signature, termed RadiotypeDx to connote similarity to OncotypeDx in the systemic therapy space, has been externally validated in an independent cohort. Although encouraging, this signature also requires external validation as part of either a prospective randomized trial or a "prospective retrospective" analysis from previously conducted randomized phase III trials for it to be adopted into clinical practice.

13.2.3 DECIPHER

Gene expression and genomic signatures are also being developed to predict the likelihood of response to definitive, adjuvant, and salvage radiation therapy for prostate cancer. Although multiple tests already exist to predict either need for surgical treatment (OncotypeDx for Prostate) (Klein et al. 2014) or risk of metastasis after surgery (Decipher) (Erho et al. 2013), none of these tests were designed specifically to predict the response of prostate tumors to radiotherapy. To address this unmet need, the 22-gene–based Decipher Gene Classifier has been extended to post-prostatectomy to evaluate the effect of this gene-based signature on effectiveness of adjuvant radiation. In these studies the data suggests that patients with low Decipher scores are best treated with salvage radiation, whereas those with high Decipher DC scores benefit little from radiation treatment and are more appropriately treated with adjuvant systemic treatment (Erho et al. 2013; Klein et al. 2014; Den et al. 2015).

13.2.4 POST-OPERATIVE RADIATION THERAPY OUTCOMES SCORE

Whereas the Decipher score was an extension of a previous genomic risk-classifier signature, more recent work described the development of a gene signature that was used to specifically identify patients likely to benefit from adjuvant or salvage radiation treatment after prostatectomy (Zhao et al. 2016). In this study, investigators describe the development of a 24-gene signature predictive of response to post-operative radiation treatment. The expression of these 24 genes was the foundation of the post-operative radiation therapy outcomes score (PORTOS) that independently predicted response to adjuvant or salvage radiation, and the PORTOS score is correlated with distant metastasis-free survival. Although validated in external retrospective cohorts, the PORTOS remains invalidated in a prospective, randomized trial, though efforts are underway to run such a genomic-risk stratified trial. Additional genomic classifiers and signatures predictive of response to radiation are currently being developed for lung, rectal, anal, glioblastoma, and head and neck cancers (Ataman et al. 2004; Huse et al. 2011; Stoyanova et al. 2016; Tang et al. 2017; Visser et al. 2017). As with signatures associated with breast and prostate cancers, these will need to be externally validated as part of prospective randomized phase III trials in order to be adopted into clinical practice, and it remains to be determined whether a more global "pan-radiation" response signature can be developed and validated.

13.2.5 RADIATION SENSITIVITY INDEX

As noted earlier, efforts to develop a "pan-cancer" genomic signature of radiation response have been reported. In these initial studies, cell line sensitivity to ionizing radiation was evaluated across the NCI-60 panel of cancer cell lines (Torres-Roca et al. 2005; Eschrich et al. 2009). Genes associated with intrinsic radiosensitivity (measured as surviving fraction at 2 Gy, SF2) at the RNA level were then identified.

Network analysis was used to identify 10 hub genes from which a radio sensitivity index (RSI) was derived. The group assessed the performance of RSI in various disease types with varying levels of success (Eschrich et al. 2009, 2012; Ahmed et al. 2015; Strom et al. 2015). Importantly, RSI was shown to predict benefit from adjuvant radiotherapy in breast cancer patients and has progressed to prospective evaluation in a clinical trial.

A similar approach was used involving the NCI-60 cell line panel to screen radiation response-associated genes (Amundson et al. 2008) with the discovery of genes whose basal expression differed between the radiosensitive and radioresistant cell lines, with a portion associated with survival following irradiation. Interestingly, genes induced (by RNA expression profiling) by radiation were remarkably consistent between tumor types and were a function of *TP53* status, suggesting an underlying conserved set of genes responsible for responding to genotoxic stress. However, there was surprisingly no overlap between the genes identified in these studies, suggesting that the response to radiation treatment is complex and differs under basal and genotoxic conditions.

13.2.6 GENOMIC-ADJUSTED RADIATION DOSE

Work on the RSI was recently expanded in an effort to develop a genomic-adjusted radiation dose (GARD) framework as the basis for future radiation trial design (Scott et al. 2017). To develop GARD, the investigators incorporated their previously developed RSI with the linear-quadratic model to derive a genomic-adjusted radiation dose that was used to calculate the GARD-value for over 8000 tumor samples. In this retrospective analysis, GARD appeared to predict clinical outcome in several cancer types. While limited by an incomplete evaluation in patients who had not received radiation treatment (to evaluate whether GARD has more than just prognostic value) or external validation, this approach holds promise and is worthy of continued investigation. It also underscores the potential utility of a genomic-based radiation response signature that may be used to individualize dose according to tumor radiosensitivity.

In addition to the development of GARD, a more comprehensive assessment was performed by investigators who sought to understand the genetic basis of DNA damage response after radiotherapy (Yard et al. 2016). In this study, the radiation response of over 500 cell lines demonstrated that radiosensitivity is characterized by significant genetic variation across and within lineages. In addition to genes whose expression associated with response, somatic copy number alterations and gene mutations that correlate with the radiation survival were identified. This work offers arguably a more comprehensive and complex picture of the genomic basis of radiation response and may serve as a foundation for future signature development to predict response of various tumors to radiation treatment across all cancers.

13.2.7 HYPOXIA SIGNATURES

Numerous hypoxia signatures have been derived that are prognostic in patients undergoing radiotherapy. These signatures are not specific for radiation given that they are also prognostic in surgically treated patients. However, they help to provide the expertise in developing transcriptomic signatures for radiation oncology and progressing these biomarkers to clinical implementation. Importantly, some of the RNA-based signatures were shown to predict benefit for the addition of hypoxia-targeted treatment to radiotherapy in head and neck (Toustrup et al. 2012; Eustace et al. 2013) and bladder (Yang et al. 2017) cancer. Two signatures are being qualified as biomarkers prospectively in clinical trials. The current thinking is that the signatures transfer poorly across tumor types and signatures must be developed for different cancers.

13.3 PREDICTION OF NORMAL TISSUE TOXICITY

13.3.1 THE RADIOGENOMICS CONSORTIUM

In addition to the development of assays predicting tumor response to radiation, a goal of research in the field of radiogenomics is to identify the genomic markers associated with the development of adverse outcomes resulting from cancer radiotherapy (Rosenstein 2017). It is anticipated that discovery of these genomic markers will serve as the foundation of an assay to predict the relatively susceptibility of patients for the development of radiation-induced toxicities. In addition, the discovery of the markers and the genes within which they reside will increase understanding of the molecular pathways through which these effects arise, possibly leading to the development of agents to either prevent or mitigate these complications that often arise following radiotherapy. However, in order to accomplish this work and definitively discover and validate the critical genomic markers, access to the radiotherapy treatment information and long-term longitudinal follow-up data reporting details as to adverse outcomes must be obtained for large numbers of patients. In order to enable the creation of large cohorts of patients who received radiotherapy, the Radiogenomics Consortium (RGC) was created in 2009, which is a cancer epidemiology consortium through the Epidemiology and Genomics Research Program of the NCI of the NIH (West and Rosenstein 2010). The RGC now has 217 member investigators located at 123 medical centers in 30 countries. The aim of the RGC is to help link investigators with common interests to pursue collaborative studies with large samples sizes so as to increase the statistical power of this research. Although the RGC has successfully assembled large cohorts to perform adequately powered studies (Rosenstein 2017), data harmonization remains a problem when multiple cohorts involve patients treated with a variety of radiotherapy techniques and evaluated using multiple grading systems. Nevertheless, a number of large studies have been accomplished in which substantial amounts of radiotherapy data have been gathered for studies that typically comprise over a thousand patients.

13.3.2 BIG DATA PROJECTS IN RADIOGENOMICS

Several large-scale candidate gene studies have been performed investigating either one or a small number of single nucleotide polymorphisms (SNPs). These RGC big data projects enabled the identification of genetic variants associated with radiotherapy toxicity where effect sizes are small. One study involved analysis of genotyping data for the rs1801516 SNP in the *ATM* gene for roughly 5000 patients treated for either breast or prostate cancer with radiotherapy for which an association between acute and late toxicities was identified with ORs of 1.5 and 1.2, respectively (Andreassen et al. 2016). A separate project included four cohorts that comprised more than 2000 breast cancer patients that received radiotherapy and were genotyped for SNPs related to the TGFβ pathway and associations reported for breast induration, telangiectasia, and overall toxicity (Talbot et al. 2012). Associations with toxicities were identified for the *TNF* SNPs rs1800629 and rs2857595, which is located 25.7 kb from rs1800629 and resides in the intergenic region between *NCR3* and *AIF1*. An additional study employed a two-stage design to identify associations between SNPs in genes whose products are involved with responses to oxidative stress with toxicities following breast cancer radiotherapy of roughly 2600 women diagnosed with breast cancer (Seibold et al. 2015). The rs2682585 SNP in *XRCC1* was found to be associated with risk for skin toxicities (odds ratio [OR] = 0.77, $p = 0.02$).

The availability of large cohorts has also permitted the performance of genomewide association studies (GWAS) in which typically several million SNPs are tested and imputed for association with various adverse effects following radiotherapy (Manolio 2010, 2013; Manolio et al. 2013). The use of large sample sizes is of particular importance for GWAS because there is a substantial requirement for a multiple hypothesis correction. Thus, only those SNPs with a p-value $<5 \times 10^{-8}$ are considered to have met genomewide statistical significance.

Among the studies that have been performed are a three-stage GWAS that included 1742 men treated with radiotherapy for prostate cancer in which the standardized total average toxicity (STAT) score (Barnett et al. 2012), which represents an evaluation of the combined toxicities for several adverse outcomes following radiation treatment (Fachal et al. 2014). The approach used in this work was to first perform a GWAS of 741 men following which several hundred of the most closely linked SNPs were then screened in a second cohort consisting of 633 men. Finally, the SNPs with the smallest p-values were tested for association with STAT score in a third cohort of 388 men. From this work, a locus including the *TANC1* gene was associated with STAT score for late effects with an odds ratio of roughly six (combined $p = 4.64 \times 10^{-11}$).

A large GWAS was performed including 1217 women who received adjuvant breast radiotherapy and 633 men treated with radiotherapy for prostate cancer. Genotype associations with both overall and individual endpoints of adverse outcomes were tested and replication of potentially associated SNPs performed in three independent cohorts. Two groups consisted of men that received treatment for prostate cancer comprising 516 and 862 patients. A third cohort consisted of 355 breast cancer radiotherapy patients.

Through this study, a SNP was identified (rs2788612) located in the *KCND3* gene that was strongly associated with late rectal incontinence (relative risk [RR] = 9.91, $p = 1.05 \times 10^{-12}$) (Barnett et al. 2014). In addition, the quantile–quantile (Q–Q) plots from this study revealed a larger number of associations at the $p < 5 \times 10^{-7}$ level than would be expected by chance for the breast cancer patient. This result suggests that many common genetic variants are associated with risk for development of adverse effects following breast cancer radiotherapy.

Several adverse outcomes were investigated in a GWAS that comprised a group of roughly 800 men diagnosed with prostate cancer that were treated with radiotherapy (Kerns et al. 2012, 2013a, 2013b). Through this work, an 8-SNP haplotype block located in the *INFK* gene was discovered displaying an association with change in American Urological Association Symptom Score (AUASS). The strongest relationship was for SNP rs17779457 (combined $p = 6.5 \times 10^{-7}$). In addition, SNP rs13035033 that resides in *MYO3B*, was associated with urinary straining ($p = 5.0 \times 10^{-9}$). Twelve SNPs were also identified that reside within or in close proximity to genes associated with the development of erectile dysfunction following radiation treatment (p-values 2.1×10^{-5} to 6.2×10^{-4}). Another study focused on rectal bleeding as the principal outcome and identified two SNPs, rs7120482 (5.4×10^{-8}) and rs17630638 ($p = 6.9 \times 10^{-7}$), which lie upstream of the *SLC36A4* gene.

The RGC conducted a single stage meta-analysis of 1564 prostate cancer patients (Kerns et al. 2016). Two SNPs were identified: rs17599026 on 5q31.2 associated with urinary frequency (OR = 3.12, $p = 4.16 \times 10^{-8}$) and rs7720298 on 5p15.2 with decreased urine stream (OR = 2.71, $p = 3.21 \times 10^{-8}$). The meta-analysis showed that heterogeneous radiotherapy cohorts can be combined to identify genetic variants associated with complications following treatment with radiation.

It is anticipated that the cost for DNA sequencing will continue to drop due primarily to the development of new methods for next-generation sequencing methodologies. This should facilitate the identification of rare variants associated with the development of adverse outcomes following radiotherapy.

13.3.3 LARGE-SCALE BIG DATA RADIOGENOMICS PROJECTS CURRENTLY IN PROGRESS

Three large studies are currently in progress whose main goal is to discover new SNPs and validate previously identified genetic biomarkers predictive of susceptibility for the development of adverse effects resulting from radiotherapy. The first project involves roughly 6000 men treated for prostate cancer, which encompasses multiple cohorts created by RGC investigators. DNA samples from all of these men have been genotyped using a GWAS chip and detailed clinical data are available with a minimum of 2 years of follow-up. The goals of this project are to (1) discover new SNP associations and validate previously identified SNPs linked with the development of adverse outcomes resulting from radiotherapy; (2) build clinically useful multi-SNP models that incorporate dosimetric and clinical factors to predict susceptibility for the development of toxicities following radiotherapy; and (3) develop a low-cost, high performance assay and companion risk assessment tool to predict risk for development of complications resulting from treatment with radiation. Related to this aim, research is being conducted that is supported by the NIH Small Business Innovation program (Prasanna et al. 2015) to help rapidly translate the findings from this project into a multiplexed Luminex bead assay ready for implementation in the clinic and routine medical care.

The second large multicenter study developed by RGC members is the Validating Predictive Models and Biomarkers of Radiotherapy Toxicity to Reduce Side-effects and Improve Quality-of-Life in Cancer Survivors (REQUITE) project (West et al. 2014). The REQUITE project addresses the challenge of data heterogeneity that, as for other big data projects, requires harmonization of the different outcome measures and confounding variables used in multiple cohorts. This study does not stipulate the radiotherapy protocols to be used but involves standardized case report forms across centers and countries to ensure data in identical categories are collected. The objectives of the REQUITE project are to (1) perform a multicenter, observational cohort study in which epidemiologic, treatment, longitudinal toxicity and quality-of-life data are collected from approximately 5000 patients treated with radiotherapy for either breast, prostate or lung cancer; (2) produce a centralized biobank in which DNA is isolated from patients enrolled in the observational study and create a centralized data management system for secure collection, integration, mining, sharing and archiving of all project data; (3) validate published SNP biomarkers of radiosensitivity and discover new variants associated with specific forms of adverse effects following radiotherapy;

(4) validate clinical/dosimetric predictors of radiotherapy toxicity and incorporate SNP biomarker data; (5) design interventional trials to reduce long-term adverse cancer treatment effects; (6) deliver interventional trial protocols using validated models incorporating biomarkers to identify patient subpopulations likely to benefit from interventions; and (7) serve as a resource exploitable for future studies exploring relationships between adverse effects resulting from radiotherapy and the genetics of radiosensitivity using developing technologies such as next-generation sequencing. A key aspect of the REQUITE project is the centralized database that includes pretreatment Digital Imaging and Communications in Medicine (DICOM) and Dose-Volume Histogram (DVH) files.

A third study involves three large cohorts comprising roughly 4500 breast cancer patients treated with radiotherapy for which blood samples and detailed clinical information are available. These samples and data are available from three large groups of patients: (1) 1500 patients treated under a series of breast cancer clinical protocols performed at New York University School of Medicine (Formenti et al. 2007; Constantine et al. 2008; Raza et al. 2012; Cooper et al. 2016); (2) approximately 2000 breast cancer patients enrolled though the REQUITE project; and (3) approximately 1000 women who received breast cancer treatment through participation in radiation therapy oncology group (RTOG) 1005, the goal of which was to determine whether an accelerated course of hypofractionated whole breast irradiation, including a concomitant boost to the tumor bed in 15 fractions following lumpectomy, is non-inferior in local control to a regimen of standard whole breast irradiation with a sequential boost following lumpectomy for early-stage breast cancer patients (Freedman et al. 2013).

13.4 APPROACHES TO RADIOGENOMICS USING MACHINE LEARNING

Rapid recent advances in biotechnologies have enhanced our ability to collect large amounts of data profiling the complex processes underlying healthy and diseased organisms. These data include clinical data collected from patients, genomic and genetic data, and measurements of environmental exposures to which populations are exposed. Clearly, manual analysis of such large amounts of data is infeasible, especially because the manifestations of the processes being studied in these data are generally unclear. For the latter reason, it is also often insufficient to use traditional parametric statistical techniques to analyze these data. In such a situation, techniques from the area of machine learning (Bishop 2006; Tan et al. 2006) hold great potential for deriving actionable knowledge from such large complex data sets (i.e., big data). These techniques generally make minimal assumptions about the forms of the processes, the data distributions, and other such factors. These techniques also often pay attention to the scalability aspects of the analysis, such as time and memory requirements, which enable their application to big data more feasible. For these reasons, machine learning has been used to address several biomedical problems effectively (Baldi and Brunak 2001; Larranaga et al. 2006; Tarca et al. 2007).

Machine learning techniques typically fall under two categories, namely unsupervised and supervised learning. In the former, the machine learning analyses are restricted to data points or objects described in terms of their descriptors, features, or attributes, such as the expression values or levels of a gene across multiple patient samples or experimental conditions. The most common form of unsupervised learning is data clustering (Jain et al. 1999; Xu and Wunsch 2010), where the feature values of data objects are used to group these objects into clusters, such that the intra-cluster distance or similarity between data objects in the same cluster is minimized or maximized respectively. Simultaneously, the opposite is expected for intra-cluster distance or similarity between data objects in different clusters. Previous computational biology analyses have used clustering to group functionally related genes based on their expression profiles and other data types (Eisen et al. 1998; Brohee and van Helden 2006) and discovering/validating disease subtypes (Bhattacharjee et al. 2001; de Souto et al. 2008).

Supervised learning includes machine learning analyses that utilize not only the features of data points, but also additional information (supervision) available about them. Examples of such information include class labels of data points (e.g., diseased or healthy), functional annotations of genes and disease phenotypes (e.g., survival rates) of patients, among many other possibilities. A common analysis of this type is predictive modeling, which includes analyses like classification and regression. In a typical workflow used for predictive modeling, a learning algorithm is applied to a training data set, where class labels are available for the data objects described in terms of attributes or features, to induce a predictive model. This model captures the relationship between feature values and class labels in a form determined by the choice of the learning algorithm. This model is then applied to unlabeled data objects, possibly constituting one or more test sets, to deduce their class labels. Using methodologies such as cross-validation or training-holdout splits, and standardized measures like area under the receiver operating characteristic (ROC) curve (AUC), precision-recall-F-measure, and mean squared error, the performance of predictive models can be objectively evaluated. For details of the concepts above, we refer the reader to standard textbooks on the topic (Tan et al. 2006; Kuhn and Johnson 2013).

Predictive modeling is integral to precision or personalized medicine, as several problems in the latter area entail making predictions of various types from large complex biomedical data sets (Hamburg and Collins 2010; Pencina and Peterson 2016). These problems include recommending suitable therapeutics or preventive measures, predicting disease phenotypes and discovering effective drug combinations. Furthermore, systematic methodologies can be employed to objectively evaluate the performance and hence the expected generalization capabilities of these models. We discuss some basic and advanced applications of the above methods, including predictive modeling ones, to radiogenomics in the following subsections.

13.4.1 SINGLE-SNP ASSOCIATION TEST

Radiogenomics studies have made numerous efforts to identify genetic variants, primarily SNPs, associated with radiation-induced toxicities using GWAS data (Rosenstein 2017). As outlined above, in recent years, substantial progress has been achieved in the discovery and validation of SNPs predictive of individual radiosensitivity risk after radiotherapy in various cancers. However, to identify these biomarkers, many GWAS analyses have used a simple statistical approach with a single association test followed by Bonferroni correction for multiple-testing correction. This conservative statistical approach decreases the possibility that false-positive biomarkers are identified, assuming that individual tests are independent without taking into account interaction effects of biomarkers in the model building process. However, at the same time, this method can increase the probability of falsely rejecting existing associations of SNPs with outcomes. Failures in validation of biomarkers in several GWAS seem to be in part attributable to this issue.

13.4.2 MACHINE LEARNING-BASED PREDICTIVE MODEL

As an alternative to overcome the drawback of the single association test, machine learning–based modeling approaches have been employed, highlighting the use of a large number of SNPs, even including those that do not individually reach genomewide statistical significance (Manor and Segal 2013; Wei et al. 2013; Botta et al. 2014; Nguyen et al. 2015). An advantage associated with use of machine learning methods is that they have the capability of identifying a set of important features (SNPs, in this context) that capture hidden patterns in a given data set and associations with outcomes and to design predictive models such that the discriminatory power for prediction of outcomes is maximized, considering interaction effects among SNPs. However, big data, consisting of large numbers of SNPs and clinical information, can easily become overwhelming and impede the design of feasible, unbiased, and reliable computational methods. To this end, effective predictive models that are computationally achievable and, at the same time, perform well have been designed by the GWAS community. A hypothesis associated with use of these multivariate models is that complex phenotypes are likely the result of interactions of many biomarkers, most of which contribute only a small differential in risk (Zhang et al. 2008).

Two of the main machine learning methods employed in GWAS analysis are the random forest model and the least absolute shrinkage and selection operator (LASSO), and they are particularly effective for studies with a large number of features such as GWAS (Sabourin et al. 2015; Papachristou et al. 2016). Despite this advantage, in the area of radiogenomics, very few studies have used machine learning methods to investigate biological associations of SNPs with outcomes (De Ruyck et al. 2013; Oh et al. 2017).

13.4.3 RANDOM FOREST MODEL

Random forest is an ensemble machine learning method, consisting of a collection of decision trees, which provide a unique capability for the design of relatively unbiased predictive models with an improved predictive accuracy compared with other machine learning approaches (Botta et al. 2014; Nguyen et al. 2015). Random forest avoids the problem of overfitting using two random selection processes: (1) a bootstrap data set is used in constructing each decision tree and (2) during the tree creation, a random subset of features is selected for each node split. These can prevent individual SNPs from dominating the estimate and the final decision is reached by averaging over many trees, thus capturing complex interaction structures in the data while being resistant to the bias of any single tree. This property has made the random model forest increasingly popular, in particular for GWAS despite its drawback that the model size is large compared with penalized models such as LASSO.

13.4.4 PENALIZED MODEL

LASSO is a regression analysis technique for predictive modeling and feature selection based on regularization with a penalty term that controls the sparsity of the predictive model (Papachristou et al. 2016). Therefore, LASSO is an attractive method in handling high-dimensional data. The sparsity in LASSO is controlled by a nonnegative regularization parameter that is chosen such that the model's cross validation error is minimized, thereby constraining only a subset of selective features to have nonzero coefficients, and for other features with exactly zero coefficients. This sparsity can enhance the interpretability of the model and facilitate the biological interpretation for biomarkers (with nonzero coefficients) chosen in the model. In contrast, the "black-box" nature of the random forest model makes it impracticable to gain insight into how the random forest model reaches its prediction across many trees. An expectation-maximization (EM) LASSO model was designed to predict dysphagia in head and neck cancer following radiotherapy using clinical variables and 19 SNPs from 13 DNA damage repair genes (De Ruyck et al. 2013).

13.4.5 PRECONDITIONING RANDOM FOREST REGRESSION

A variant of the random forest model, termed preconditioning random forest regression, was designed based on a preconditioning idea in which the original binary outcomes were converted into continuous pseudo-outcomes (preconditioned outcomes) that were used in the random forest-based modeling (Paul et al. 2008; Oh et al. 2017). The preconditioned outcomes were produced using logistic regression coupled with principal components in a way that the preconditioned outcomes are highly correlated with the original outcomes as well as a set of important SNPs resulting in principal components. A study using preconditioning random forest regression on a radiogenomic data set demonstrated that this approach outperformed conventional random forest and LASSO in predicting the risk of developing late rectal bleeding (AUC = 0.7) and erectile dysfunction (AUC = 0.62) in prostate cancer following radiotherapy (Figure 13.1) (Oh et al. 2017).

13.4.6 BIOLOGICAL DATA ANALYSIS

To enhance our understanding of biological mechanisms associated with radiation-induced toxicities, it is important to identify key biological processes or pathways relevant to tissue-specific complications. The random forest model has the capability of providing the extent of individual SNP contribution to the

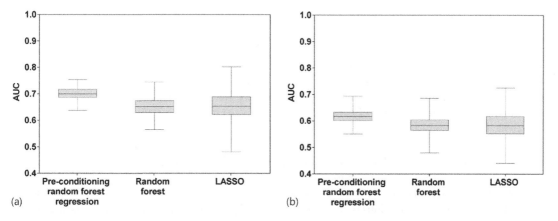

Figure 13.1 Performance comparison of preconditioning random forest regression with alternative methods on the validation data set for (a) rectal bleeding and (b) erectile dysfunction.

Big data in radiation oncology

predictive model by measuring the degradation of predictive power when each SNP is randomly permutated. In the analysis of radiogenomic data, SNPs were ranked on the basis of the importance score and their nearby genes were found. With an assumption that SNPs with larger importance scores are more likely to be causally related to outcomes than those with smaller importance scores, a gene ontology enrichment analysis was performed for top-ranked genes. The biological processes resulting from different sets of genes were examined to find which set of genes produces the most plausible biological pathways linked with outcomes. As a result, ion-transport related functions were associated with rectal bleeding, whereas heart and muscle contraction– and blood circulation–related biological processes were found to be associated with erectile dysfunction (Oh et al. 2017). Interestingly, when SNPs were filtered using a threshold of chi-square p-value of 0.001, no common SNPs between the two endpoints (rectal bleeding and erectile dysfunction) were found. These results suggest that biological mechanisms related to tissue-specific complications may differ and are involved in a set of different SNPs or genes.

13.4.7 PREDICTIVE MODEL OPTIMIZATION BASED ON BIOLOGICAL PROCESSES

Once plausible biological processes are identified, the model-building process can be performed using the corresponding SNPs to optimize the predictive capability of the designed model because the likelihood that these SNPs are putative and relevant to outcomes is increased. When predictive models were rebuilt using these SNPs, predictive power for late rectal bleeding was slightly improved, with an AUC of 0.71, whereas there was greater improvement with an AUC of 0.65 for erectile dysfunction (Oh et al. 2017). These results suggest that the biological analysis using important SNPs measured by computational techniques can identify more plausible biological processes and that SNP filtering based on the biological relevance analysis can improve performance of the predictive model.

13.5 OTHER APPLICATIONS OF MACHINE LEARNING IN BIOMEDICAL SCIENCES

As indicated above, big data and machine learning methods have not been extensively applied in radiogenomics. However, several areas of biomedical sciences have been significantly positively impacted by the application of these methods to complex, generally large, data sets. We describe some of these areas and applications below, with the hope that they will motivate similar applications in radiogenomics.

13.5.1 PROTEIN FUNCTION PREDICTION

Proteins perform some of the most important cellular functions, such as transcription (transcriptional factors), metabolism (enzymes), and signaling (kinases). Thus, knowledge of the functions of proteins can significantly aid an understanding of complex cellular processes, as well as applications such as the development of new drugs. Although numerous genomes have been sequenced, and their resultant proteins identified, the rate of functional annotation of these proteins has not kept pace (Radivojac et al. 2013; Jiang et al. 2016). In addition to the limitations of our knowledge as to how proteins function in different organisms and context, the primary reason for this critical gap is that laboratory experiments are generally time, labor, and, hence, cost intensive.

Thus, several computational approaches have been developed to narrow the space of possible hypotheses about potential protein function, followed by experimental/literature-based validation, thus expediting the overall process. The first natural approach was to use sequence homology assessment tools, such as the Basic Local Alignment Search Tool (BLAST) and Position-Specific Iterative BLAST (PSI-BLAST) (Altschul et al. 1990, 1997), to transfer functional annotations to unannotated proteins from proteins having similar amino acid sequences. However, other studies demonstrated that this approach does not always yield accurate results due to the multidomain structure of proteins and the insufficiency of sequence homology to reflect the effects of the evolutionary process of gene duplication (Gerlt and Babbitt 2000; Whisstock and Lesk 2003). Thus, a much wider spectrum of data types was leveraged to expand the types, specificity, and accuracy of protein functions that can be predicted. Appropriate data analysis methods, including those from machine learning (e.g., clustering, classification, and network analysis), were employed to infer protein function form these data types. Table 13.1 lists the most well-investigated data types, as well as the

Table 13.1 **Types of biological data and approaches used for predicting protein function**

DATA TYPE	APPROACHES USED FOR PROTEIN FUNCTION PREDICTION
Primary structure (amino acid sequences)	• Clustering and matching based on sequence similarity measures • Classification based on sequence elements such as motifs and domains • Classification based on biological, physical and chemical features derived from sequences
Secondary and tertiary structure	• Inference based on structural similarity • Inference based on structural motifs • Classification based on features of the protein's three-dimensional surface (e.g., binding sites, active sites)
Evolution and genomic context	• Proximity of genes and their orthologs in multiple genomes (e.g., operons in prokaryotic genomecs) • Gene or protein fusion • Co-occurrence of genes in multiple genomes (phylogenetic profiles) • Inference on the basis of evolutionary events recorded in phylogenetic trees
Gene expression data	• Clustering into functionally coherent groups • Classification using gene expression profiles as features • Classification of time-series/longitudinal gene expression data
Protein–protein interaction networks	• Transfer of function(s) from neighboring proteins in network • Global transfer of function(s) across the whole network • Clustering of proteins into densely connected regions
Biomedical literature/text	• Inference using frequency statistics of biomedical terms • Classification and clustering of scientific documents • Natural language processing
Multiple types of data	• Aggregation of function predictions from diverse data types • Inference of functional associations between genes and proteins • Integration of higher-order representations of individual data types (e.g., kernels)

Source: Pandey, G. et al., Computational approaches for protein function prediction: A survey, Technical Reports of the Department of Computer Science and Engineering, University of Minnesota, 2006; Lee, D. et al., *Nat. Rev. Mol. Cell. Biol.*, 8, 995–1005, 2007; Sharan, R. et al., *Mol. Syst. Biol.*, 3, 88, 2007; Kamath, C. et al., Scientific data analysis, in Shoshani, A. and D. Rotem (Eds.), *Scientific Data Management*, Chapman and Hall/CRC, London, UK, 2009.

Note: The content was inspired by the discussion of protein function prediction.

most prominent data analysis approaches, used to predict protein function. For details of these data types and approaches, we refer the reader to extensive reviews (Pandey et al. 2006; Lee et al. 2007; Sharan et al. 2007) and reports of recent large-scale assessments (Radivojac et al. 2013; Jiang et al. 2016).

13.5.2 DISCOVERY OF GENETIC INTERACTIONS

Genetic interactions are inferred when the cumulative phenotypic effect of mutations in two or more genes is significantly different from that expected if the effects of mutations in each individual gene were independent of each other (Mani et al. 2008; Boucher and Jenna 2013). These interactions are important for delineating functional relationships among genes and their corresponding proteins, as well as elucidating complex biological processes and diseases. A variety of genetic interactions have been described, but one of the most well-studied types is synthetic lethality, which describes how combinations of mutations confer (cellular) lethality while individual ones do not (Mani et al. 2008). One of the most exciting biomedical applications of genetic interactions lies in the use of synthetic lethality to selectively target cancer cells (Kaelin 2005; McLornan et al. 2014), as exemplified by the pharmacological PARP inhibition in BRCA-mutated (and deficient) tumors (Weil and Chen 2011). Such opportunities are therapeutically promising

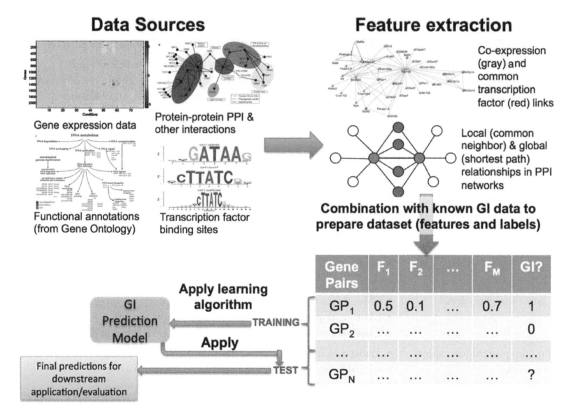

Figure 13.2 Overview of a commonly used approach to predict genetic interactions. Here, a generally large number and variety of features are extracted from diverse data sources, examples of both of which are shown in the top panel. The feature data are combined with known genetic interaction data from public databases like BioGRID, leading to a feature + label table/matrix. Some of the gene pairs in this table, whose genetic interaction status is known, are used as training examples, from which a genetic interaction prediction model is learned using an appropriate algorithm. Finally, the model is applied to test gene pairs to make predictions of their genetic interaction status, which can be used for downstream evaluations or applications. (From Madhukar, N.S. et al., *Front. Bioeng. Biotechnol.*, 3, 172, 2015. With permission.)

because targeting a gene that is synthetic lethal to a cancer-associated mutation may preferentially kill cancer cells and spare normal ones. However, the availability of large numbers of genetic interactions screened in simpler organisms, such as yeast (Costanzo et al. 2010) opened the doors for computational approaches that aim to fill this gap by predicting potential genetic interactions in various organisms and contexts (e.g., tumors) that could be experimentally validated. Figure 13.2 illustrates a commonly used approach employing genomic data integration and machine learning for predicting genetic interactions (Pandey et al. 2010). This approach was applied to conduct large-scale prediction of synthetic lethal genetic interactions in yeast and to analyze which data sources or features were the most predictive for genetic interaction status of a gene pair. These sources proved to be Gene Ontology functional annotations and protein–protein interactions, which were not surprising because genetic interactions are known to indicate functional relationships and are correlated with other cellular interactions (Kelley and Ideker 2005; Boucher and Jenna 2013). Other studies have used alternative hypotheses and data sources, such as evolutionary patterns, high-throughput phenotyping, and network structure, for this prediction (Madhukar et al. 2015).

13.5.3 BIOMARKER DISCOVERY AND PERSONALIZED MEDICINE

One of the most clinically relevant applications of computational methods developed has been the analysis of large complex biomedical data sets to discover disease biomarkers that can aid personalized diagnosis, prognosis, therapy planning, and other tasks (Frank and Hargreaves 2003; Yang et al. 2009; Hamburg and Collins 2010). These biomarkers might be (changes in) individual molecules or entities such as mutations in BRCA1 and BRCA2 genes for breast cancer risk prediction (Ford et al. 1998), or groups of them, such as MammaPrint (van't Veer et al. 2002) and Oncotype DX (Paik et al. 2004) gene expression panels for the same task.

Computational methods are particularly capable of discovering the latter types of biomarkers from diverse types of biomedical data, such as genetic–protein interaction networks, SNP panels, gene expression profiles, and proteomics and metabolomics data (McDermott et al. 2013). Machine learning methods have played a productive role in this area due to their ability to sift through noisy high-dimensional biomedical data sets to find useful and actionable biomarkers. A machine learning subarea that is particularly relevant to biomarker discovery is feature selection (Saeys et al. 2007), which includes techniques for selecting a subset of all the features in a data set that can help accomplish the target data analysis task better than all the features taken together. This is typically accomplished by eliminating or reducing the effect of redundant or irrelevant features based on some objective criteria. In this sense, the identification of differentially expressed genes or SNPs significantly associated with a phenotype can be thought of as traditional univariate feature selection tasks. However, if the end goal is the prediction of a clinically relevant phenotype, a far more powerful approach is to embed the feature selection operation within the predictive modeling process (Ma and Huang 2008). This enables the simultaneous selection and evaluation of feature sets that can *collectively* be used to eventually learn an accurate predictive model for the target phenotype. Indeed, the MammaPrint gene expression panel for breast cancer risk prediction was derived using a similar approach (van't Veer et al. 2002).

13.6 SUMMARY AND FUTURE DIRECTIONS

The research outlined in this chapter highlights recent advances and progress in the development of genomic signatures and "omics"-based tests for the prediction of tumor and normal tissue response to radiation treatment. Future efforts aimed at validation current signatures and incorporating additional information about tumor cell signaling, metabolism, the immune response, and imaging characteristics will further refine the predictive power, sensitivity, and, ultimately, the clinical utility of these tests. In addition, continued efforts by the Radiogenomics Consortium to unravel the genes and SNPs associated with normal tissue toxicity will further aid in the personalization of radiation treatment delivery. Further, the development of centralized databases and data collection standardization similar to that achieved for the REQUITE project will substantially enhance the identification of biomarkers predictive of responses and outcomes resulting from cancer radiotherapy. Thus, the anticipated continued progress over the coming years in the field of radiogenomics will result in the incorporation of these genomic-based tests into clinical decision thereby enhancing the ability to personalize treatment decisions for cancer patients and enhancing precision radiation oncology.

REFERENCES

Ahmed, K.A. et al. (2015) The radiosensitivity index predicts for overall survival in glioblastoma. *Oncotarget* 6(33):34414–34422.

Albain, K.S. et al. (2010) Prognostic and predictive value of the 21-gene recurrence score assay in postmenopausal women with node-positive, oestrogen-receptor-positive breast cancer on chemotherapy: A retrospective analysis of a randomised trial. *Lancet Oncol.* 11(1):55–65.

Altschul, S.F. et al. (1990) Basic local alignment search tool. *J. Mol. Biol.* 215:403–410.

Altschul, S.F. et al. (1997) Gapped BLAST and PSI-BLAST: A new generation of protein database search programs. *Nucleic Acids Res.* 25:3389–3402.

Amundson, S.A. et al. (2008) Integrating global gene expression and radiation survival parameters across the 60 cell lines of the National Cancer Institute Anticancer Drug Screen. *Cancer Res.* 68(2):415–424.

Andreassen, C.N. et al. (2016) Individual patient data meta-analysis shows a significant association between the ATM rs1801516 SNP and toxicity after radiotherapy in 5456 breast and prostate cancer patients. *Radiother. Oncol.* 121(3):431–439.

ASCO (2017) Assays and predictive markers. American Society of Clinical Oncology; Available from: https://www.asco.org/practice-guidelines/quality-guidelines/guidelines/assays-and-predictive-markers.

Ataman, O.U. et al. (2004) Molecular biomarkers and site of first recurrence after radiotherapy for head and neck cancer. *Eur. J. Cancer* 40(18):2734–2741.

Baldi, P. and S. Brunak (2001) *Bioinformatics: The Machine learning Approach*. Cambridge, MA: MIT Press.

Barnett, G.C. et al. (2012) Standardized total average toxicity score: A scale- and grade-independent measure of late radiotherapy toxicity to facilitate pooling of data from different studies. *Int. J. Radiat. Oncol. Biol. Phys.* 82(3):1065–1074.

Barnett G.C. et al. (2014) A genome wide association study (GWAS) providing evidence of an association between common genetic variants and late radiotherapy toxicity. *Radiother. Oncol.* 111(2):178–185.

Begg, A.C. et al. (1999) The value of pretreatment cell kinetic parameters as predictors for radiotherapy outcome in head and neck cancer: A multicenter analysis. *Radiother. Oncol.* 50(1):13–23.

Bhattacharjee, A. et al. (2001) Classification of human lung carcinomas by mRNA expression profiling reveals distinct adenocarcinoma subclasses. *Proc. Natl. Acad. Sci. (U.S.A.)* 98:13790–13795.

Bishop, C.M. (2006) *Pattern Recognition and Machine Learning*. New York: Springer.

Botta, V. et al. (2014) Exploiting SNP correlations within random forest for genome-wide association studies. *PLoS One* 9:e93379.

Boucher, B. and S. Jenna (2013) Genetic interaction networks: better understand to better predict. *Front. Genet.* 4:290.

Brohee, S. and J. van Helden (2006) Evaluation of clustering algorithms for protein-protein interaction networks. *BMC Bioinformatics* 7:488.

Burnet, N.G. et al. (1992) Prediction of normal-tissue tolerance to radiotherapy from in-vitro cellular radiation sensitivity. *Lancet* 339(8809):1570–1571.

Constantine, C. et al. (2008) Feasibility of accelerated whole-breast radiation in the treatment of patients with ductal carcinoma in situ of the breast. *Clin. Breast Cancer* 8(3):269–274.

Cooper, B.T. et al. (2016) Prospective randomized trial of prone accelerated intensity modulated breast radiation therapy with a daily versus weekly boost to the tumor bed. *Int. J. Radiat. Oncol. Biol. Phys.* 95(2):571–578.

Costanzo, M. et al. (2010) The genetic landscape of a cell. *Science* 327:425–431.

De Ruyck, K. et al. (2013) A predictive model for dysphagia following IMRT for head and neck cancer: Introduction of the EMLasso technique. *Radiother. Oncol.* 107(3):295–299.

de Souto, M.C. et al. (2008) Clustering cancer gene expression data: A comparative study. *BMC Bioinformatics* 9:497.

Den, R.B. et al. (2015) Genomic classifier identifies men with adverse pathology after radical prostatectomy who benefit from adjuvant radiation therapy. *J. Clin. Oncol.* 33(8):944–951.

Dowsett, M. et al. (2013) Comparison of PAM50 risk of recurrence score with oncotype DX and IHC4 for predicting risk of distant recurrence after endocrine therapy. *J. Clin. Oncol.* 31(22):2783–2790.

Eisen, M.B. et al. (1998) Cluster analysis and display of genome-wide expression patterns. *Proc. Natl. Acad. Sci. (U.S.A.)* 95:14863–14868.

Erho, N. et al. (2013) Discovery and validation of a prostate cancer genomic classifier that predicts early metastasis following radical prostatectomy. *PLoS One* 8(6):e66855.

Eschrich, S.A. et al. (2009) A gene expression model of intrinsic tumor radiosensitivity: Prediction of response and prognosis after chemoradiation. *Int. J. Radiat. Oncol. Biol. Phys.* 75(2):489–496.

Eschrich, S.A. et al. (2012) Validation of a radiosensitivity molecular signature in breast cancer. *Clin. Cancer Res.* 18(18):5134–5143.

Eustace, A. et al. (2013) A 26-gene hypoxia signature predicts benefit from hypoxia-modifying therapy in laryngeal cancer but not bladder cancer. *Clin. Cancer Res.* 19(17):4879–4888.

Fachal, L. et al. (2014) A three-stage genome-wide association study identifies a susceptibility locus for late radiotherapy toxicity at 2q24.1. *Nat. Genet.* 46(8):891–894.

Fertil, B. and E.P. Malaise (1981) Inherent cellular radiosensitivity as a basic concept for human tumor radiotherapy. *Int. J. Radiat. Oncol. Biol. Phys.* 7(5):621–629.

Ford, D. et al. (1998) Genetic heterogeneity and penetrance analysis of the BRCA1 and BRCA2 genes in breast cancer families. The Breast Cancer Linkage Consortium. *Am. J. Hum. Genet.* 62:676–689.

Formenti, S.C. et al. (2007) Phase I-II trial of prone accelerated intensity modulated radiation therapy to the breast to optimally spare normal tissue. *J. Clin. Oncol.* 25(16):2236–2242.

Frank, R. and R. Hargreaves (2003) Clinical biomarkers in drug discovery and development. *Nat. Rev. Drug Discov.* 2:566–580.

Freedman, G.M. et al. (2013) Accelerated fractionation with a concurrent boost for early stage breast cancer. *Radiother. Oncol.* 106(1):15–20.

Geara, F.B. et al. (1993) Prospective comparison of in vitro normal cell radiosensitivity and normal tissue reactions in radiotherapy patients. *Int. J. Radiat. Oncol. Biol. Phys.* 27(5):1173–1179.

Gerlt, J.A. and P.C. Babbitt (2000) Can sequence determine function? *Genome Biol.* 1(5):REVIEWS0005.

Glas, A.M. et al. (2006) Converting a breast cancer microarray signature into a high-throughput diagnostic test. *BMC Genomics* 7:278.

Hamburg, M.A. and F.S. Collins (2010) The path to personalized medicine. *N. Engl. J. Med.* 363:301–304.

Huse, J.T. et al. (2011) Molecular subclassification of diffuse gliomas: Seeing order in the chaos. *Glia* 59(8):1190–1199.

Irlam-Jones, J.J. et al. (2016) Expression of miR-210 in relation to other measures of hypoxia and prediction of benefit from hypoxia modification in patients with bladder cancer. *Br. J. Cancer* 115(5):571–578.

Jain, A.K. et al. (1999) Data clustering: A review. *ACM Comput. Surv.* 31:264–323.

Jiang, Y. et al. (2016) An expanded evaluation of protein function prediction methods shows an improvement in accuracy. *Genome Biol.* 17:184.

Johansen, J. et al. (1994) Evidence for a positive correlation between in vitro radiosensitivity of normal human skin fibroblasts and the occurrence of subcutaneous fibrosis after radiotherapy. *Int. J. Radiat. Biol.* 66(4):407–412.

Kaelin, W.G. Jr. et al. (2005) The concept of synthetic lethality in the context of anticancer therapy. *Nat. Rev. Cancer* 5:689–698.

Kamath, C. et al. (2009) Scientific data analysis. In Shoshani, A. and D. Rotem (Eds.), *Scientific Data Management*. London, UK: Chapman & Hall/CRC Press.

Kelley, R. and T. Ideker (2005) Systematic interpretation of genetic interactions using protein networks. *Nat. Biotech.* 23:561–566.

Kerns, S.L. et al. (2013a) A 2-stage genome-wide association study to identify single nucleotide polymorphisms associated with development of erectile dysfunction following radiation therapy for prostate cancer. *Int. J. Radiat. Oncol. Biol. Phys.* 85(1):e21–e28.

Kerns, S.L. et al. (2013b) Genome-wide association study identifies a region on chromosome 11q14.3 associated with late rectal bleeding following radiation therapy for prostate cancer. *Radiother. Oncol.* 107(3):372–376.

Kerns, S.L. et al. (2016) Meta-analysis of genome wide association studies identifies genetic markers of late toxicity following radiotherapy for prostate cancer. *EBioMedicine* 10:150–163.

Klein, E.A. et al. (2014) A 17-gene Assay to predict prostate cancer aggressiveness in the context of Gleason grade heterogeneity, tumor multifocality, and biopsy undersampling. *Eur. Urol.* 66(3):550–560.

Knauer, M. et al. (2010) The predictive value of the 70-gene signature for adjuvant chemotherapy in early breast cancer. *Breast Cancer Res. Treat.* 120(3):655–661.

Kreike, B. et al. (2006) Gene expression profiles of primary breast carcinomas from patients at high risk for local recurrence after breast-conserving therapy. *Clin. Cancer Res.* 12(19):5705–5712.

Kuhn, M. and K. Johnson (2013) *Applied Predictive Modeling*. New York: Springer.

Lalonde, E. et al. (2016) Translating a prognostic DNA genomic classifier into the clinic: Retrospective validation in 563 localized prostate tumors. *Eur. Urol.* 72(1):22–31.

Larranaga, P. et al. (2006) Machine learning in bioinformatics. *Brief Bioinform.* 7:86–112.

Lee, D. et al. (2007) Predicting protein function from sequence and structure. *Nat. Rev. Mol. Cell. Biol.* 8:995–1005.

Ma, S. and J. Huang (2008) Penalized feature selection and classification in bioinformatics. *Brief Bioinform.* 9:392–403.

Madhukar, N.S. et al. (2015) Prediction of genetic interactions using machine learning and network properties. *Front. Bioeng. Biotechnol.* 3:172.

Mani, R. et al. (2008) Defining genetic interaction. *Proc. Natl. Acad. Sci. (U.S.A.)* 105:3461–3466.

Manolio, T.A. (2010) Genomewide association studies and assessment of the risk of disease. *N. Engl. J. Med.* 363(2):166–176.

Manolio, T.A. (2013) Bringing genome-wide association findings into clinical use. *Nat. Rev. Genet.* 14(8):549–558.

Manolio, T.A. et al. (2013) Implementing genomic medicine in the clinic: The future is here. *Genet. Med.* 15(4):258–267.

Manor, O. and E. Segal (2013) Predicting disease risk using bootstrap ranking and classification algorithms. *PLoS Comput. Biol.* 9(8):e1003200.

McDermott, J.E. et al. (2013) Challenges in biomarker discovery: Combining expert insights with statistical analysis of complex omics data. *Expert Opin. Med. Diagn.* 7:37–51.

McLornan, D.P. et al. (2014) Applying synthetic lethality for the selective targeting of cancer. *N. Engl. J. Med.* 371:1725–1735.

Network, N.C.C. (2017) National Comprehensive Cancer Network Guidelines for Breast Cancer. [cited May 15, 2017].

Nguyen, T.T. et al. (2015) Genome-wide association data classification and SNPs selection using two-stage quality-based random forests. *BMC Genomics* 16 Suppl 2:S5.

Niméus-Malmström, E. et al. (2008) Gene expression profiling in primary breast cancer distinguishes patients developing local recurrence after breast-conservation surgery, with or without postoperative radiotherapy. *Breast Cancer Res.* 10(2):R34–R34.

Nuyten, D.S. et al. (2006) Predicting a local recurrence after breast-conserving therapy by gene expression profiling. *Breast Cancer Res.* 8(5):R62.

Oh, J.H. et al. (2017) Computational methods using genome-wide association studies to predict radiotherapy complications and to identify correlative molecular processes. *Sci. Rep.* 7:43381.

Paik, S. et al. (2004) A multigene assay to predict recurrence of tamoxifen-treated, node-negative breast cancer. *N. Engl. J. Med.* 351(27):2817–2826.

Paik, S. et al. (2006) Gene expression and benefit of chemotherapy in women with node-negative, estrogen receptor-positive breast cancer. *J. Clin. Oncol.* 24(23): 3726–3734.

Pandey, G. et al. (2006) Computational approaches for protein function prediction: A survey. Technical Reports of the Department of Computer Science and Engineering, University of Minnesota Minneapolis, MN.

Pandey, G. et al. (2010) An integrative multi-network and multi-classifier approach to predict genetic interactions. *PLoS Comput. Biol.* 6(9):e1000928.

Papachristou, C. et al. (2016) A LASSO penalized regression approach for genome-wide association analyses using related individuals: Application to the Genetic Analysis Workshop 19 simulated data. *BMC Proc.* 10(Suppl 7):221–226.

Paul, D. et al. (2008) "Pre-conditioning" for feature selection and regression in high-dimensional problems. *Ann. Stat.* 36:1595–1618.

Pencina, M.J. and E.D. Peterson (2016) Moving from clinical trials to precision medicine: The role for predictive modeling. *JAMA* 315:1713–1714.

Piening, B.D. et al. (2009) A radiation-derived gene expression signature predicts clinical outcome for breast cancer patients. *Radiat. Res.* 171(2):141–154.

Prasanna, P.G. et al. (2015) Radioprotectors and radiomitigators for improving radiation therapy: The Small Business Innovation Research (SBIR) Gateway for Accelerating Clinical Translation. *Radiat. Res.* 184(3):235–248.

Radivojac, P. et al. (2013) A large-scale evaluation of computational protein function prediction. *Nat. Methods* 10:221–227.

Raza, S. et al. (2012) Comparison of acute and late toxicity of two regimens of 3- and 5-week concomitant boost prone IMRT to standard 6-week breast radiotherapy. *Front. Oncol.* 2:44.

Rosenstein, B.S. (2017) Radiogenomics: Identification of genomic predictors for radiation toxicity. *Semin. Radiat. Oncol.* 27(4):300–309.

Sabourin, J. et al. (2015) Fine-mapping additive and dominant SNP effects using group-LASSO and fractional resample model averaging. *Genet. Epidemiol.* 39(2):77–88.

Saeys, Y. et al. (2007) A review of feature selection techniques in bioinformatics. *Bioinformatics* 23:2507–2517.

Scott, J.G. et al. (2017) A genome-based model for adjusting radiotherapy dose (GARD): A retrospective, cohort-based study. *Lancet Oncol.* 18(2):202–211.

Seibold, P. et al. (2015) XRCC1 polymorphism associated with late toxicity after radiation therapy in breast cancer patients. *Int. J. Radiat. Oncol. Biol. Phys.* 92(5):1084–1092.

Servant, N. et al. (2012) Search for a gene expression signature of breast cancer local recurrence in young women. *Clin. Cancer Res.* 18(6):1704–1715.

Sharan, R. et al. (2007) Network-based prediction of protein function. *Mol. Syst. Biol.* 3:88.

Sparano, J.A. et al. (2015) Prospective validation of a 21-gene expression assay in breast cancer. *N. Engl. J. Med.* 373(21):2005–2014.

Speers, C. et al. (2015) Development and validation of a novel radiosensitivity signature in human breast cancer. *Clin. Cancer Res.* 21(16):3667–3677.

Steel, G.G. (1993) *Basic Clinical Radiobiology for Radiation Oncologists*. London, UK: Edward Arnold.

Stoyanova, R. et al. (2016) Prostate cancer radiomics and the promise of radiogenomics. *Transl. Cancer Res.* 5(4):432–447.

Strom, T. et al. (2015) Radiosensitivity index predicts for survival with adjuvant radiation in resectable pancreatic cancer. *Radiother. Oncol.* 117(1):159–164.

Talbot, C.J. et al. (2012) A replicated association between polymorphisms near TNFalpha and risk for adverse reactions to radiotherapy. *Br. J. Cancer* 107(4):748–753.

Tan, P.-N. et al. (2006) *Introduction to Data Mining*. London, UK: Pearson Education India.

Tang, G. et al. (2011) Comparison of the prognostic and predictive utilities of the 21-gene Recurrence Score assay and Adjuvant! for women with node-negative, ER-positive breast cancer: Results from NSABP B-14 and NSABP B-20. *Breast Cancer Res. Treat.* 127(1):133–142.

Tang, H. et al. (2017) Comprehensive evaluation of published gene expression prognostic signatures for biomarker-based lung cancer clinical studies. *Ann. Oncol.* 28(4):733–740.

Tarca, A.L. et al. (2007) Machine learning and its applications to biology. *PLoS Comput. Biol.* 3:e116.

Thomson, D. et al. (2014) NIMRAD - A phase III trial to investigate the use of nimorazole hypoxia modification with intensity-modulated radiotherapy in head and neck cancer. *Clin. Oncol. (R. Coll. Radiol.)* 26(6):344–347.

Torres-Roca, J.F. et al. (2005) Prediction of radiation sensitivity using a gene expression classifier. *Cancer Res.* 65(16):7169–7176.

Toustrup, K. et al. (2016) Validation of a 15-gene hypoxia classifier in head and neck cancer for prospective use in clinical trials. *Acta Oncol.* 55(9–10):1091–1098.

van de Vijver, M.J. et al. (2002) A gene-expression signature as a predictor of survival in breast cancer. *N. Engl. J. Med.* 347(25):1999–2009.

van't Veer, L.J. et al. (2002) Gene expression profiling predicts clinical outcome of breast cancer. *Nature* 415(6871):530–536.

Visser, E. et al. (2017) Prognostic gene expression profiling in esophageal cancer: A systematic review. *Oncotarget* 8(3):5566–5577.

Wei, Z. et al. (2013) Large sample size, wide variant spectrum, and advanced machine-learning technique boost risk prediction for inflammatory bowel disease. *Am. J. Hum. Genet.* 92(6):1008–1012.

Weichselbaum, R.R. et al. (2008) An interferon-related gene signature for DNA damage resistance is a predictive marker for chemotherapy and radiation for breast cancer. *Proc. Natl. Acad. Sci. (U.S.A.)* 105(47):18490–18495.

Weil, M.K. and A.P. Chen (2011) PARP inhibitor treatment in ovarian and breast cancer. *Curr. Probl. Cancer* 35:7–50.

West, C. and B.S. Rosenstein (2010) Establishment of a radiogenomics consortium. *Radiother. Oncol.* 94(1):117–118.

West, C.M. et al. (1991) Prediction of cervical carcinoma response to radiotherapy. *Lancet* 338(8770):818.

West, C. et al. (2014) The REQUITE project: Validating predictive models and biomarkers of radiotherapy toxicity to reduce side-effects and improve quality of life in cancer survivors. *Clin. Oncol. (R. Coll. Radiol.)* 26(12):739–742.

Whisstock, J.C. and A.M Lesk (2003) Prediction of protein function from protein sequence and structure. *Q. Rev. Biophys.* 36:307–340.

Xu, R. and D.C. Wunsch 2nd (2010). Clustering algorithms in biomedical research: A review. *IEEE Rev. Biomed. Eng.* 3:120–154.

Yang, Y. et al. (2009) Target discovery from data mining approaches. *Drug Discov. Today* 14:147–154.

Yang, L. et al. (2017) A gene signature for selecting benefit from hypoxia modification of radiotherapy for high risk bladder cancer patients. *Clin. Cancer Res.* 23(16):4761–4768.

Yard, B.D. et al. (2016) A genetic basis for the variation in the vulnerability of cancer to DNA damage. *Nat. Commun.* 7:11428.

Zhang, Z. et al. (2008) An ensemble learning approach jointly modeling main and interaction effects in genetic association studies. *Genet. Epidemiol.* 32(4):285–300.

Zhao, S.G. et al. (2016) Development and validation of a 24-gene predictor of response to postoperative radiotherapy in prostate cancer: A matched, retrospective analysis. *Lancet Oncol.* 17(11):1612–1620.

14 Radiomics and quantitative imaging

Dennis Mackin and Laurence E. Court

Contents

14.1	Introduction	219
14.2	Image Acquisition	221
	14.2.1 Patient Cohort Homogeneity	221
	14.2.2 Region of Interest Size	221
	14.2.3 Image Data Homogeneity	221
	14.2.4 Data Collection	222
14.3	Image Segmentation	222
14.4	Image Preprocessing	223
14.5	Quantitative Feature Extraction	223
	14.5.1 Agnostic Features	224
	14.5.2 Derived Features	224
	14.5.3 Engineered Features	225
14.6	Model Building	227
	14.6.1 Explore or Validate?	227
	14.6.2 Explain or Predict?	229
	14.6.3 Prognostic or Predictive?	229
	14.6.4 Feature Selection	229
	14.6.4.1 Robustness Tests	230
	14.6.4.2 Redundancy Tests	230
	14.6.4.3 Feature Selection Methods	230
	14.6.5 Model Algorithms	231
	14.6.6 Controlling for False Discovery	232
14.7	Radiomics Results	233
14.8	Future Work	234
References		235

14.1 INTRODUCTION

Radiomics is the high-throughput extraction of many quantitative image features with the goal of creating mineable data (Kumar et al. 2012; Lambin et al. 2012). Thus, it is a fundamentally big data approach to quantitative imaging. Although radiomics is relative new field, quantitative imaging is not. A method for the computer-aided diagnosis (CAD) of primary bone tumors was reported in Lodwick et al. (1963). In the same year, Lodwick et al. discussed methods for coding Roentgen images for computer-aided analysis of lung tumors (Figure 14.1). By the 1990s, CAD was being applied to

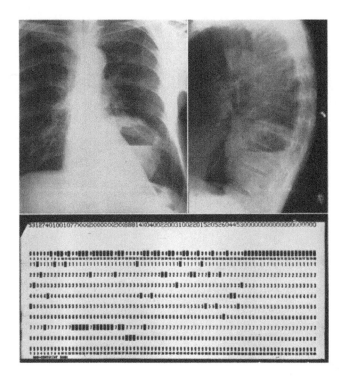

Figure 14.1 A tabulating card from 1963, appropriately punched for the analysis of a tumor in the lower left lobe of the lung. (From Lodwick, G. et al., *Radiology*, 81, 1963.)

new imaging modalities such as CT, MRI, ultrasonography, and mammography (Chan et al. 1990; Dawant et al. 1993; Vyborny and Giger 1994; Yamamoto et al. 1994; Rasheed et al. 1995). However, the effectiveness of the techniques was limited by the quality of the images and availability of computer hardware and software.

In contrast to prior quantitative imaging approaches designed to target specific information, radiomics is a more general approach that attempts to maximize the information extracted from standard-of-care images and make it available for data mining (Gillies et al. 2015). Advances in computer hardware such as multicore computers and graphics processing units have made it possible to process thousands of images in seconds. Image processing software is widely and freely available. Further, high-quality medical images are now routine, especially in oncology. Even though these advances alone are enough to stimulate renewed interest in quantitative imaging, radiomics is more than faster quantitative imaging: radiomics aims to personalized medicine by using quantitative imaging, much like genomics aims to personalize medicine by using genetic markers. However, unlike genomics, which is limited to the information obtained at the biopsy site, radiomics can sample the entire tumor (Gillies et al. 2015). Further, radiomics has the potential to identify and quantify gene signatures expressed as intra-tumoral heterogeneity, a process referred to as radiogenomics (Rutman and Kuo 2009) (discussed in Chapter 13, "Radiogenomics"). This relationship is a central hypothesis of radiomics: imaging features can probe the phenotype of tissues, and this phenotype is related to the underlying genotype (Lambin et al. 2012).

The radiomics workflow can be divided into four stages (Figure 14.2). (1) The workflow starts with the collection of medical images; (2) the regions of interest (ROIs) are segmented; (3) the radiomics features are extracted, and (4) the radiomics features are used to build models for prediction and inference. This chapter describes each of these stages in more detail, presents some compelling results from radiomics, and concludes with a discussion of the future of radiomics.

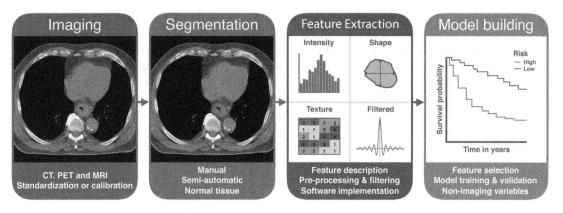

Figure 14.2 The four steps in radiomics analysis. (Adapted from Larue, R.T. et al., *Br. J. Radiol.*, 90, 2017 under Creative Commons Attribution 4.0 License.)

14.2 IMAGE ACQUISITION

The first step in the radiomics workflow begins with identifying the characteristics of the patient group to be studied. Those characteristics vary depending on the task that is being investigated. Key considerations in this task include the homogeneity of the cohort to be studied, the size of the ROI(s), the homogeneity of the imaging data, and the size of the cohort. Developing a radiomics project requires careful evaluation of all of these factors, and the possible effects of any variations on the validity of the project results must be carefully considered as well.

14.2.1 PATIENT COHORT HOMOGENEITY

Depending on the radiomics task, the need for homogeneity in the patient cohort should be carefully assessed. For example, will the results be meaningful if patients with stage I–IV disease are included in the cohort? Similarly, what would be the possible effect of including patients who were treated with different regimens (e.g., radiotherapy vs. surgery)? The importance of these questions depends on the project, but care must be taken to avoid diluting the impact of the project results in an attempt to increase patient numbers.

14.2.2 REGION OF INTEREST SIZE

The most important question here is whether the ROI is too small for radiomics image features to make sense: If the ROI is too small, the image features may be excessively influenced by image noise. No consistent guidelines have been established with regard to the smallest ROI that can be used in radiomics studies, and the choice is affected by the imaging modality (including voxel size) and the features being used. Some authors have suggested 5 cm³ as a suitable cutoff for CT; for PET, which has much larger voxel sizes than CT, Brooks and Grigsby showed that selecting ROIs smaller than 45 cm³ can influence comparisons of intra-tumoral uptake heterogeneity metrics (Brooks and Grigsby 2014). Researchers should consider volume sizes for their own studies, including an assessment of the relationship between volume and calculated feature values for small features.

14.2.3 IMAGE DATA HOMOGENEITY

Some data sets have significant heterogeneity in imaging parameters. For CT imaging, examples include pixel size, slice spacing, reconstruction kernel, and others, some of which can significantly affect the calculated image features, especially pixel size (which can be corrected for), reconstruction kernel (Kumar et al. 2012; Mackin et al. 2015, 2017b), and motion management approach (Fave et al. 2015). Other factors have a less significant influence, such as the tube current and its corresponding effect on the radiation dose to the patient (Mackin et al. 2018). However, many imaging parameters have yet to be adequately investigated; one example is the timing of the contrast injection used for contrast-enhanced CT. Preliminary findings suggest that this is not a major source of uncertainty in radiomics studies (Yang et al. 2016), but more data are still needed. The story is similar, although generally less well studied, for other imaging modalities such as PET and MRI. Information about the imaging protocols used to acquire the images from the patient cohort is listed in the digital imaging and communications in medicine (DICOM) header and is easily accessed. Although many radiomics studies have not considered the effects of the image acquisition conditions on their modeling studies (indicating that some features are reasonably robust to the acquisition parameters), it seems likely that consideration of the image acquisition parameters may improve radiomics models.

14.2.4 DATA COLLECTION

The imaging data for an individual study may come from a single modality (e.g., CT), but data from several modalities (e.g., both non-contrast CT and contrast-enhanced CT, or CT and PET) or data collected at several time points (sometimes known as delta-radiomics [Fave et al. 2017]) are often used. The considerations summarized above underscore the importance of carefully controlling the quality of the data. In addition, small patient cohorts increase both type I (incorrectly detecting a difference) and type II (not detecting an actual difference) error rates. To minimize these issues, Chalkidou et al. (2015) suggested that linear models (e.g., multiple regressions) require a minimum of 10–15 observations per variable.

In some instances, it is possible to collect enough data from a single institution (or, even better, a single imaging device). However, in many cases the big data needed for such studies often necessitates the use of data from several institutions, perhaps supplemented with data from the Cancer Imaging Archive (Kirby et al. 2015), an open-access database that facilitates sharing image data. The Cancer Imaging Archive has numerous image sets covering a range of anatomic sites and includes several data sets that have been collected specifically for radiomics studies (Gevaert et al. 2012; Aerts et al. 2014). These image sets are an excellent source of data for testing algorithms or for validating models, but their use should be considered in the context of the various uncertainties described above.

Design of a prospective study allows much more opportunity for harmonizing data acquisition, including the use of consistent pixel size and comparable reconstruction kernels. Such an approach is likely to produce much more reproducible radiomics results than use of retrospective data.

14.3 IMAGE SEGMENTATION

The next step in the radiomics workflow after collection of the image data set and verifying the quality of the images is segmentation of the ROI. The main considerations here are the choice of which ROI is going to be segmented and how the segmentation is to be done.

Many studies involve use of manually segmented ROIs. Because radiotherapy images are usually manually delineated for creating the treatment plan, they are particularly attractive for radiomics studies. However, manually segmented ROIs are well known to have high interuser variability, especially for some modalities (White et al. 2009; Persson et al. 2012), which could affect the radiomics image features. In addition, segmentation for creating the treatment plan is not necessarily the same as segmentation for a radiomics study. Some researchers use images that have been segmented by several physicians to identify image features that are robust to variability in the manual segmentation process (Aerts et al. 2014). Another way to reduce the impact of interuser variability is to use algorithms such as simultaneous truth and performance-level estimation (Warfield et al. 2004) to create consensus contours from segmentations generated by experts or from different auto-segmentation algorithms. Interuser variability can be also reduced by the use of semi- or fully automated segmentation tools (Parmar et al. 2014), with the awareness that these tools can fail; we strongly recommend that the results always be visually checked. We also caution that any automatic algorithm should be carefully evaluated because automation can actually degrade the reproducibility of image features in some instances. An example of this was shown by Hunter and colleagues, who found that the use of inappropriate thresholds to segment lung tumors could reduce the information content of the extracted image features (Hunter et al. 2013).

In addition to the reliability of the segmentation, another consideration is the time needed to segment the ROI. This is particularly important when considering big data and the segmentation of hundreds or thousands of patient images (each with tens to hundreds of image slices). Automated segmentation tools can significantly reduce the time required for segmentation.

The details of the region to be segmented are also important; an example would be whether to delineate the entire tumor (or other structure) or only portions of the ROI such as a single slice (Ganeshan et al. 2012a). Segmenting a single slice or fixed-size ROI significantly improves efficiency when manual segmentation is used, but the extracted ROI may not represent the entire tumor. The effect of segmenting a single slice or fixed-size ROI on the extracted radiomics image features varies widely depending on the image feature, but can be significant (Fave et al. 2015). Other approaches to identifying the ROI, improve workflow, and reduce interuser variability are using a fixed-size ROI, perhaps a spherical ROI (Bang et al. 2015) and including the maximum circle within manually segmented regions (Echegaray et al. 2015).

14.4 IMAGE PREPROCESSING

Before quantitative features are extracted, images may require preprocessing such as resampling, band pass and low-pass filtering, and image gray-level resampling. The choice of the preprocessing method depends on the problem to be solved. Images with inconsistent voxel sizes are problematic in many retrospective radiomics studies. For example, in one study of 74 patients with non-small cell lung cancer (NSCLC), the pixel spacing varied from 0.59 to 0.93 mm, and the slice thickness varied from 3 to 6 mm (Kumar et al. 2012). Differences in the voxel sizes directly affect the values of the texture features, which incorporate the spatial relationships as discrete voxels. Resampling the images to a uniform voxel size has been shown to improve the robustness of the features (Vallières et al. 2015; Yip et al. 2017; Mackin et al. 2017a; Shafiq-ul-Hassan et al. 2017). The resampling method used does not seem to be important; Larue and colleagues (2017b) found little difference between the results of cubic, linear, and nearest-neighbor resampling.

Slightly more controversial is the application of low-pass or band-pass filter to reduce the effects of image noise. Although low-pass filtering (also called *smearing*) removes high-frequency information from images, at least one study has shown that it reduces feature variability (Mackin et al. 2017a). To emphasize features of a particular size, band-pass filters may be applied, often in combination with the Laplace operator (Laplacian) (Ganeshan et al. 2012a; He et al. 2016; Yasaka et al. 2017).

Because texture features are calculated from the spatial relationships of image voxel intensities, preprocessing the images to reduce the number of gray levels can reduce the sparseness of the underlying matrices and improve the significance of the images. Because the number of gray levels is customarily made to an integer power of 2, gray-level resampling is also called bit-depth resampling, and it is common to resample the intensities to 3 to 10 bits for 8 to 1024 gray levels (Yip and Aerts 2016; Fave et al. 2017; Larue et al. 2017b). An alternative to fixing the number of gray levels in the images is to fix the resolution of underlying histograms. Leijenaar and others (2015b) demonstrated that applying these two seemingly similar approaches to PET images produced surprisingly different rankings of patients with NSCLC.

14.5 QUANTITATIVE FEATURE EXTRACTION

The primary difference between radiomics and other forms of quantitative imaging is the approach. Radiomics was originally defined as a method involving the "automated high-throughput extraction of large amounts (200+) of quantitative features of medical images" (Lambin et al. 2012). Instead of starting with a few preselected and targeted features, radiomics starts with hundreds of features and then applies statistical methods to find the most effective features for classifying, predicting, or estimating the quantity of interest. Although some studies discuss the interpretation of features after they have been determined to be prognostic (Van Dijk et al. 2017), interpretability is not a requirement in radiomics. Instead, the primary goal for most studies is to produce the most accurate and generalizable models.

14.5.1 AGNOSTIC FEATURES

Several categories of features, adopted from computer vision, have become widely used in radiomics studies. These features are referred to as "agnostic" because they are based on mathematical formulas rather than the descriptive lexicon of radiology (Gillies et al. 2015). First-order intensity histogram features are the most basic of the agnostic features. These quantities are calculated directly from image voxel intensities and include common statistics such as the mean, median, standard deviation, and the range. Other commonly studied first-order features include skewness, kurtosis, energy, entropy, and uniformity. A second group of features is the aptly named shape features, which include volume, surface area, roundness, spherical disproportion, and the number of objects.

Texture features use formulations that combine the intensities and the spatial relationships of voxels. Three commonly studied texture features come from computer vision and have been studied and applied for more than 25 years. First, gray-level co-occurrence matrix features were introduced by Haralick and others (1973) to quantify texture characteristics such as homogeneity, linear structure, contrast, and complexity. These "Haralick features," as they are commonly referred to, emphasize the differences between tone (areas of low-intensity variation) and texture (areas of high-intensity variation). Second, neighborhood gray-tone difference matrix features are derived from a one-dimensional (1D) matrix in which the differences between intensity values and the intensity value of adjacent voxels is measured. The features attempt to represent image characteristics that can be perceived by humans (Amadasun and King 1989). The third commonly studied class of texture features extracts information from the runs, that is, consecutive adjacent voxels with the same intensity value (Galloway 1975; Chu et al. 1990; Tang 1998). These three feature groups each calculate features from the counts in a matrix, the gray-level co-occurrence matrix, neighborhood gray-tone difference matrix, and gray-level run-length matrix. Therefore, when these types of features are also used to plot the matrices, good practice dictates plotting the matrices and verifying that their distribution of values is reasonable.

Radiomics features and especially their software implementations vary from study to study. Fortunately, efforts are underway to produce a standard set of radiomics features. The Image Biomarker Standardization Initiative, a growing collaboration of at least 40 researchers from 25 institutions, has compiled a document specifying a common set of definitions and nomenclature for morphology (29 features), statistical (18), intensity histogram (23), intensity-volume histogram (5), gray-level co-occurrence (25), gray-level run length (16), gray-level size zone (16), neighborhood gray-tone difference (5), gray-level distance zone (16), and neighboring gray-level dependence (17), for a total of 170 features. Also provided are descriptions and mathematical definitions for each of these features and feature groups, and feature values for a digital phantom and several patient ROIs to be used as a baseline for comparison with radiomics software (Zwanenburg et al. 2016).

Many commercial and open-source software packages for extracting radiomics features are available, and several review articles have summarized their strengths and weaknesses (Fave et al. 2015; Larue et al. 2017a). Open-source radiomics software packages include Imaging Biomarker Explorer (IBEX) (Zhang et al. 2015), PyRadiomics (van Griethuysen et al. 2017), Radiomics Enabler (Seymour and Payoux 2018), Computational Environment for Radiological Research (CERR) (Deasy et al. 2003), and ePAD (Schaer et al. 2018).

14.5.2 DERIVED FEATURES

Derivative features are standard features that have been modified to make them more effective for modeling a specific phenotype. For example, to emphasize features of a particular size, Ganeshan et al. applied Laplacian of Gaussians filters (LoG) of multiple widths. The Laplacian, or second derivative of the spatial coordinates, is applied to emphasize regions of rapid change in the images. The Gaussian is applied to smooth the image and reduce the contributions of image structures smaller than the width of the Gaussian (Ganeshan et al. 2009). LoG filters have been shown to be widely applicable and effective for extracting characteristics of tumors of the lung (Ganeshan et al. 2012a), liver (Ganeshan et al. 2007, 2011), colon (Ganeshan et al. 2011; Ganeshan and Miles 2013), and esophagus (Ganeshan et al. 2012b). Yip and colleagues used LoG-derived texture features to evaluate changes in heterogeneity in tumors of the esophagus (Yip et al. 2015) (Figure 14.3).

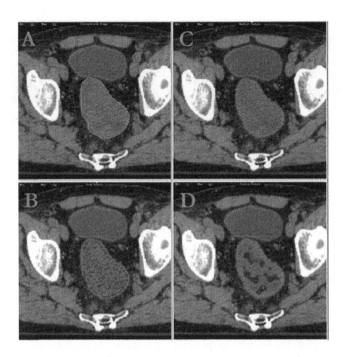

Figure 14.3 CT image of colorectal lesion: (a) unfiltered and with (b) fine, (c) medium, and (d) coarse Laplacian of Gaussians textures superimposed. (From Ganeshan, B. and Miles, K.A., *Cancer Imaging*, 13, 2013. With permission.)

Calculating the features on the 3D wavelet decompositions of images produces another class of derived features. The eight permutations of the decompositions—high and low frequencies for each of three spatial dimensions—greatly increase the number of available features (Aerts et al. 2014). Several studies have found wavelet features to be prognostic, including one that used wavelet-based features in a naïve Bayes classifier to classify tumor histology (Wu et al. 2016) and one that included high-low-high gray-level non-uniformity in their radiomics signature associated with overall survival in lung cancer and head-and-neck cancer (Aerts et al. 2014).

14.5.3 ENGINEERED FEATURES

What makes some machine learning projects succeed when others fail? According to Pedro Domingos, a Professor of Computer Science at the University of Washington, Seattle, "Easily the most important factor is the features used. So there is ultimately no replacement for the smarts you put into feature engineering" (Domingos 2012). Feature engineering is the deliberate attempt to incorporate domain knowledge into quantitative features used for machine learning. Thus, to some extent feature engineering is antithetical to radiomics, which seeks to develop pipelines that automate the process of biomarker development from a large suite of agnostic features (Gillies et al. 2015). However, a single engineered feature from a domain expert often beats a suite of agnostic features in predictive performance. So even as radiomics researchers strive to develop completely automated radiomics workflow pipelines, they should bear this advice from Domingos in mind and apply their domain expertise to develop custom features to complement the agnostic ones.

An interesting and successful example of feature engineering is found in a study of usual interstitial pneumonia by Depeursinge and colleagues (2015). By using a combination of oriented Riesz wavelets, they generated features modeling normal, ground glass, reticulation, and honeycomb patterns observed in CT images of lung tissue (Figure 14.4). They next used the feature to train support vector machines (SVMs) to classify the tissue as classic or atypical usual interstitial pneumonia. The first of the 3 SVMs was trained using the Riesz wavelets, the second using intensity features, and the third using gray-level co-occurrence

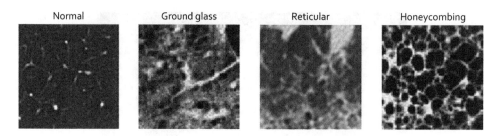

| Normal | Ground glass | Reticular | Honeycombing |

Figure 14.4 Common parenchymal appearances of interstitial pneumonia in CT. (From Depeursinge, A. et al., *Inv. Radiology*, 50, 261–267, 2015. With permission.)

matrix features. The classifications of the three SVMs were compared with the consensus classification of two thoracic radiologists by using the area under the curve (AUC) of the receiver operating characteristic (ROC) curve. The Riesz wavelet features SVM strongly agreed with the consensus classification (AUC = 0.81); the agreement between the consensus classification and the two standard radiomics feature SVMs was not much better than chance (AUC = 0.54 for intensity SVM and AUC = 0.60 for gray-level co-occurrence matrix SVM; Figure 14.5).

Oakden-Rayner and others (2017) engineered features to match biomarkers from the radiological literature. They applied the Agatston et al. (1990) method to pixel intensities to quantify cardiac and aortic calcification. They also engineered features to quantify cardiac and aortic calcification and pulmonary emphysema and bone mineral density. In univariate logistic regression models of 5-year survival, the engineered features targeting bone were the most likely to be significant.

Some of the most effective engineered features are those that quantify and facilitate easy extraction of image characteristics that are visible to the human eye. For example, Fried et al. (2015) developed the feature *solidity* to quantify the proximity between primary lesions and involved nodes in stage III

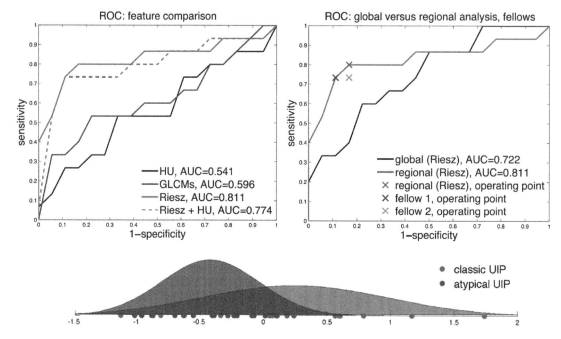

Figure 14.5 Comparison of models built using histogram, gray-level co-occurrence matrix (GLCM), and engineered Riesz wavelet features. The engineered features produce a much larger area under the curve (AUC). The density plots (bottom) of the computer score for classic (red) and atypical (blue) interstitial pneumonia. HU, histogram features; ROC, receiver operating characteristic. (From Depeursinge, A. et al., *Inv. Radiology*, 50, 261–267, 2015. With permission.)

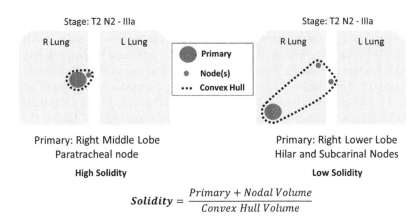

Figure 14.6 Illustration of the engineered feature solidity. (From Fried, D.V. et al., *Radiology*, 278, 214–222. With permission.)

NSCLC (Figure 14.6). They found that solidity and the co-occurrence matrix energy extracted from ^{18}F-fluorodeoxyglucose (FDG)–PET images could be combined with conventional clinical prognostic factors to improve the stratification of patients into overall survival risk groups compared with models that included only conventional clinical prognostic factors (log-rank $p = 0.18$ vs. $p = 0.0001$).

14.6 MODEL BUILDING

Once the image features have been extracted, radiomics model building should follow the best practices identified by the statistical and machine learning communities. Often the initial challenge is deciding what type of model to build. Three questions can help to guide the model building process: (1) Will the study explore or validate? (2) Is the model's purpose to explain or to predict? and (3) Is the model prognostic or predictive? The answers to these questions help determine what type of models are most appropriate and what results should be reported. This section reviews how some previous radiomics studies have answered these questions.

14.6.1 EXPLORE OR VALIDATE?

In data analysis, observation of the data can used for exploratory model building or model validation, but not both (Wickham and Grolemund 2016). Exploratory studies include a full analysis with feature selection, model building, and model hyperparameter optimization. Validation studies, on the other hand, test a model predetermined by a prior study using new observations of data. Exploratory studies generate hypotheses. Validation studies test them. One example of an exploratory study was reported by Wu and colleagues, who examined 440 features for potential association with histologic subtypes in a group of 350 patients with lung cancer (Wu et al. 2016). After considering 24 feature-selection methods and three classifiers, they found that a 5-feature naïve Bayes classifier produced the largest AUC, that is, 0.72. Although this model is promising, it must be considered exploratory until it is validated with unseen data. Another exploratory study looked for correlations between radiomics features and gene expression models; that radiogenomics model found 77 significant correlations from 54 image features (Gevaert et al. 2014). A third exploratory study found a significant difference between the features extracted from T2-weighted MR images of the inner-ear labyrinths of patients with Ménière's disease compared with control subjects. The authors conclude that their methods, once validated, could be used clinically as a biomarker of the disease.

Validation studies are not as common as exploratory studies, which is unfortunate because external validation is a necessary step for demonstrating model effectiveness (Taylor et al. 2008). To better demonstrate that a model generalizes well, many researchers believe validation data should be independent and externally collected (Yip and Aerts 2016; Summers 2017). Moon and colleagues (2012a) describe three categories for validation data: geographic, domain, and temporal. For geographic validation, the data set is collected at a different institution, often in a different country. For example, a CT radiomics signature developed exclusively from a group of patients treated in the Netherlands was validated with a group of patients treated in North America. This signature was originally developed for patients with NSCLC and was then applied to two groups of patients with oropharyngeal squamous cell carcinoma patients. Harrell's c-index was 0.628 for the North American cohort as compared with 0.686 and 0.685 for the two oropharyngeal cancer cohorts from the Netherlands (Leijenaar et al. 2015a). Domain validation is the term for when the training and validation patient groups have different types of disease. For example, the co-occurrence matrix energy from FDG-PET, which had previously been found to be prognostic for overall survival in NSCLC, was combined with clinical factors and found to be prognostic for overall survival in patients with oligometastatic NSCLC as well (Jensen et al. 2017). In temporal validation, a time point is chosen for splitting a data set, with the more recent data typically designated for validation (Moons et al. 2012a). This approach was used to develop a radiomics-based nomogram for predicting lymph node metastasis in colorectal cancer. For that study, the training and validation data sets comprised 326 patients treated from January 2007 through April 2010 and 200 patients treated from May 2010 through December 2011. The nomogram generalized well to this validation cohort, with c-indexes of 0.74 for the training set and 0.79 for the validation set.

In some cases, exploration and validation have been successfully combined into a single study. Aerts and colleagues developed a radiomics survival model based on a cohort of 422 patients with NSCLC treated at the Maastricht Radiation Oncology (MAASTRO) Clinic and then validated the model with 225 patients with NSCLC treated at Radboud University Nijmegen Medical Centre. They then tested the same radiomics signature using two groups of patients with head-and-neck squamous cell carcinoma, 136 treated at the MAASTRO Clinic and 95 treated at the VU University Medical Center Amsterdam. These three external data sets provided both geographic and domain validation of the model (Aerts et al. 2014). Another study involved developing and validating a radiomics signature for predicting the presence of human papillomavirus in patients with oropharyngeal cancer. Patients were randomly assigned to either the training ($n = 628$) or validation ($n = 150$) data sets (Leijenaar et al. 2018). Notably, both of these studies had large numbers of patients, and data sets of this size are rare in radiomics studies.

How much data is needed for the exploratory and validation portions of the study? In simulation studies, Peduzzi and others established a rule of 10:1 for the ratio of the number of examples of an outcome or class in the data to the number of predictor variables in logistic regression and Cox proportional hazards models (Peduzzi et al. 1995, 1996). Therefore, the number of observations needed in the training sample is related to the number of truly predictive features in the model. Feature selection can be used to reduce the number of features based on the amount of training data, but if used for this purpose the selection method should be unsupervised, that is, it should not include the response variable (also referred to as the dependent variable). Few studies have examined how many observations are needed for a validation sample to produce a stable estimate of a model's accuracy. Steyerberg and others developed a logistic regression model to predict bacterial infection from 555 patients presenting with fever (120 of whom had bacterial infection). They repeatedly split the patients into training ($n = 376$) and validation ($n = 179$) samples. For each iteration, they repeated the process of optimizing the logistic regression model and then calculating the AUC for the validation sample. For this relatively large validation sample, the standard deviation of the AUC was 0.05 (Steyerberg et al. 2001). Radiomics studies can and should provide estimates of stability for external validation by using bootstrap procedures.

Because a large number of samples is needed for both exploratory model building and validation, splitting the sample for validation is "not appropriate unless the number of subjects is huge" (Harrell 2015). The splits must be large enough that the model is not weakened and validation is stable, which occurs when the sample is large enough that overfitting is not a problem (Steyerberg and Harrell 2016). Rather than setting aside a sizeable portion of the data for validation, most radiomics studies should use all of the data for exploration, and they should report the bootstrapped or cross-validated results (Steyerberg et al. 2001; Simon et al. 2011; Harrell 2015). A more natural split for data is according to medical center. Steyerberg and Harrell (2016) proposed an "internal–external" cross-validation approach in which models are built using data from multiple centers, with the data from one center left out for validation. The process would be repeated until the data from each center has been used for validation once.

14.6.2 EXPLAIN OR PREDICT?

In model building and in radiomics, there is often a trade-off between the accuracy of the prediction and the interpretability of the model. To understand this trade-off, it helps to first consider the three categories of models common in radiomics: regression, classification, or survival analysis. Regression models predict a continuous outcome value. Classification models assign data observations to one or more groups. Survival analysis predicts the time until an event, such as a local recurrence of a tumor or death, from data that are censored. In censored data, the time of the event of interest is unknown for some observations. Each of these three types of model is associated with a commonly used statistical model: the linear model for regression, logistic regression model for classification, and Cox proportional hazard model for survival analysis. The popularity of these models is due in a large part to the interpretability of the model coefficients. The coefficients in linear models are the expected change in the predicted value for a unit change in the corresponding data feature. For logistic regression, the coefficients estimate the change in the log odds ratio of classes for a unit change of the date features. The coefficients in a Cox proportional hazards model scale the likelihood of the occurrence of the event, or "hazard." These models can be explained to caregivers and patients, perhaps making them more likely to gain acceptance in the clinic (Caruana et al. 2015; Breiman 2001b).

What is the cost of this interpretability? In a study of 17 classifier families with 179 software implementations on 121 data sets, logistic regression was never among the top 20 in accuracy (Fernández-Delgado et al. 2014). In a study evaluating machine learning methods for predicting mortality in 2287 patients with pneumonia (1601 for training, 686 set aside for validation), logistic regression performance was significantly worse than an artificial neural network (ANN) (Cooper et al. 2005). However, in two studies of classifier performance radiomics data sets of 464 patients with lung cancer and 101 patients with head-and-neck cancer, the prediction performance of logistic regression was similar to the best-performing methods (Parmar et al. 2015a, 2015b). Thus the cost of the interpretable models may be smaller for smaller data sets of hundreds rather than thousands of observations.

A thoughtful discussion of this trade-off is provided by Leo Breiman, inventor of the random forest ensemble method (Breiman 2001b). He admits that "forests are A+ predictors... on interpretability, they rate an F," but he also points out an inherent contradiction in the interpretation of statistical models. When multiple models provide nearly equivalent fits to the data and prediction accuracy, which of the interpretable models should be believed is not clear. Choosing one of many models for interpretation can lead to questionable scientific conclusions (Breiman 2001b).

14.6.3 PROGNOSTIC OR PREDICTIVE?

Describing a radiomics study as prognostic or predictive may seem like a semantic or even pedantic distinction. Prognostic studies model the patient's overall outcome; predictive studies model the patient's response to treatment (Oldenhuis et al. 2008). A goal of radiomics, however, is to produce biomarkers that can be used for personalized medicine (Lambin et al. 2012; Aerts et al. 2014), and prognostic studies are not sufficient to achieve this goal. Radiomics studies need to be predictive. Despite dozens of prognostic radiomics results (e.g., Aerts et al. 2014; Huang et al. 2016; Liang et al. 2016), predictive results have been few. Ypsilantis et al. (2015) developed a model using a convolutional neural network to predict which patients with esophageal cancer would benefit from neoadjuvant chemotherapy. Their model achieved accuracy of 73.4% ± 5.3%, and significantly outperformed models built using random forest, support vector machines, or gradient boosting algorithms. In a prognostic study, Fried et al. showed that tumor solidity and the gray-level co-occurrence matrix energy from FDG-PET were prognostic for overall survival in patients with stage III NSCLC. In a follow-up study, they showed that the same two features could predict which patients would benefit from radiation dose increases (Fried et al. 2015, 2016).

14.6.4 FEATURE SELECTION

In radiomics studies, the number of features, m, available for model building is often greater than the number of observations in the data, n, called the $m > n$ problem. As previously mentioned, numerous studies have suggested that the ratio of data observations to features used for model building should be at least 10:1

(Peduzzi et al. 1995; Moons et al. 2012b; Chalkidou et al. 2015; Harrell 2015). Therefore, until image data sets containing thousands of observations become available, feature selection will be an important step in radiomics analyses. The challenge for radiomics studies is to select only the most informative of the features for model building to avoid overfitting the training data. In radiomics, selecting the best features begins with eliminating the worst. Features that not robust or are redundant are removed before a feature selection method from machine learning is applied to select the most informative features.

14.6.4.1 Robustness tests

Robustness tests are applied as part of feature selection to ensure that radiomics features describe the underlying phenotype of tissue rather than patient–setup uncertainties, variability in contouring, image-acquisition parameters, or statistical noise. Robustness tests fall into two classes: repeatability and reproducibility. *Repeatability* measures the effects on a feature's precision when the imaging is repeated under identical conditions and is assessed in test/retest studies in which an imaging procedure is repeated on a patient after some time has elapsed (anywhere from 15 min to several weeks) (Leijenaar et al. 2013; Hunter et al. 2013; Balagurunathan et al. 2014; Zhao et al. 2016). For example, a study of 328 features on 25 test/retest 4D-CT scans compared the repeatability of the average, end-of-exhalation, and breath-hold scans. The features extracted from the average scans were the most repeatable. The concordance correlation coefficient was greater than 0.9 for 93% of the features compared with just 73% for the end-of-exhalation scans and 61% for the breath-hold scans. Other test/retest studies have compared the repeatability of segmentation methods (van Velden et al. 2016) and radiomics features extracted from FDG-PET images (Leijenaar et al. 2013; van Velden et al. 2016). *Reproducibility* measures the effects on a feature's precision when there is a change in the imaging conditions, such as scan location, scanner manufacturer, image acquisition parameters, or tumor contouring (Hatt et al. 2017; Sullivan et al. 2015; Bartlett and Frost 2008). Reproducibility studies have shown that features are not robust to changes in reconstruction kernel and slice thickness in CT (Lu et al. 2016; He et al. 2016; Zhao et al. 2016) or changes in the dynamic range or matrix size in MRI (Molina et al. 2017).

Other studies have applied these results by removing the non-robust features. In their exploratory radiogenomic study of glioblastoma multiforme, Gevaert and colleagues (2014) selected 18 of 55 features that were robust (interclass correlation coefficient >0.6) to digital algorithmic modification and test/retest analysis of the ROI segmentation. Hatt et al. (2015) eliminated features that were not robust to partial-volume, segmentation, and reconstruction parameters in a study of the relationship between the metabolically active tumor volume and texture features.

14.6.4.2 Redundancy tests

Many radiomics features are highly correlated with each other. Much of the correlation results from similarities in the feature algorithms, especially for features in the group. Correlation can also be caused by features describing similar characteristics of the tissue or limitations in the imaging (Hatt et al. 2017). Removing redundant or collinear features from a model can reduce the uncertainty of the parameters and improve the prediction accuracy on external data sets (Kiers and Smilde 2007). Just as important is that redundant variables do not provide additional information. Removing them can reduce the time needed for model building and improve estimates of the false discovery rate. For these reasons, most statistical software packages can easily remove correlated features. The *caret* and *CoreLearn* packages in R (Kuhn 2008; Robnik-Sikonja et al. 2015; R Core Team 2015) identify pairs of features with correlations exceeding an arbitrary threshold M. The feature from the pair that has the largest average correlation is removed. Both the Pearson and Spearman rank correlation coefficients are often used to measure the correlations. For most radiomics studies, the Spearman and Kendall rank correlation measures are preferred to the Pearson correlation because the variable correlation may be nonlinear (Hatt et al. 2017). Unfortunately, little guidance is available regarding the appropriate value for the cutoff threshold M, although values from 0.8 to 0.95 have been reported (Wu et al. 2016; Aerts et al. 2016; Desbordes et al. 2017; Li et al. 2017a; Van Dijk et al. 2017; Kirienko et al. 2018; Lee et al. 2018).

14.6.4.3 Feature selection methods

There are three classes of feature selection methods: wrapper, filter, and embedded (Guyon and Elisseeff 2003). Wrapper methods use the predictive model algorithms to select best combinations of features.

In most cases, an exhaustive search testing all possible models is impractical, so other wrapper algorithms have been proposed. One commonly applied method is sequential forward selection. Sequential forward selection begins by testing each feature to determine the best univariate model. Building on the best univariate model, sequential forward selection next tests the remaining features to determine the best bivariate model, and it continues to expand the model until it ceases to improve. Sequential backward selection reverses this process, starting with a model including all of the features and then iteratively removing the least important ones. Although sequential forward selection and sequential backward selection are intuitive, they will not necessarily select the best model and feature combination. When used with regression models, sequential forward selection and sequential backward selection overestimate the magnitude of the coefficients and underestimate their uncertainties (Harrell 2015). Other wrapper methods such as genetic algorithms and simulated annealing address these limitations, albeit at the cost of increased computation (Saeys et al. 2007).

Filter methods assess the characteristics of the data to determine the relative importance of the features. Examples of filters include univariate ranking algorithms such as the Wilcoxon rank-sum, t-test, and Euclidean distance. Multivariate filters, which take interactions between features into account, include the uncorrelated shrunken centroid (Yeung and Bumgarner 2003), the minimum redundancy-maximum relevance (Ding and Peng 2005), and the Markov blanket (Koller and Sahami; 1996, Saeys et al. 2007).

Embedded algorithms include feature selection in the model-building process. Logistic regression with the least absolute shrinkage and selection operator (LASSO) (Tibshirani 1996) and the random forest (Breiman 2001a) are perhaps the two most commonly used classifiers in radiomics, and they both embed feature selection. The LASSO is a regularization method that shrinks the regression coefficients to reduce model overfitting. By shrinking some of the regression coefficients to zero, the LASSO performs feature selection as it regularizes (Muthukrishnan and Rohini 2016). The random forests classifier chooses a random sample or "bag" of data for each tree, and for each split in the tree, it chooses from a random subset of features. As the data are fit, the random forest calculates the classifier's performance by using the out-of-bag sample and provides metrics for the importance of each feature. Thus, random forest can be used without feature selection, and it can also be used to provide feature selection for other classifiers.

Parmar and colleagues (2015a) evaluated the performance and stability of 14 filter-based feature selection methods specifically for radiomics. These methods were tested for selecting among 440 radiomics features for use in overall survival models built from 464 patients with lung cancer. Each feature-selection method was paired with 12 classification methods. Feature selection using the Wilcoxon (1945) signed rank test demonstrated the highest predictive performance and was among the most stable. Haury and colleagues reported a comparison of 32 feature-selection methods on gene expression data sets. They found that simpler univariate methods such as those based on the Student's t-test and Wilcoxon signed rank test generally performed as well as more complex wrapper and ensemble methods (Haury et al. 2011). Although numerous additional studies of feature selection have been done (Guyon and Elisseeff 2003; Friedman et al. 2009; Grömping 2009; Hall and Smith 1999; Genuer et al. 2010), none are definitive. The optimal feature selection method will depend on the needs of each study.

14.6.5 MODEL ALGORITHMS

The three primary model building tasks are classification, regression, and survival analysis. Many of the popular classification algorithms, including random forest, support vector machines, and neural networks, can also provide regression results. The classifiers can also be used to stratify patients into risk groups for survival analysis. Thus, in this section we provide an overview of the some of the most popular and powerful classifiers.

Dozens of classifiers have been proposed in the statistical and machine learning literature, and many have been applied to radiomics. Parmar and others (2015a) attempted to determine the best of 12 classifiers for predicting the overall survival class of 464 patients with lung cancer from radiomics features. Although random forest had the largest AUC (66% ± 0.03%), several other classifiers produced similar results. In an attempt to answer the question, "Do we really need so many classifiers?" Fernández-Delgado and colleagues (2014) evaluated 179 classifiers and once again found that "random forest is clearly the best family of classifiers." Despite the confusion caused by so many choices, some guidelines do exist for choosing a classifier.

Logistic regression models are interpretable, fast to train, and when combined with the LASSO, have built-in feature selection and regularization. As noted in the previous section, the LASSO is a regularization term added to the cost function of regression models. It shrinks some coefficients to reduce overfitting to the training data; it also performs feature selection by setting other coefficients to 0. LASSO regression is popular, as evidenced by the more than 20,000 citations of the study introducing it (Tibshirani 1996). On the other hand, logistic regression, even with the LASSO, is often less accurate than machine learning classifiers.

Although single-decision trees have strengths and weakness similar to those of logistic regression, random forest, which is an ensemble of dozens or hundreds of decision trees, is one of the more widely used machine learning classifiers (Breiman 2001a). Random forests are easy to train because they generally produce good results with default tuning parameters, and training of the individual trees can be parallelized on multiple CPU cores. Because they combine the results of many trees, random forests gain prediction accuracy while losing interpretability.

A commonly used alternative to the random forest is the SVM. SVMs attempt to find the hyperplane that maximizes the margins between the classes. The data points that determine the margin are the support vectors, and these data points determine the effectiveness of the classification. Like random forests, SVMs trade interpretability for performance. Unlike random forest, the effectiveness of SVMs depends heavily on the tuning parameters. Properly choosing and tuning the kernels used for the SVM can greatly improve its performance. Because SVMs rely on the support vectors to determine the maximum margin hyperplane, they are subject to overfitting (Vapnik et al. 1995).

Boosting algorithms, such as Adaboost and gradient boosting, use an iterative learning approach in which the observations in the training sample are weighted in the cost function used to optimize the model. The weights for misclassified events are increased, or "boosted," in subsequent iterations. In some cases, these learners greatly outperform random forests and SVMs. Adaboost, short for "adaptive boosting," was the first boosting algorithm (Freund and Schapire 1997). More recently, gradient boosting algorithms have been shown to be among the most effective classifiers (Mason et al. 2000; Taieb and Hyndman 2014).

ANNs were an early attempt at artificial intelligence that attempted to simulate biological learning by using layers of interconnected and weighted "neurons." The networks consist of input and output layers with one or more hidden layers in between. The hidden layers receive inputs, perform a calculation, and then pass the result to the next layer. Training an ANN requires a computationally expensive process of iteratively reweighting the neurons in each layer to improve the fit to the data. To avoid overfitting, early stopping or regularization is applied. As one of the older machine learning algorithms, ANNs initially fell out of favor as the popularity of the simpler and faster-to-train methods such as random forest and SVM grew. Because of their unavoidable complexity, ANNs are not commonly used in radiomics. However, recent advances, especially the advent of graphics processing unit computing and "deep learning," have led to resurgence in the interest in ANNs. Rather than having one or two hidden layers, deep learning networks have as many layers as needed to optimize the model (Bengio 2009; Lecun et al. 2015). For some tasks, especially image recognition, deep learning greatly outperforms other algorithms. Further, deep learning methods can learn features on their own, which is likely to have significant implications for radiomics. One notable study may provide a glimpse into the future. In that study, Ypsilantis and colleagues (2015) compared machine learning models trained using conventional radiomics features extracted from FDG-PET images to a convolution neural network, a type of ANN, which automatically extracts variability in the radiotracer uptake directly from three image slices at a time. The convolution neural network approach achieved an accuracy of $73.4\% \pm 5.3\%$ for predicting whether a patient would respond to neoadjuvant chemotherapy compared with $66.7\% \pm 5.2\%$ for gradient boosting, the best performing of the conventional machine learning models.

14.6.6 CONTROLLING FOR FALSE DISCOVERY

The large number of features in radiomics produces a large number of possible prediction models, and a correspondingly high likelihood of finding models to be significant when they are not, referred to as type I errors and false discoveries. Radiomics studies commonly correct for the possibility of false discoveries by using one of three methods: Holm-Bonferroni, Benjamini-Hochberg, or permutation.

The Holm-Bonferroni method determines the family-wise error rate, which is the probability of making at least one false discovery (type I error) in a group or "family" of inferences (Holm 1979; Bland and Altman 1995). This method corrects the significance level of a test, α, by dividing it by the number of inferences in the family, M. For example, when testing 100 features for univariate significance when modeling an outcome (e.g., local control of a lesion) with a test significance level of 0.05, the adjusted significance level, $\alpha_{adj} = \frac{\alpha}{100} = 0.0005$. Although apparently simple, the Holm-Bonferroni method implicitly assumes that each of the tests are statistically independent, and because this assumption often false in radiomics, this method may lead to type II errors, in which significant models are falsely rejected (Perneger 1998).

The Benjamini-Hochberg method is a less conservative approach that calculates the false discovery rate rather than family-wise error rate. For the Benjamini-Hochberg method, the p values for each of the hypotheses tested are put into order, $p_1, p_2, p_k, ..., p_M$, and the largest k such that $p_k < \frac{k}{M}\alpha$ is identified. The null hypothesis is rejected for all tests with $p < p_k$ (Benjamini and Hochberg 1995; Friedman et al. 2009). Consider the same example of testing 100 features for significance, with $\alpha = 0.05$, in a univariate model predicting an outcome. If the 5th and 6th smallest p values are 0.001 and 0.0004, then p_5 is

$$p_5 = 0.001 < \frac{5}{100}0.05 = 0.0025 \text{ and } p_6 \text{ is}$$

$$p_6 = 0.004 \nless \frac{6}{100}0.05 = 0.003.$$

Only the five hypotheses with the smallest p values would be rejected.

The third approach, permutation, is also straightforward but computationally expensive. For this method, the outcome labels in the data are randomly permuted, and the model-building process is repeated a large number of times, perhaps 10,000 times. The entire model-building process including variable selection should be repeated with each permutation. The estimate of the significance of the original model is provided by the fraction of the repeated models with less significant test results. The permutation method does not assume that models being tested are independent. Further, because the permutation method tests for spurious correlations, it penalizes less for large data sets and more for large feature sets. This method is described in detail by Friedman and colleagues (2009), and an example of its use is given by Fried and colleagues (2015).

All three of these approaches can help to frame the importance of results in exploratory studies. However, we reiterate that the key to controlling against false discovery in radiomics is the external validation of results with unseen data. These methods for controlling the false discovery rate provide guidance as to which models are worthy of external validation.

14.7 RADIOMICS RESULTS

More than 400 radiomics studies have been published since 2014. Despite all of these results, radiomics has yet to make an impact on clinical decision making. A problem for most studies is that they lack big data sets. Another problem is that the results are statistically significant without being clinically significant. Study reports should make it clear how radiomics results can help to personalize clinical decision making. For example, an exploratory study from Fried and colleagues attempted to identify subgroups of patients with stage III NSCLC who would benefit from radiation dose escalation. Their results indicated that increases in dose from a baseline of 60–70 Gy to 74 Gy would increase the overall survival of patients with high values for both gray-level co-occurrence matrix energy and the engineered feature solidity extracted from FDG PET images. Further, increasing the dose for patients with low values for these two features was associated with poorer overall survival (Fried et al. 2016). This result is promising but requires external validation.

It is important that radiomics biomarkers be supported by the appropriate statistical tests. But to gain wide acceptance as clinical decision support tools, radiomics biomarkers must be believable as well. Believability is obtained through interpretability of models, correlation with other biomarkers, and external validation by an independent research groups using independently collected data. Leger and colleagues

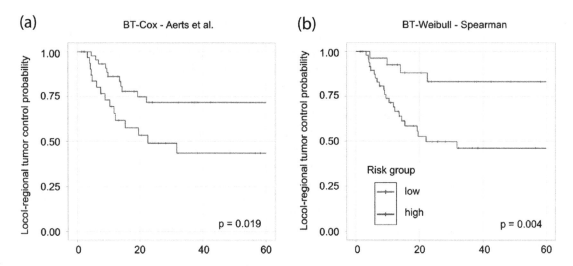

Figure 14.7 (a) The boosted trees–Cox (BT-Cox) model using a previously defined radiomics signature from Aerts et al. compared with (b) a boosted trees–Weibull (BT-Weibull) model using features selected by a Spearman rank correlation filter. Both models significantly stratify the patients into high- and low-risk groups. (Adapted from Leger, S. et al., *Sci. Rep.*, 2018 under Creative Commons Attribution 4.0 License.)

used pretreatment CT images from a multicenter cohort of 293 patients with head-and-neck squamous cell carcinoma, applied a radiomics signature identified by Aerts and colleagues, and compared the models developed using this prior signature with a new signature developed using 213 of the observations in their data (Leger et al. 2017). The highest predictive performance for local–regional control using the prior radiomics signature was achieved using a boosted trees–Cox model (Figure 14.7). A boosted trees–Weibel model using features selected by a Spearman rank correlation–based filter achieved better performance, but the comparable results from the old and new approaches makes them both more believable (Figure 14.7).

14.8 FUTURE WORK

The central hypothesis of radiomics is that quantitative features measured from routine medical images are related to the tumor phenotype and provide important information for personalized medicine. This chapter has described many successful examples of the explicit extraction of quantitative features ("radiomics features") from medical images, as well as the subsequent modeling processes to relate these features to clinical endpoints. As noted, the modeling can be traditional statistical modeling (e.g., survival regression models, penalized regressions), or it can use machine learning approaches (e.g., random forest). Although many publications have highlighted successes in radiomics research, these successes have yet to make their way into clinical decision making, meaning there is still work to be done to improve the accuracy and reliability of radiomics models. One way this is likely to be achieved is with better control of the input data, reducing sources of uncertainty (e.g., using controlled imaging protocols). Another way is the use of alternative image-analysis techniques.

A likely candidate for an alternative approach is that of deep learning. Over the past few years, deep learning has emerged as an important tool that can "learn" directly from image data, without the need to explicitly extract image features (Kontos et al. 2017; Choi 2018; Hosny et al. 2018). Avoiding the image segmentation step has several advantages, including time savings and the removal of a process that is plagued with uncertainty (including interuser uncertainties and the fact that often the "truth" is unknown). Avoiding the features-extraction step has similar advantages, potentially avoiding additional errors that this step may introduce. Deep learning approaches, including convolution neural networks, have been shown to be quite useful for a range of tasks, including risk assessment, classification, and many other medical image analysis tasks (Cheng et al. 2016; Li et al. 2017b; Lao et al. 2017; Korfiatis et al. 2017; Cha et al. 2017; Kang et al. 2017). Their success for segmenting ROIs (Lustberg et al. 2018) further suggests that hybrid approaches combining deep learning and more traditional radiomics approaches may benefit from the advantages of both techniques.

Although it is too early to fully assess the influence of deep learning on radiomics, it will be huge. The remarkable success and rapid growth of deep learning approaches mean that these approaches will inevitably and quickly become routine tools in radiomics research. The next few years in the field of radiomics will be exciting and hopefully will lead to the successful implementation of deep learning tools in clinical decision making.

REFERENCES

Aerts, H. J. W. L., Velazquez, E. R., Leijenaar, R. T. H., Parmar, C., Grossmann, P., Carvalho, S., Bussink, J. et al. 2014. Decoding tumour phenotype by noninvasive imaging using a quantitative radiomics approach. *Nature Communications,* 5, 1–8.

Aerts, H. J., Grossmann, P., Tan, Y., Oxnard, G. G., Rizvi, N., Schwartz, L. H. & Zhao, B. 2016. Defining a radiomic response phenotype: A pilot study using targeted therapy in NSCLC. *Scientific Reports,* 6.

Agatston, A. S., Janowitz, W. R., Hildner, F. J., Zusmer, N. R., Viamonte, M. & Detrano, R. 1990. Quantification of coronary artery calcium using ultrafast computed tomography. *Journal of the American College of Cardiology,* 15, 827–832.

Amadasun, M. & King, R. 1989. Textural features corresponding to textural properties. *Systems, Man and Cybernetics, IEEE Transactions on,* 19, 1264–1274.

Balagurunathan, Y., Kumar, V., Gu, Y., Kim, J., Wang, H., Liu, Y., Goldgof, D. B., Hall, L. O., Korn, R. & Zhao, B. 2014. Test–retest reproducibility analysis of lung CT image features. *Journal of Digital Imaging,* 27, 805–823.

Bang, J.-I., HA, S., Kang, S.-B., Lee, K.-W., Lee, H.-S., Kim, J.-S., Oh, H.-K., Lee, H.-Y. & Kim, S. E. 2015. Prediction of neoadjuvant radiation chemotherapy response and survival using pretreatment [18F]FDG PET/CT scans in locally advanced rectal cancer. *European Journal of Nuclear Medicine and Molecular Imaging,* 43, 422–431.

Bartlett, J. & Frost, C. 2008. Reliability, repeatability and reproducibility: Analysis of measurement errors in continuous variables. *Ultrasound in Obstetrics & Gynecology,* 31, 466–475.

Bengio, Y. 2009. Learning deep architectures for AI. *Foundations and Trends® in Machine Learning,* 2, 1–127.

Benjamini, Y. & Hochberg, Y. 1995. Controlling the false discovery rate: A practical and powerful approach to multiple testing. *Journal of the Royal Statistical Society. Series B (Methodological),* 289–300.

Bland, J. M. & Altman, D. G. 1995. Multiple significance tests: The Bonferroni method. *Britsh Medical Journal,* 310, 170.

Breiman, L. 2001a. Random forests. *Machine Learning,* 45, 5–32.

Breiman, L. 2001b. Statistical modeling: The two cultures (with comments and a rejoinder by the author). *Statistical Science,* 16, 199–231.

Brooks, F. J. & Grigsby, P. W. 2014. The effect of small tumor volumes on studies of intratumoral heterogeneity of tracer uptake. *Journal of Nuclear Medicine,* 55, 37–42.

Caruana, R., Lou, Y., Gehrke, J., Koch, P., Sturm, M. & Elhadad, N. Intelligible models for healthcare: Predicting pneumonia risk and hospital 30-day readmission. *Proceedings of the 21th ACM SIGKDD International Conference on Knowledge Discovery and Data Mining,* 2015. ACM, 1721–1730.

Cha, K. H., Hadjiiski, L., Chan, H. P., Weizer, A. Z., Alva, A., Cohan, R. H., Caoili, E. M., Paramagul, C. & Samala, R. K. 2017. Bladder cancer treatment response assessment in CT using radiomics with deep-learning. *Scientific Reports,* 7, 8738.

Chalkidou, A., O'Doherty, M. J. & Marsden, P. K. 2015. False discovery rates in PET and CT studies with texture features: A systematic review. *PLoS One,* 10, e0124165.

Chan, H.-P., Doi, K., Vyborny, C. J., Schmidt, R. A., Metz, C. E., Lam, K. L., Ogura, T., Wu, Y. & Macmahon, H. 1990. Improvement in radiologists' detection of clustered microcalcifications on mammograms. The potential of computer-aided diagnosis. *Investigative Radiology,* 25, 1102–1110.

Cheng, J. Z., Ni, D., Chou, Y. H., Qin, J., Tiu, C. M., Chang, Y. C., Huang, C. S., Shen, D. & Chen, C. M. 2016. Computer-aided diagnosis with deep learning architecture: Applications to breast lesions in US images and pulmonary nodules in CT scans. *Scientific Reports,* 6, 24454.

Choi, J. Y. 2018. Radiomics and deep learning in clinical imaging: What should we do? *European Journal of Nuclear Medicine and Molecular Imaging,* 52, 89–90.

Chu, A., Sehgal, C. M. & Greenleaf, J. F. 1990. Use of gray value distribution of run lengths for texture analysis. *Pattern Recognition Letters,* 11, 415–419.

Cooper, G. F., Abraham, V., Aliferis, C. F., Aronis, J. M., Buchanan, B. G., Caruana, R., Fine, M. J., Janosky, J. E., Livingston, G. & Mitchell, T. 2005. Predicting dire outcomes of patients with community acquired pneumonia. *Journal of Biomedical Informatics,* 38, 347–366.

Dawant, B. M., Zijdenbos, A. P. & Margolin, R. A. 1993. Correction of intensity variations in MR images for computer-aided tissue classification. *IEEE Transactions on Medical Imaging,* 12, 770–781.

Deasy, J. O., Blanco, A. I. & Clark, V. H. 2003. CERR: A computational environment for radiotherapy research. *Medical Physics,* 30, 979–985.

Depeursinge, A., Chin, A. S., Leung, A. N., Terrone, D., Bristow, M., Rosen, G. & Rubin, D. L. 2015. Automated classification of usual interstitial pneumonia using regional volumetric texture analysis in high-resolution computed tomography. *Investigative Radiology,* 50, 261–267.

Desbordes, P., Ruan, S., Modzelewski, R., Pineau, P., Vauclin, S., Gouel, P., Michel, P., DI Fiore, F., Vera, P. & Gardin, I. 2017. Predictive value of initial FDG-PET features for treatment response and survival in esophageal cancer patients treated with chemo-radiation therapy using a random forest classifier. *PLoS One,* 12, e0173208.

Ding, C. & Peng, H. 2005. Minimum redundancy feature selection from microarray gene expression data. *Journal of Bioinformatics and Computational Biology,* 3, 185–205.

Domingos, P. 2012. A few useful things to know about machine learning. *Communications of the ACM,* 55, 78–87.

Echegaray, S., Gevaert, O., Shah, R., Kamaya, A., Louie, J., Kothary, N. & Napel, S. 2015. Core samples for radiomics features that are insensitive to tumor segmentation: Method and pilot study using CT images of hepatocellular carcinoma. *Journal of Medical Imaging,* 2, 041011.

Fave, X., Cook, M., Frederick, A., Zhang, L. F., Yang, J. Z., Fried, D., Stingo, F. & Court, L. 2015. Preliminary investigation into sources of uncertainty in quantitative imaging features. *Computerized Medical Imaging and Graphics,* 44, 54–61.

Fave, X., Zhang, L., Yang, J., Mackin, D., Balter, P., Gomez, D., Followill, D., Jones, A. K., Stingo, F. & Liao, Z. 2017. Delta-radiomics features for the prediction of patient outcomes in non–small cell lung cancer. *Scientific Reports,* 7, 588.

Fernández-Delgado, M., Cernadas, E., Barro, S. & Amorim, D. 2014. Do we need hundreds of classifiers to solve real world classification problems. *Journal of Machine Learning Research,* 15, 3133–3181.

Freund, Y. & Schapire, R. E. 1997. A decision-theoretic generalization of on-line learning and an application to boosting. *Journal of Computer and System Sciences,* 55, 119–139.

Fried, D. V., Mawlawi, O., Zhang, L., Fave, X., Zhou, S., Ibbott, G., Liao, Z. & Court, L. E. 2015. Stage III non–small cell lung cancer: Prognostic value of FDG PET quantitative imaging features combined with clinical prognostic factors. *Radiology,* 278, 214–222.

Fried, D. V., Mawlawi, O., Zhang, L., Fave, X., Zhou, S., Ibbott, G., Liao, Z. & Court, L. E. 2016. Potential use of 18 F-fluorodeoxyglucose positron emission tomography–Based quantitative imaging features for guiding dose escalation in stage III non–small cell lung cancer. *International Journal of Radiation Oncology* Biology* Physics,* 94, 368–376.

Friedman, J., Hastie, T. & Tibshirani, R. 2009. *The Elements of Statistical Learning,* Second ed. Springer series in statistics Springer, Berlin, Germany.

Galloway, M. M. 1975. Texture analysis using gray level run lengths. *Computer Graphics and Image Processing,* 4, 172–179.

Ganeshan, B. & Miles, K. A. 2013. Quantifying tumour heterogeneity with CT. *Cancer Imaging,* 13, 140.

Ganeshan, B., Burnand, K., Young, R., Chatwin, C. & Miles, K. 2011. Dynamic contrast-enhanced texture analysis of the liver: Initial assessment in colorectal cancer. *Investigative Radiology,* 46, 160–168.

Ganeshan, B., Miles, K. A., Young, R. & Chatwin, C. 2007. Hepatic entropy and uniformity: Additional parameters that can potentially increase the effectiveness of contrast enhancement during abdominal CT. *Clinical Radiology,* 62, 761–768.

Ganeshan, B., Miles, K. A., Young, R. C. & Chatwin, C. R. 2009. Texture analysis in non-contrast enhanced CT: Impact of malignancy on texture in apparently disease-free areas of the liver. *European Journal of Radiology,* 70, 101–110.

Ganeshan, B., Panayiotou, E., Burnand, K., Dizdarevic, S. & Miles, K. 2012a. Tumour heterogeneity in non-small cell lung carcinoma assessed by CT texture analysis: A potential marker of survival. *European Radiology,* 22, 796–802.

Ganeshan, B., Skogen, K., Pressney, I., Coutroubis, D. & Miles, K. 2012b. Tumour heterogeneity in oesophageal cancer assessed by CT texture analysis: Preliminary evidence of an association with tumour metabolism, stage, and survival. *Clinical Radiology,* 67, 157–164.

Genuer, R., Poggi, J.-M. & Tuleau-Malot, C. 2010. Variable selection using random forests. *Pattern Recognition Letters,* 31, 2225–2236.

Gevaert, O., Mitchell, L. A., Achrol, A. S., XU, J., Echegaray, S., Steinberg, G. K., Cheshier, S. H., Napel, S., Zaharchuk, G. & Plevritis, S. K. 2014. Glioblastoma multiforme: Exploratory radiogenomic analysis by using quantitative image features. *Radiology,* 273, 168–174.

Gevaert, O., Xu, J., Hoang, C. D., Leung, A. N., Xu, Y., Quon, A., Rubin, D. L., Napel, S. & Plevritis, S. K. 2012. Non–small cell lung cancer: Identifying prognostic imaging biomarkers by leveraging public gene expression microarray data—Methods and preliminary results. *Radiology,* 264, 387–396.

Gillies, R. J., Kinahan, P. E. & Hricak, H. 2015. Radiomics: Images are more than pictures, they are data. *Radiology,* 278, 563–577.

Grömping, U. 2009. Variable importance assessment in regression: linear regression versus random forest. *The American Statistician,* 63, 308–319.

Guyon, I. & Elisseeff, A. 2003. An introduction to variable and feature selection. *Journal of Machine Learning Research,* 3, 1157–1182.

Hall, M. A. & Smith, L. A. 1999. Feature selection for machine learning: Comparing a correlation-based filter approach to the wrapper. *FLAIRS Conference,* 235–239.

Haralick, R. M., Shanmugam, K. & Dinstein, I. H. 1973. Textural features for image classification. *IEEE Transactions on Systems, Man and Cybernetics,* 610–621.

Harrell, F. 2015. *Regression Modeling Strategies: With Applications to Linear Models, Logistic and Ordinal Regression, and Survival Analysis,* Springer, Cham, Switzerland.

Hatt, M., Majdoub, M., Vallières, M., Tixier, F., Le Rest, C. C., Groheux, D., Hindié, E., Martineau, A., Pradier, O. & Hustinx, R. 2015. 18F-FDG PET uptake characterization through texture analysis: Investigating the complementary nature of heterogeneity and functional tumor volume in a multi–cancer site patient cohort. *Journal of Nuclear Medicine,* 56, 38–44.

Hatt, M., Tixier, F., Pierce, L., Kinahan, P. E., Le Rest, C. C. & Visvikis, D. 2017. Characterization of PET/CT images using texture analysis: The past, the present… any future? *European Journal of Nuclear Medicine and Molecular Imaging,* 44, 151–165.

Haury, A.-C., Gestraud, P. & Vert, J.-P. 2011. The influence of feature selection methods on accuracy, stability and interpretability of molecular signatures. *PloS One,* 6, e28210.

He, L., Huang, Y., Ma, Z., Liang, C., Liang, C. & Liu, Z. 2016. Effects of contrast-enhancement, reconstruction slice thickness and convolution kernel on the diagnostic performance of radiomics signature in solitary pulmonary nodule. *Scientific Reports,* 6.

Holm, S. 1979. A simple sequentially rejective multiple test procedure. *Scandinavian Journal of Statistics,* 65–70.

Hosny, A., Parmar, C., Quackenbush, J., Schwartz, L. H. & Aerts, H. 2018. Artificial intelligence in radiology. *Nature Reviews Cancer,* 18, 500–510.

Huang, Y., Liu, Z., He, L., Chen, X., Pan, D., Ma, Z., Liang, C., Tian, J. & Liang, C. 2016. Radiomics signature: A potential biomarker for the prediction of disease-free survival in early-stage (I or II) non-small cell lung cancer. *Radiology,* 281, 947–957.

Hunter, L. A., Krafft, S., Stingo, F., Choi, H., Martel, M. K. & Kry, S. F. 2013. High quality machine-robust image features: Identification in nonsmall cell lung cancer computed tomography images. *Medical Physics,* 40, 121916.

Jensen, G. L., Yost, C. M., Mackin, D. S., Fried, D. V., Zhou, S. & Gomez, D. R. 2017. Prognostic value of combining a quantitative image feature from positron emission tomography with clinical factors in oligometastatic non-small cell lung cancer. *Radiotherapy and Oncology* 126, 362–367.

Kang, G., Liu, K., Hou, B. & Zhang, N. 2017. 3D multi-view convolutional neural networks for lung nodule classification. *PLoS One,* 12, e0188290.

Kiers, H. A. & Smilde, A. K. 2007. A comparison of various methods for multivariate regression with highly collinear variables. *Statistical Methods and Applications,* 16, 193–228.

Kirby, J., Tarbox, L., Freymann, J., Jaffe, C. & Prior, F. 2015. TU-AB-BRA-03: The cancer imaging archive: Supporting radiomic and imaging genomic research with open-access data sets. *Medical Physics,* 42, 3587–3587.

Kirienko, M., Cozzi, L., Antunovic, L., Lozza, L., Fogliata, A., Voulaz, E., Rossi, A., Chiti, A. & Sollini, M. 2018. Prediction of disease-free survival by the PET/CT radiomic signature in non-small cell lung cancer patients undergoing surgery. *European Journal of Nuclear Medicine and Molecular Imaging,* 45, 207–217.

Koller, D., and Sahami, M. 1996. Toward optimal feature selection. In *Machine Learning: Proceedings of the Thirteenth International Conference,* 284–292. Morgan Kaufmann.

Kontos, D., Summers, R. M. & Giger, M. 2017. Special section guest editorial: Radiomics and deep learning. *J Med Imaging (Bellingham),* 4, 041301.

Korfiatis, P., Kline, T. L., Lachance, D. H., Parney, I. F., Buckner, J. C. & Erickson, B. J. 2017. Residual deep convolutional neural network predicts MGMT methylation status. *Journal of Digital Imaging,* 30, 622–628.

Kuhn, M. 2008. Building predictive models in R using the caret package. *Journal of Statistical Software,* 28, 1–26.

Kumar, V., Gu, Y., Basu, S., Berglund, A., Eschrich, S. A., Schabath, M. B., Forster, K., Aerts, H. J., Dekker, A. & Fenstermacher, D. 2012. Radiomics: The process and the challenges. *Magnetic Resonance Imaging,* 30, 1234–1248.

Lambin, P., Rios-Velazquez, E., Leijenaar, R., Carvalho, S., Van Stiphout, R. G., Granton, P., Zegers, C. M., Gillies, R., Boellard, R. & Dekker, A. 2012. Radiomics: Extracting more information from medical images using advanced feature analysis. *European Journal of Cancer,* 48, 441–446.

Lao, J., Chen, Y., Li, Z. C., Li, Q., Zhang, J., Liu, J. & Zhai, G. 2017. A deep learning-based radiomics model for prediction of survival in glioblastoma multiforme. *Scientific Reports,* 7, 10353.

Larue, R. T., Defraene, G., De Ruysscher, D., Lambin, P. & Van Elmpt, W. 2017a. Quantitative radiomics studies for tissue characterization: A review of technology and methodological procedures. *The British Journal of Radiology,* 90, 20160665.

Larue, R. T., Van Timmeren, J. E., De Jong, E. E., Feliciani, G., Leijenaar, R. T., Schreurs, W. M., Sosef, M. N., Raat, F. H., Van Der Zande, F. H. & Das, M. 2017b. Influence of gray level discretization on radiomic feature stability for different CT scanners, tube currents and slice thicknesses: A comprehensive phantom study. *Acta Oncologica,* 56, 1544–1553.

Lecun, Y., Bengio, Y. & Hinton, G. 2015. Deep learning. *Nature,* 521, 436–444.

Lee, J., Li, B., Cui, Y., Sun, X., Wu, J., Zhu, H., Yu, J., Gensheimer, M. F., Loo JR, B. W. & Diehn, M. 2018. A quantitative CT imaging signature predicts survival and complements established prognosticators in stage I non-small cell lung cancer. *International Journal of Radiation Oncology* Biology* Physics.*

Leger, S., Zwanenburg, A., Pilz, K., Lohaus, F., Linge, A., Zöphel, K., Kotzerke, J., Schreiber, A., Tinhofer, I. & Budach, V. 2017. A comparative study of machine learning methods for time-to-event survival data for radiomics risk modelling. *Scientific Reports,* 7, 13206.

Leijenaar, R. T., Bogowicz, M., Jochems, A., Hoebers, F. J., Wesseling, F. W., Huang, S. H., Chan, B., Waldron, J. N., O'sullivan, B. & Rietveld, D. 2018. Development and validation of a radiomic signature to predict HPV (p16) status from standard CT imaging: A multicenter study. *The British Journal of Radiology,* 91, 20170498.

Leijenaar, R. T., Carvalho, S., Hoebers, F. J., Aerts, H. J., Van Elmpt, W. J., Huang, S. H., Chan, B., Waldron, J. N., O'sullivan, B. & Lambin, P. 2015a. External validation of a prognostic CT-based radiomic signature in oropharyngeal squamous cell carcinoma. *Acta Oncologica,* 54, 1423–1429.

Leijenaar, R. T., Carvalho, S., Velazquez, E. R., Van Elmpt, W. J., Parmar, C., Hoekstra, O. S., Hoekstra, C. J., Boellaard, R., Dekker, A. L. & Gillies, R. J. 2013. Stability of FDG-PET Radiomics features: An integrated analysis of test-retest and inter-observer variability. *Acta Oncologica,* 52, 1391–1397.

Leijenaar, R. T., Nalbantov, G., Carvalho, S., Van Elmpt, W. J., Troost, E. G., Boellaard, R., Aerts, H. J., Gillies, R. J. & Lambin, P. 2015b. The effect of SUV discretization in quantitative FDG-PET Radiomics: The need for standardized methodology in tumor texture analysis. *Scientific Reports,* 5, 11075.

Li, Q., Kim, J., Balagurunathan, Y., Liu, Y., Latifi, K., Stringfield, O., Garcia, A., Moros, E. G., Dilling, T. J. & Schabath, M. B. 2017a. Imaging features from pre-treatment CT scans are associated with clinical outcomes in non-small-cell lung cancer patients treated with stereotactic body radiotherapy. *Medical Physics,* 44, 4341–4349.

Li, Z., Wang, Y., Yu, J., Guo, Y. & Cao, W. 2017b. Deep Learning based Radiomics (DLR) and its usage in noninvasive IDH1 prediction for low grade glioma. *Scientific Reports,* 7, 5467.

Liang, C., Huang, Y., He, L., Chen, X., Ma, Z., Dong, D., Tian, J., Liang, C. & Liu, Z. 2016. The development and validation of a CT-based radiomics signature for the preoperative discrimination of stage I-II and stage III-IV colorectal cancer. *Oncotarget,* 7, 31401.

Lodwick, G. S., Haun, C. L., Smith, W. E., Keller, R. F. & Robertson, E. D. 1963. Computer diagnosis of primary bone tumors: A preliminary report. *Radiology,* 80, 273–275.

Lu, L., Ehmke, R. C., Schwartz, L. H. & Zhao, B. 2016. Assessing agreement between radiomic features computed for multiple CT imaging settings. *PLoS One,* 11, e0166550.

Lustberg, T., Van Soest, J., Gooding, M., Peressutti, D., Aljabar, P., Van Der Stoep, J., VAN Elmpt, W. & Dekker, A. 2018. Clinical evaluation of atlas and deep learning based automatic contouring for lung cancer. *Radiotherapy and Oncology,* 126, 312–317.

Mackin, D., Fave, X., Zhang, L., Fried, D., Yang, J., Taylor, B., Rodriguez-Rivera, E., Dodge, C., Jones, A. K. & Court, L. 2015. Measuring computed tomography scanner variability of radiomics features. *Investigative Radiology,* 50, 757–765.

Mackin, D., Fave, X., Zhang, L., Yang, J., Jones, A. K. & Ng, C. S. 2017a. Harmonizing the pixel size in retrospective computed tomography radiomics studies. *PLoS One,* 12, e0178524.

Mackin, D., Fave, X., Zhang, L., Yang, J., Jones, A. K., Ng, C. S. & Court, L. 2017b. Harmonizing the pixel size in retrospective computed tomography radiomics studies. *PLoS One,* 12, e0178524.

Mackin, D., Ger, R., Dodge, C., Fave, X., Chi, P. C., Zhang, L., Yang, J., Bache, S., Dodge, C., Jones, A. K. & Court, L. 2018. Effect of tube current on computed tomography radiomic features. *Scientific Reports,* 8, 2354.

Mason, L., Baxter, J., Bartlett, P. L. & Frean, M. R. 2000. Boosting algorithms as gradient descent. *Advances in Neural Information Processing Systems,* 512–518.

Molina, D., Pérez-Beteta, J., Martínez-González, A., Martino, J., Velasquez, C., Arana, E. & Pérez-García, V. M. 2017. Lack of robustness of textural measures obtained from 3D brain tumor MRIs impose a need for standardization. *PLoS One,* 12, e0178843.

Moons, K. G., Kengne, A. P., Grobbee, D. E., Royston, P., Vergouwe, Y., Altman, D. G. & Woodward, M. 2012a. Risk prediction models: II. External validation, model updating, and impact assessment. *Heart.* doi:10.1136/heartjnl-2011-301247.

Moons, K. G., Kengne, A. P., Woodward, M., Royston, P., Vergouwe, Y., Altman, D. G. & Grobbee, D. E. 2012b. Risk prediction models: I. Development, internal validation, and assessing the incremental value of a new (bio) marker. *Heart,* 98, 683–690.

Muthukrishnan, R. & Rohini, R. 2016. Lasso: A feature selection technique in predictive modeling for machine learning. *Advances in Computer Applications (ICACA), IEEE International Conference on,* 18–20.

Oakden-Rayner, L., Carneiro, G., Bessen, T., Nascimento, J. C., Bradley, A. P. & Palmer, L. J. 2017. Precision Radiology: Predicting longevity using feature engineering and deep learning methods in a radiomics framework. *Scientific Reports,* 7.

Oldenhuis, C., Oosting, S., Gietema, J. & De Vries, E. 2008. Prognostic versus predictive value of biomarkers in oncology. *European Journal of Cancer,* 44, 946–953.

Parmar, C., Grossmann, P., Bussink, J., Lambin, P. & Aerts, H. J. 2015a. Machine learning methods for quantitative radiomic biomarkers. *Scientific Reports,* 5, 13087.

Parmar, C., Grossmann, P., Rietveld, D., Rietbergen, M. M., Lambin, P. & Aerts, H. J. 2015b. Radiomic machine-learning classifiers for prognostic biomarkers of head and neck cancer. *Frontiers in Oncology,* 5, 272.

Parmar, C., Rios Velazquez, E., Leijenaar, R., Jermoumi, M., Carvalho, S., Mak, R. H., Mitra, S. et al. 2014. Robust Radiomics feature quantification using semiautomatic volumetric segmentation. *PLoS One,* 9, e102107.

Peduzzi, P., Concato, J., Feinstein, A. R. & Holford, T. R. 1995. Importance of events per independent variable in proportional hazards regression analysis II. Accuracy and precision of regression estimates. *Journal of Clinical Epidemiology,* 48, 1503–1510.

Peduzzi, P., Concato, J., Kemper, E., Holford, T. R. & Feinstein, A. R. 1996. A simulation study of the number of events per variable in logistic regression analysis. *Journal of Clinical Epidemiology,* 49, 1373–1379.

Perneger, T. V. 1998. What's wrong with Bonferroni adjustments. *British Medical Journal,* 316, 1236.

Persson, G. F., Nygaard, D. E., Hollensen, C., Munck Af Rosenschold, P., Mouritsen, L. S., Due, A. K., Berthelsen, A. K. et al. 2012. Interobserver delineation variation in lung tumour stereotactic body radiotherapy. *The British Journal of Radiology,* 85, e654–e660.

R Core Team 2015. R: A Language and Environment for Statistical Computing. R Foundation for Statistical computing.

Rasheed, Q., Dhawale, P. J., Anderson, J. & Hodgson, J. M. 1995. Intracoronary ultrasound—defined plaque composition: Computer-aided plaque characterization and correlation with histologic samples obtained during directional coronary atherectomy. *American Heart Journal,* 129, 631–637.

Robnik-Šikonja, M., Savicky, P. and Alao, J.A., CORElearn: Classification, Regression and Feature Evaluation, R package version 0.9. 45 (2015). URL http://CRAN. R-project. org/package= CORElearn.

Rutman, A. M. & Kuo, M. D. 2009. Radiogenomics: Creating a link between molecular diagnostics and diagnostic imaging. *European Journal of Radiology,* 70, 232–241.

Saeys, Y., Inza, I. & Larrañaga, P. 2007. A review of feature selection techniques in bioinformatics. *Bioinformatics,* 23, 2507–2517.

Schaer, R., Cid, Y. D., Alkim, E., John, S., Rubin, D. L. & Depeursinge, A. 2018. Web-based tools for exploring the potential of quantitative imaging biomarkers in radiology: Intensity and texture analysis on the ePAD platform. *Biomedical Texture Analysis.* Elsevier.

Seymour, K. & Payoux, P. 2018. Radiomics Enabler®, an ETL (Extract-Transform-Load) for biomedical imaging in big-data projects.

Shafiq-Ul-Hassan, M., Zhang, G. G., Latifi, K., Ullah, G., Hunt, D. C., Balagurunathan, Y., Abdalah, M. A., Schabath, M. B., Goldgof, D. G. & Mackin, D. 2017. Intrinsic dependencies of CT radiomic features on voxel size and number of gray levels. *Medical Physics,* 44, 1050–1062.

Simon, R. M., Subramanian, J., Li, M.-C. & Menezes, S. 2011. Using cross-validation to evaluate predictive accuracy of survival risk classifiers based on high-dimensional data. *Briefings in Bioinformatics,* 12, 203–214.

Steyerberg, E. W. & Harrell, F. E. 2016. Prediction models need appropriate internal, internal–external, and external validation. *Journal of Clinical Epidemiology,* 69, 245–247.

Steyerberg, E. W., Harrell, F. E., Borsboom, G. J., Eijkmans, M., Vergouwe, Y. & Habbema, J. D. F. 2001. Internal validation of predictive models: efficiency of some procedures for logistic Regression analysis. *Journal of Clinical Epidemiology,* 54, 774–781.

Sullivan, D. C., Obuchowski, N. A., Kessler, L. G., Raunig, D. L., Gatsonis, C., Huang, E. P., Kondratovich, M., Mcshane, L. M., Reeves, A. P. & Barboriak, D. P. 2015. Metrology standards for quantitative imaging biomarkers. *Radiology,* 277, 813–825.

Summers, R. M. 2017. Texture analysis in radiology: Does the emperor have no clothes? *Abdominal Radiology,* 42, 342–345.

Taieb, S. B. & Hyndman, R. J. 2014. A gradient boosting approach to the Kaggle load forecasting competition. *International Journal of Forecasting,* 30, 382–394.

Tang, X. 1998. Texture information in run-length matrices. *IEEE Transactions on Image Processing,* 7, 1602–1609.

Taylor, J. M., Ankerst, D. P. & Andridge, R. R. 2008. Validation of biomarker-based risk prediction models. *Clinical Cancer Research,* 14, 5977–5983.

Tibshirani, R. 1996. Regression shrinkage and selection via the lasso. *Journal of the Royal Statistical Society. Series B (Methodological),* 267–288.

Vallières, M., Freeman, C. R., Skamene, S. R. & El Naqa, I. 2015. A radiomics model from joint FDG-PET and MRI texture features for the prediction of lung metastases in soft-tissue sarcomas of the extremities. *Physics in Medicine & Biology,* 60, 5471.

Van Dijk, L. V., Brouwer, C. L., Van Der Schaaf, A., Burgerhof, J. G., Beukinga, R. J., Langendijk, J. A., Sijtsema, N. M. & Steenbakkers, R. J. 2017. CT image biomarkers to improve patient-specific prediction of radiation-induced xerostomia and sticky saliva. *Radiotherapy and Oncology,* 122, 185–191.

Van Griethuysen, J. J., Fedorov, A., Parmar, C., Hosny, A., Aucoin, N., Narayan, V., Beets-Tan, R. G., Fillion-Robin, J.-C., Pieper, S. & Aerts, H. J. 2017. Computational radiomics system to decode the radiographic phenotype. *Cancer Research,* 77, e104–e107.

Van Velden, F. H., Kramer, G. M., Frings, V., Nissen, I. A., Mulder, E. R., De Langen, A. J., Hoekstra, O. S., Smit, E. F. & Boellaard, R. 2016. Repeatability of radiomic features in non-small-cell lung cancer [18F] FDG-PET/CT studies: Impact of reconstruction and delineation. *Molecular Imaging and Biology,* 18, 788–795.

Vapnik, V., Guyon, I. & Hastie, T. 1995. Support vector machines. *Machine Learning,* 20, 273–297.

Vyborny, C. J. & Giger, M. L. 1994. Computer vision and artificial intelligence in mammography. *American Journal of Roentgenology,* 162, 699–708.

Warfield, S. K., Zou, K. H. & Wells, W. M. 2004. Simultaneous truth and performance level estimation (STAPLE): An algorithm for the validation of image segmentation. *IEEE Transactions on Medical Imaging,* 23, 903–921.

White, E. A., Brock, K. K., Jaffray, D. A. & Catton, C. N. 2009. Inter-observer variability of prostate delineation on cone beam computerised tomography images. *Clinical Oncology,* 21, 32–38.

Wickham, H. & Grolemund, G. 2016. *R for Data Science.* Sebastopol, CA: O'Reilly. http://r4ds.had.co.nz.

Wilcoxon, F. 1945. Individual comparisons by ranking methods. *Biometrics Bulletin,* 1, 80–83.

Wu, W., Parmar, C., Grossmann, P., Quackenbush, J., Lambin, P., Bussink, J., Mak, R. & Aerts, H. J. 2016. Exploratory study to identify radiomics classifiers for lung cancer histology. *Frontiers in Oncology,* 6.

Yamamoto, S., Tanaka, I., Senda, M., Tateno, Y., Iinuma, T., Matsumoto, T. & Matsumoto, M. 1994. Image processing for computer-aided diagnosis of lung cancer by CT (LSCT). *Systems and Computers in Japan,* 25, 67-80.

Yang, J., Zhang, L., Fave, X. J., Fried, D. V., Stingo, F. C., Ng, C. S. & Court, L. E. 2016. Uncertainty analysis of quantitative imaging features extracted from contrast-enhanced CT in lung tumors. *Computerized Medical Imaging and Graphics,* 48, 1–8.

Yasaka, K., Akai, H., Mackin, D., Court, L., Moros, E., Ohtomo, K. & kiryu, S. 2017. Precision of quantitative computed tomography texture analysis using image filtering: A phantom study for scanner variability. *Medicine,* 96.

Yeung, K. Y. & Bumgarner, R. E. 2003. Multiclass classification of microarray data with repeated measurements: application to cancer. *Genome Biology,* 4, R83.

Yip, C., Davnall, F., Kozarski, R., Landau, D., Cook, G., Ross, P., Mason, R. & Goh, V. 2015. Assessment of changes in tumor heterogeneity following neoadjuvant chemotherapy in primary esophageal cancer. *Diseases of the Esophagus,* 28, 172–179.

Yip, S. S. & Aerts, H. J. 2016. Applications and limitations of radiomics. *Physics in Medicine and Biology,* 61, R150.

Yip, S. S., Parmar, C., Kim, J., Huynh, E., Mak, R. H. & Aerts, H. J. 2017. Impact of experimental design on PET radiomics in predicting somatic mutation status. *European Journal of Radiology,* 97, 8–15.

Ypsilantis, P.-P., Siddique, M., Sohn, H.-M., Davies, A., Cook, G., Goh, V. & Montana, G. 2015. Predicting response to neoadjuvant chemotherapy with PET imaging using convolutional neural networks. *PLoS One,* 10, e0137036.

Zhang, L., Fried, D. V., Fave, X. J., Hunter, L. A. & Yang, J. 2015. IBEX: An open infrastructure software platform to facilitate collaborative work in radiomics. *Medical Physics,* 42, 1341–1353.

Zhao, B., Tan, Y., Tsai, W.-Y., Qi, J., Xie, C., Lu, L. & Schwartz, L. H. 2016. Reproducibility of radiomics for deciphering tumor phenotype with imaging. *Scientific Reports,* 6.

Zwanenburg, A., Leger, S., Vallières, M. & Löck, S. 2016. Image biomarker standardisation initiative-feature definitions. *arXiv preprint arXiv:1612.07003.*

15

Radiotherapy outcomes modeling in the big data era

Joseph O. Deasy, Aditya P. Apte, Maria Thor, Jeho Jeong, Aditi Iyer,
Jung Hun Oh, and Andrew Jackson

Contents

15.1	Introduction	242
15.2	Is Radiotherapy Obsolete?	242
15.3	Why Do We Need Models for Normal Tissue Toxicity and Tumor Control?	242
15.4	The Data Crisis in Radiotherapy, and in Medicine, More Generally	243
15.5	Toward Quantitative Health Care: Does Your Imaging Data Tell a Meaningful Story?	244
15.6	How Can the Field of Radiotherapy Overcome the Problem of Low Model Generalizability?	245
15.7	What Data Is Relevant for Predictive Model Development?	246
15.8	Tumor Control Probability Models	246
	15.8.1 Isoeffect Fractionation Regimes	246
	15.8.2 Radiobiological Assumptions in Mechanistic TCP Models	246
	15.8.3 Tumor Response–Evolution Models	246
	15.8.4 Radiobiological Assays	247
	15.8.5 Imaging to Refine Tumor Control Models	248
	15.8.6 PET Measurements of Hypoxia	248
15.9	Normal Tissue Complication Probability Models: A Brief Overview	248
	15.9.1 Assays to Predict Normal Tissue Response. We Briefly Note Some Interesting Results	249
	15.9.2 Genomewide Association Studies	249
15.10	Methodological Issues	250
	15.10.1 Bias and Confounding Factors in Determinations of TCP and NTCP Models	250
	15.10.2 Predictive Model Development	250
	15.10.3 Data Handling with CERR	250
	15.10.4 A Tool to Envision Dose–Response Curves Based on Patient Specific Treatment Plans	251
15.11	Data Cohorts for Predictive Modeling	251
	15.11.1 The Problem of Data Pooling	252
	15.11.2 The Need for Consistent Data Collection	253
	15.11.3 Randomized Clinical Trials (RCTs) versus Observational Cohorts	253
	15.11.4 Pooling Data to Improve Normal Tissue Toxicity Prediction Studies	254
15.12	Patient-Reported Outcomes as a Basis for Big Data Studies in Radiation Oncology	254
15.13	Modeling Methodology	255
	15.13.1 Data Splitting for Validation	255
	15.13.2 Dose–Volume Histogram Parameters	255
	15.13.3 Clinical Variables	255
	15.13.4 How Robust Are the Modeling Results?	256
	15.13.5 Things Were Simpler in the Past	256
	15.13.6 Spatial Dose Patterns for NTCP Modeling	256
	15.13.7 Image-Based Predictors for NTCP Models	256

15.13.8   The Untapped Value of Longitudinal Imaging for TCP Modeling   256
15.13.9   The Rise of the Machines   256
15.13.10   Some Common Ways to Stumble in TCP and NTCP Modeling   257
15.13.11   Some Rules of Thumb from Experience   257
15.14   Summary   257
Acknowledgments   258
References   258

15.1 INTRODUCTION

"There is a lot to be gained by a much better understanding of the responses of normal tissues (and tumors) to a whole range of dose–volume distributions."—Michael Goitein (2007)

15.2 IS RADIOTHERAPY OBSOLETE?

The question naturally arises: why work so hard at incremental improvements in radiotherapy? Is all this effort going to pay off? In fact, physicists, physicians, and other collaborators continue to develop innovations, large and small, that contribute to more precise radiotherapy delivery, whereas the literature continues to grow on dose response (Buettner et al. 2012; Lee and Fang 2013; Appelt et al. 2014; Benadjaoud et al. 2014; Cella et al. 2015; Christianen et al. 2016; Dean et al. 2016; Brodin et al. 2018). Arguably, radiotherapy in many cases should become the preferred first-line treatment for many patients with localized disease who would today be treated with surgery.

Unlike surgery and chemotherapy, including targeted drugs, radiotherapy delivers dose distributions that can be highly focused and sculpted into the desired patient-specific pattern. Although recent immunotherapy progress is astounding (Gubin et al. 2014; Patel and Kurzrock 2015), it is less effective against large tumors, possibly due to increased clonal heterogeneity as well as increasingly poor vasculature with increasing mass (Li et al. 2010). Radiotherapy, in theory, deals with large disease masses by simply increasing the dose, and is relatively less sensitive to clonal and microenvironmental heterogeneity, although these are still important factors. Even highly hypoxic cells can be reliably sterilized with large enough doses. Compared with surgery, radiotherapy has a much lower risk of death as a treatment side effect. For example, a recent pan-Canadian review determined in-hospital surgical mortality rates, averaged over 9 years, of 3.7% for pancreatic cancer, 5.6% for esophagectomy, 3% for liver resection, 2.3% for lung cancer, and 0.9% for ovarian cancer. Major acute complication rates were typically 20%–40% (Finley et al. 2016).

Similarly, a recently published comparison of 30-day mortality for early-stage lung cancer patients treated either with SBRT or with surgery showed a significantly higher 30- and 90-day mortality rate following surgery; this difference increased with patient age (Stokes et al. 2018). Even when disease has metastasized, radiation therapy (RT) can be used at reduced, less-toxic doses to achieve high rates of sterilization of any occult deposits.

From a financial view, many recent advances in cancer care, such as immunotherapy and CAR-T cell therapy, come with extremely high patient prices, creating "financial toxicity," despite insurance, especially in the United States. In contrast, *optimized radiotherapy*, although often needing some incremental extra effort to deploy, represents a way to improve outcomes, even if only incrementally, without dramatically increasing the cost of care.

15.3 WHY DO WE NEED MODELS FOR NORMAL TISSUE TOXICITY AND TUMOR CONTROL?

The central hypothesis of this chapter is that radiotherapy could be optimized for a significant fraction of the patient population by developing individualized dose–response curves that integrate dose distribution aspects, biological assays, imaging, and other patient-specific predictors. This has been the long-term dream of many pioneers in radiotherapy research, including Lionel Cohen, Jack Fowler, Anders Brahme, Michael Goitein, and many others.

Even accepting these arguments in favor of radiotherapy, state-of-the-art oncology already utilizes a deep knowledge of tumor and normal tissue biology. Molecular panels to grade breast cancer aggressiveness were introduced nearly 20 years ago and are now relatively mature (Ferté et al. 2010). A rich variety of tumor data is becoming increasingly available to help manage patients, including sequencing, microRNA expression profiles, RNA profile levels (RNAseq), circulating tumor cells (automatically recognized through surface antigens) and levels of extracellular circulating tumor DNA (Garcia-Murillas et al. 2017).

Other kinds of information besides the treatment plan are becoming more important. Recently, the rapidly expanding field of *radiomics*, which explores the tumor "*radiophenotype*," is supplying information relevant to radiotherapy outcomes, including the risk of metastatic progression (Coroller et al. 2015; Vallières et al. 2015, 2017). Hence, in the long run, radiomics should inform patient management and further enhance the value of a local radiotherapy that is able to achieve local control without surgery.

Due to disease and anatomic variability between patients, the prescription dose (and dose distribution coverage) to achieve a high probability of local control with an acceptably low risk of morbidity, is patient-specific. However, tumor control probability (TCP) and normal tissue complication probability (NTCP) models could be used to individually optimize treatment planning despite these variations. Specifically, potential uses for accurate TCP and NTCP models include: mathematical guidance (as an objective function or constraint) for intensity modulated radiotherapy (IMRT) or intensity modulated proton therapy (IMPT); better selection of patients for dose de-escalation, or selection of patients for dose intensification; and selection of optimal dose–fractionation schemes (more fractions if tumor reoxygenation is needed or normal tissue volume irradiated is inconsistent with hypofractionation).

Regarding the use of models to determine the best treatment technique, a recent Lancet review article (van Loon et al. 2012) on the evaluation of new technology in radiotherapy is particularly insightful on the potential use of predictive models: "…dose–distribution models are mainly useful to identify tumour types and locations that are most likely to benefit from the new technology and to provide an estimate of the predicted size of this benefit. Data from these models can provide helpful information about the design of future clinical trials, such as for calculation of sample size or definition of the research population."

A weakness of current TCP and NTCP models that include volume dependence is the lack of validated fractionation dependence. However, we are fortunately working in a "golden age" of progress in fractionation, with a rich variety of clinical trials ongoing or completed that can provide crucial data to underpin modeling (e.g., see the recently published meta-analysis on prostate fractionation results by Vogelius and Bentzen [2017]).

For some historical perspective, quantitative analysis of normal tissue effects in the clinic (QUANTEC), was in fact primarily a qualitative effort and was quantitative only in the sense that graphs and tables summarizing literature results were produced as input to multi-expert recommendations of dose–volume thresholds to avoid toxicity (Marks et al. 2010). Although this was a very successful step past a reliance on the under-supported tables of Emami et al. (1991), the QUANTEC group recognized that, for the vast majority of endpoints, 3-dimensional (3D) planning data in radiotherapy had not yet been effectively exploited or analyze (Jackson et al. 2010). As described in this chapter, we are now poised as a community to go far beyond the QUANTEC paradigm; some reviews of outcome data based on fully 3D data sets that have recently come online, with more in the pipeline.

15.4 THE DATA CRISIS IN RADIOTHERAPY, AND IN MEDICINE, MORE GENERALLY

Excitement about big data is well justified, given the explosion of information that can be applied to understanding outcomes. Grappling with big data means engaging in all aspects of the pipeline, as one review puts it: "Creating value from Big Data is a multistep process: Acquisition, information extraction and cleaning, data integration, modeling and analysis, and interpretation and deployment" (Jagadish et al. 2014). Furthermore, we should learn from the wide variety of patients we actually treat, not just those who fit clinical trial guidelines, which is easy to state, hard to implement in practice (Institute of Medicine and Roundtable on Evidence-Based Medicine 2007; Hsu et al. 2017). Many patients do not fit clinical trial guidelines due to comorbidities or other reasons. We therefore briefly discuss obstacles to learning from routine clinical data from a clinical informatics perspective.

It is widely accepted that patient records in institutional Electronic Health Records (EHRs), also known as Electronic Medical Records (EMRs), even at academic institutions, are typically not accurate enough to support academic publication or knowledge generation without the further step of curation (Birtwhistle and Williamson 2015; Ewing et al. 2015). Of course, this greatly drives up the cost of any knowledge generation. This, therefore, represents a key unaddressed problem with the use of clinical data. We have observed the following as a (nearly) general law of clinical informatics: *The accuracy of data entered by busy professionals varies inversely to the value personally placed on the data.* If data entry for that specific item is considered a "nuisance" (even subconsciously), the data is unlikely to be reliable. Consequently, better input processes are needed along with methods that identify/filter out data that are likely unreliable.

Filtering of medical data is an option, but the challenge will be to throw away a minimal amount of good data with the bad. Unfortunately, simple outlier detection filters on data input (although useful) are unlikely to solve the problem because the most interesting cases will have unexpected characteristics that may be outliers with regard to some statistical population criteria. It seems likely to us that machine learning methods will be needed to identify clusters of characteristics associated with "bad entries." Despite these obstacles, it seems likely that data entry reliability will improve considerably over the next decade based on real-time analyses of inputs and use feedback. We are therefore likely to be at the beginning of the era of successful efforts to develop predictive models from real-world data (i.e., the EHR/EMR).

As one example that pushes far beyond the current paradigm, the European Surveillance of Congenital Anomalies (EUROCAT) effort of the MAASTRO group relies on imaging and clinical data being collected locally. Local instances are (literally) networked globally (Deist et al. 2017). Software is installed locally that automatically gathers picture archiving and communication system (PACS) images and other information. This data is kept on local servers either to test and validate models developed elsewhere or to serve as anonymous data to guide optimization algorithms developing predictive models. For example, a predictive model for life-expectancy following a diagnosis of lung cancer could be constructed via an algorithm that computes a local gradient in the local repository (Jochems et al. 2016). This can be done without "unmasking" or extraction of the data itself, thereby preserving a relatively complete form of data protection. The EUROCAT project has already had significant success in recruiting partners. A more complete description can be found at http://www.eurocat.info.

15.5 TOWARD QUANTITATIVE HEALTH CARE: DOES YOUR IMAGING DATA TELL A MEANINGFUL STORY?

Unfortunately, the problem of usable data for real-time learning goes far beyond reliable data entry. A much larger issue is that a great deal of information is required to ensure that data (e.g., images, treatment records) could be reused at a later date. In the typical data-driven study, there are important facts about how the data were collected that are key to reliable analyses and conclusions (e.g., were images taken before or after shifts? or taken due to suspicion of the need for replanning? ... with what technical parameters?) We are here referring here to data use far beyond what is required for a given protocol. Today, data typically is usually stored *without* the context and linkage to other data so that the patient story—including the story of how and why the data was collected—cannot be authoritatively reviewed through the data records. Some knowledge from clinical personnel is usually needed to drive an analysis. Access to this knowledge decays over time. Effectively, data effectively "rots" due to the inability to access the context of the data (e.g., that postdoctoral fellow left, no one remembers why we took the data this way).

Building a data workflow and repository that preserve enough context and data linkage to establish the unambiguous "data story" is a difficult challenge in clinical informatics. Fortunately, this challenge matches up to the ongoing effort to better clinical data records in other facets of oncology and medicine more generally. One approach to increasing the value of "data story" effort would be to develop a health care–specific language that could be used to link together the processes of radiation oncology into a "story." Such a language could potentially be used to graphically represent the events of a patient's trajectory through the health care system. For example, object types could include ACTIONS (e.g., a CT scan for treatment planning, plan was approved by a physician), DECISIONS (e.g., "Patient setup was approved"),

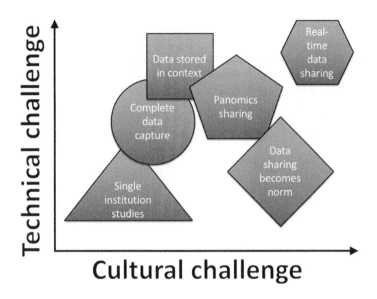

Figure 15.1 Achieving data collection and sharing of different types will require overcoming technical as well as cultural challenges. The word "cultural" here stands for resistance to change and required buy-in of stakeholders who may not initially understand the value proposition.

INFORMATION (e.g., "hemoglobin test result is xx.xx"), and MONITORING (e.g., "Warning: contours not approved"). A graphical interface with various icons could be developed that would link in real time to data sources of each object type. Such a system, or a similar idea, would be a welcome advance toward data archiving that preserves context, story, and linkage. One could imagine an interface using simple icons to indicate images, treatments, lab results, clinical decisions, and even monitoring processes, with links between them. Unfortunately, many of the low-level trigger events that are required to write an unambiguous "data story" are not currently available. One can see part of what will be required (including many new vendor notification interfaces) through the "event net" real-time framework being developed at the University of Michigan (Hadley et al. 2016).

Achieving a "quantitative oncology" infrastructure would routinely guarantee that a significant subset of data used for clinical decision making was accurate and complete enough to serve as input to scientific analyses, thereby supporting the vision of a "learning healthcare system" (Mayo et al. 2016).

Summarizing this vision of future informatics needs, Figure 15.1 schematically shows the relative difficulty involved in managing different types of data needed for outcomes analyses. In many cases, the key challenges are expected to be cultural and institutional rather than purely technical.

15.6 HOW CAN THE FIELD OF RADIOTHERAPY OVERCOME THE PROBLEM OF LOW MODEL GENERALIZABILITY?

The need for data sharing to understand toxicity is actually more fundamental in radiation oncology, a technologically detailed methodology, compared with medical oncology, where the typical question of generalizability is about the efficacy of a new drug, with a similar prescription, across multiple clinical cohorts. The most important method of increasing predictive model generalizability, rarely pursued, is to combine multiple data sets sharing the same endpoint, but with differing patient, treatment, and prescription characteristics. Often, relevant data sets are effectively locked away after the initial reports are published. One new initiative to increase data reuse, supported by the *Journal of Medical Physics*, is the "Medical Physics Dataset Article." This is a new type of article that describes a data set publicly archived by the authors, including details on size, completeness, and expected applications (Williamson et al. 2017).

15.7 WHAT DATA IS RELEVANT FOR PREDICTIVE MODEL DEVELOPMENT?

To give the necessary context, we briefly review TCP and NTCP modeling. This section is only meant to review some representative models, and in particular, the data required.

15.8 TUMOR CONTROL PROBABILITY MODELS

15.8.1 ISOEFFECT FRACTIONATION REGIMES

The "Fowler equation." Simple equations to guide overall prescriptions as a function of fractionation have been popularized by Jack Fowler (e.g., using the linear-quadratic-kickoff model), who synthesized and harmonized contributions from many investigators (Fowler 2006, 2010). The fundamental equation that Jack Fowler popularized over several decades can be used to estimate iso-effectiveness dose–fractionation combinations by estimating cell kill as though the treatment were given with a large number of very small doses over the same overall treatment time (the "biologically effective dose"). A time factor reduces treatment effectiveness after a "kickoff time," typically 21–24 days from the start of radiotherapy. This so-called "LQ+time" equation has been shown to usefully organize and rank the effectiveness of clinical regimes; it is perhaps the most simplified model possible that has any clinical credibility.

15.8.2 RADIOBIOLOGICAL ASSUMPTIONS IN MECHANISTIC TCP MODELS

Newer developments in TCP modeling have sought to place the empirical "Fowler equation" approach on a more solid mechanistic footing. Going up the scale of sophistication, models have been proposed that explicitly account for the known radiobiological principles of hypoxia, proliferation, or both together, as well as glucose consumption. We briefly review the main radiobiological effects known to impact outcome.

Extensive radiobiological experiments *in vitro* have established that hypoxic tumor cells are more resistant to radiation, by up to a factor (known as the oxygen enhancement ratio) of 3 (Hill 2017). However, there is evidence that this extra resistance is smaller for chronically hypoxic cells (Chan et al. 2008). Modeling results from Jeong et al. (2017) indicate a value of 1.5–1.8 for the oxygen enhancement ratio for hypoxic cells (hypoxic, but still receiving glucose) in early-stage lung tumors. Importantly, proliferation usually requires oxygen, and so hypoxic conditions normally suppress proliferation. The impact of hypoxia, being reduced in chronic situations, and effectively reducing proliferation, is not as dominant regarding tumor sterilization as once thought.

The clinical impact of hypoxia has been observed in limited, but compelling, clinical data that we summarize here (Overgaard 2007). Rischin et al. (2006) showed that patients treated for head and neck cancer had a better rate of local control using an anti-hypoxia compound (tirapazamine), but only when the tumor was positive on PET-F-misonidazole imaging. The Scandinavian inter-institutional trials (DAHANCA) consortium have demonstrated a clear, but modest, improvement in head and neck outcomes when patients received the anti-hypoxic cell compound nimorazole (Overgaard et al. 2005). Moreover, the ARCON trial results, using a regime that incorporates carbogen breathing (an oxygen enhanced gas) into an accelerated fractionation schedule, as well as an anti-hypoxia drug, demonstrated improved local control for head and neck tumors (Kaanders et al. 2002).

15.8.3 TUMOR RESPONSE–EVOLUTION MODELS

The recent work of Jeong et al. (2013) is an example of a more sophisticated tumor state-simulation model that attempts to capture a mechanistic picture of tumor evolution in response to a radiotherapy regime. The model assumes that each voxel of the tumor has a limited vascular capacity to deliver nutrients (glucose or oxygen). As radiotherapy continues over many days, cells with access to oxygen go through mitosis, frequently dying due to chromosome aberrations (a result of DNA damage). Dead cells are cleared from the tumor, resulting in tumor regression. Following this, previously hypoxic cells can take up available oxygen and glucose at each time step. Thus, over many days, the tumor reoxygenates and the growth fraction (fraction of tumor cells that are progressing through cell cycle) rises, though possibly not above a fixed level due to tumor vascularity limitations. Mathematically, this process is described as a "state machine" approach to

TCP modeling, whereby tumor cells in a given voxel of the tumor are divided into three states: proliferating (receiving oxygen and glucose), metabolically active but not proliferating (receiving glucose but hypoxic), and starving (hypoxic and lacking glucose). The key assumption is that the vascular network supplies each voxel of tissue with a given, conserved, level of chemical nutrient support (oxygen, glucose) to those tumor cells (over a course of radiotherapy). In reality, this vascular support varies between voxels and certainly on a more microscopic level. The software program simulates a string of 15-min. intervals over which tumor cells proliferate, attempt mitosis, are sterilized by radiotherapy, resort into different compartments, and ultimately reoxygenate. Initial values of these compartments are determined mathematically by the initial growth fraction and cell loss factors (or tumor doubling time).

Results of the Jeong model typically correlate well with the results of the Fowler equation (unpublished); although the Jeong model results in different radiobiological parameters, it is calibrated into an absolute probability prediction and eliminates the need for a discontinuous "kickoff time." The Jeong model was used to fit local control/TCP results for early-stage lung cancer, relying on more than 36 patient cohorts representing more than 3000 patients, with only four fitting parameters. The fit produced radiobiological parameters with reasonable values, and an excellent fit could be obtained from single fraction cohorts to standard fractionation results (Jeong et al. 2017). At the very least, this shows that treatment for some patients can be fine-tuned to seek a near optimal dose–fractionation scheme, that neither overtreats nor undertreats.

Despite this success, later stage/larger volume disease is almost certainly a much greater challenge, due to tumor heterogeneity, which manifests itself in terms of vascular and possibly clonal and epigenetic spatial heterogeneity. Fortunately, imaging could be used to better understand variability in local/voxel conditions. Any model that seeks to project the response of a large lung tumor in detail must introduce varying initial conditions in different parts of the tumor, presumably from imaging inputs (Titz and Jeraj 2008; Crispin-Ortuzar et al. 2017).

Another interesting approach has recently been published (Prokopiou et al. 2015) that focuses on proliferating cells and ignores the nonproliferating and severely hypoxic compartments. How much complexity is actually required in such models? It will be of great interest to see how models of different complexity can be used to guide treatment and predict outcomes. We next briefly review other biological data that can be incorporated to better predict tumor control.

15.8.4 RADIOBIOLOGICAL ASSAYS

Although efforts to use biological assays as a basis for predictions or prognosis have to date mostly been unfruitful, there are a few examples of success, with more likely in the future.

The observation that different tumors—even of similar size and histopathology type—respond variably, has long been a focus of investigations (Fertil and Malaise 1985). In the late 1990s and early 2000s, it was widely believed that the key to understanding tumor response was the variability in two factors: (a) tumor radiosensitivity, assayed as the intrinsic probability of a cell retaining proliferative capacity, such as the surviving fraction of biopsied cells after receiving 2 Gy *in vitro* (West et al. 1993), and (b) hypoxia, which was known to be nonuniform in the cellular microenvironment, assayed using invasive probes (Fyles et al. 1998; Nahum et al. 2003). However, assays of the proliferative fraction of cells or the surviving fraction at 2 Gy *in vitro*, although correlating with outcome in some studies, were not found to be reproducible. Intra-tumoral radiosensitivity variation showed extremely wide tumor-to-tumor variabilities (coefficients of variation of typically 40%) (Taghian et al. 1992), which were irreconcilable with the steepness of commonly observed dose–response curves. This indicated that such biopsies could not be taken as accurately correlating with dose required for control (Deasy 1998). Hypoxia sampling is invasive and was not deemed clinically feasible in a routine setting.

One remarkably successful assay in measuring the patient-specific variation in tumor radiosensitivity is the seminal work by Torres-Rocca, Esrich, et al., who applied an innovative framework to develop a radiosensitivity predictive index (Eschrich et al. 2009). Utilizing *in vitro* radiosensitivity cell line data sets, they derived predictors of radiosensitivity, composed of linear sums of mRNA protein expression. Arguing from system biology principles, this formulation was later simplified to focus on hub-genes (whose expression integrates signals from many other gene/protein pairs). This assay signature has since been validated on several data sets across cancer types, including prostate, breast, pancreatic, and brain cancer (Strom et al. 2015;

Torres-Roca et al. 2014, 2015; Ahmed et al. 2015). Importantly, it was demonstrated that this radiosensitivity index (RSI) impacted outcomes for patients receiving radiotherapy, but not for patients undergoing surgery. RSI, therefore, is specific to radiation response, as hypothesized, and is not a general marker of biological aggressiveness. These are impressive results, demonstrating that there are predictors of radiosensitivity that are potentially useful as clinical assays. Despite this, the RSI has not come into widespread clinical use.

One hurdle is that such an assay needs to be incorporated as just one component of an estimated dose–response curve, an issue that is being addressed by the Moffitt group (Scott et al. 2017). Without knowing where the proposed treatment sits on the local control versus dose curve, the optimal treatment cannot be selected. If RSI, or something like it, eventually gets integrated into predictive dose–response models, the long-term prospects for clinical use seem positive. A second hurdle is simply the effort required to get good biopsy material and to perform the needed microarray analysis. Nonetheless, Torres-Rocca and colleagues have established the possibility that such a predictor exists. It seems likely that even better predictive indices may be possible using sequencing methods (e.g., RNAseq) that give a deeper view of tumor genomic defects as well as RNA expression levels.

15.8.5 IMAGING TO REFINE TUMOR CONTROL MODELS

Image-based metrics that likely impact on tumor radiocurability include Fluorodeoxyglucose (FDG)-PET standard uptake value (SUV) maximum values and related histogram parameters; hypoxia imaging; and MRI-derived parameters, including apparent diffusion coefficient (ADC) estimates as well as dynamic contrast perfusion related markers (Mayr et al. 2000, 2009; Huang et al. 2014). There are robust published results showing consistently that tumors with higher SUV-max values have a higher risk of local failure. Published data for head and neck cancer was synthesized by Jeong et al. (2014) using a graphical method that is equivalent to the random effects statistical methodology. They showed that the average difference in dose needed to equalize local control rates, for "FDG hot" tumors (higher than mean SUV-max) compared with "FDG cold" tumors (lower than mean SUV-max) is about 20% extra dose. This is a population averaged value, so the result for any given tumor may vary by a larger or smaller amount. Motivated by image-based analyses that show local failure is most often in the region of high SUV (Soto et al. 2008; Calais et al. 2015; Trani et al. 2015), several clinical trials have been started with the goal of establishing standard nonuniform dose escalations to high FDG-PET regions.

15.8.6 PET MEASUREMENTS OF HYPOXIA

Following the introduction of PET agents for hypoxia imaging—there are now several accepted hypoxia imaging compounds (Wack et al. 2015), although cu-ATSM has essentially been deprecated (Carlin et al. 2014; Colombié et al. 2015)—several papers and conceptual notes appeared emphasizing the opportunity to paint dose levels appropriate to different image characteristics, an approach sometimes called "dose painting" (Ling et al. 2000).

From this discussion, it should be apparent (although opinions vary) that modern quantitative imaging techniques can provide information that is highly relevant to local radiotherapy response, although we are at the beginning of learning how to integrate the information into a highly predictive model. One approach is to assume that machine learning on image features or dose distributions, from the beginning, will provide the final answer (Naqa et al. 2010). Our favored approach is to start with established scientific principles and to add information, based on the concept that science is more likely to advance if the relationship with established principles remains clear. Data driven information should be added. It is certainly possible, though unlikely in our view, that completely data driven/machine learning/radiomics results will dominate predictive models without a clear bridge to understood principles.

15.9 NORMAL TISSUE COMPLICATION PROBABILITY MODELS: A BRIEF OVERVIEW

The connection between normal tissue tolerance variation as a function of fractionation versus dose–volume factors is poorly understood. Fractionation and dose–volume factors remain almost scientifically "orthogonal" (to use a characterization I owe to a private conversation with Eli Glatstein). This remains an important area of investigation.

Regarding fractionation, in parallel to tumor response, Fowler, Thames, Withers, and others considered the same framework for normal tissue tolerance (Thames et al. 1982; Fowler 2010). Some common assumptions that have been commonly taught are questionable. It has been assumed that acutely responding normal tissues have a reduced alpha/beta parameter, based on data that considers the entire delivered dose. The disconnect is that the complication (say, acute esophagitis) typically comes along within a few weeks of the beginning of radiotherapy, well before the end of treatment. It only stands to reason that acute toxicity *must* have a significant dependence on the rate at which dose is accumulated in the weeks before enough cells die to manifest the initial symptoms. New analysis should consider this aspect. Late effects, in contrast, are not expected to have this entanglement and may be treated in a more straightforward manner regarding fractionation effects. The relevant alpha/beta parameter has been determined for many different endpoints to demonstrate significant repair, frequently being 2–5 per Gy (Joiner and van der Kogel 2016). It has often been concluded that tumor alpha/beta values are higher than normal tissue values, for radiobiological reasons that are not clear. The modeling work by Jeong et al., yielding relatively low best-fit tumor alpha/beta values, shows that this disconnect seems to come from not accounting for proliferation and reoxygenation in a realistic way. Notwithstanding these caveats, a useful summary of linear quadratic modeling results has been given by Chapman and Gillespie (2012).

Quantitative clarity regarding dose–volume effects has been much harder to achieve, partly because preclinical guidance is minimal. The QUANTEC effort has already been noted above. Although early efforts to model NTCP assumed that analytic functions of a given form would be appropriate tools, experience shows that clinical predictors and multiple dose–volume factors often need to be analyzed in the same model (Appelt et al. 2014). We will further discuss methodological issues below.

15.9.1 ASSAYS TO PREDICT NORMAL TISSUE RESPONSE. WE BRIEFLY NOTE SOME INTERESTING RESULTS

Focusing on normal tissue toxicity, van Oorschot et al. (2014) performed *ex vivo* radiation (1 Gy) of blood lymphocytes from prostate cancer patients. DNA damage repair was quantified via gamma-H2AX foci. They found that the group of patients who experienced less treatment toxicity demonstrated a greater propensity to repair DNA damage, as measured by a lower average ratio of gamma-H2AX foci at 24 h compared with 2 h posttreatment. In a similar study, Beaton et al. (2013) performed *ex vivo* irradiation on blood lymphocytes (6 Gy) for 10 prostate cancer patients who experienced toxicity and for 20 matched controls. Measurements of the resulting residual chromosome aberrations, such as dicentric chromosomes and excess fragments per cell, where measured. Chromosome aberration levels were much higher in the patient cohort that experienced toxicity. Despite these results, an inherent drawback of methods that use *ex vivo* irradiation is the expensive and time-consuming nature of the assay. This is perhaps reflected in the overall small number of such studies, and the relatively low patient numbers.

15.9.2 GENOMEWIDE ASSOCIATION STUDIES

Radiogenomics. Across medicine, there have now been many studies that seek to better understand intersubject variations of health states based on common variations in the genetic code. Such studies are commonly referred to as genomewide association studies (GWAS). In radiotherapy, this strategy applied to normal tissue effects is called "radiogenomics." GWAS studies to date have seen limited success (Wijmenga and Zhernakova 2018). In our view, this is likely due to modeling approaches that are too simple, often assuming that genetic markers (single nucleotide polymorphisms, or SNPs) individually contribute large and independent risks. In contrast, Oh et al. (2017) used nonlinear machine learning and statistical modeling methods that naturally accounted for complicated multi-SNP dependencies (Lee et al. 2018). In our view, this approach is likely to continue to be successful. The approach is general: It can be applied to any normal tissue reaction endpoint as well as a general range of medical endpoints. Unlike for *in vitro* lymphocyte assays, GWAS measurements are easy to make, via mouth swabs or blood samples, and analysis costs are now quite low, often less than $100 per subject.

15.10 METHODOLOGICAL ISSUES

15.10.1 BIAS AND CONFOUNDING FACTORS IN DETERMINATIONS OF TCP AND NTCP MODELS

Bias refers to any systematic deviation in any part of the research process that results in error in the research conclusions. This is usually distinct from recognized uncertainty in the conclusions. Applied to modeling, a highly biased result will not be generalizable to another population. The term "confounding effects," or *factors*, refers literally to a confusion regarding what is being observed and measured (Rawlins 2011). Variability in the factors that impact outcome often include performance status, the patient's general sense of well-being, comorbidities and patient chronic conditions, age, sex, smoking (e.g., New York City has low smoking rates compared with most of the United States), biological status of disease (e.g., the fraction of patients with HPV-driven tumors), and so on. Given the number and importance of such factors, it is not surprising that many published prediction models do not generalize between institutions.

15.10.2 PREDICTIVE MODEL DEVELOPMENT

It is a truism that it takes many years to bring a new development to the clinic. There are currently very few predictive models that are used in daily radiotherapy clinical practice (although the biologically effective dose/linear-quadratic [BED/LQ] model is one exception). A debate was recently published on the applicability of models resulting in so-called "biologically guided radiotherapy planning" (Deasy et al. 2015). Obstacles to the use of such models include, in roughly decreasing order of importance: (1) lack of established, standard protocols based on published results, for incorporating models, (2) lack of validated predictive models for key clinical endpoints, and (3) cultural and regulatory friction against adopting models.

Even if predictive models are available, it is currently unclear how they should be used. There have been previous proposals to customize target dose based limiting predicted normal tissue complication probabilities (Spalding et al. 2007; Fenwick et al. 2009), but such approaches remains rare. Today, establishing the standard use of such models effectively entails embracing a somewhat nonstandard treatment philosophy based on individual plan customization. Although there is a tradition of customizing radiotherapy plans in terms of dose distribution, this does not extend to the physician's prescription, which is typically standard for a given set of classifying clinical variables such as site, stage, and histopathology. One groundbreaking example of how models could be used effectively was published by a cooperative group in the Netherlands: models will be used to determine the appropriateness of the use of proton therapy versus photon therapy, via plan comparisons (Cheng et al. 2016). In the long term, it seems likely that full "radiobiologically optimized" treatment planning will become common, by driving the plan optimization using outcome models (Uzan et al. 2016; Christianen et al. 2016).

The frequent lack of validated models is gradually being ameliorated as the publication of new models based on significant data sets steadily continues. Lambin et al. (2013) recently published a very useful overview of clinical decision support systems in radiation oncology, covering many aspects of model development and validation, as well as various areas of potential clinical impact.

15.10.3 DATA HANDLING WITH CERR

Our group (led by Aditya Apte) maintains the open-source CERR software system, now renamed as the Computational Environment for Radiological Research, to emphasize increasing generality. CERR has literally hundreds graphical and batch numerical processing tools, as well as broad multimodality import capabilities useful for outcomes analyses. Details of CERR can be found on the wiki page hosted by github, at https://github.com/cerr/CERR/wiki. CERR contains tools that streamline review, structure identification, and data extraction for further analysis. CERR is an evolving system, with new features added each year. A wiki describes the different CERR subsystems and therefore a detailed list will not be given here. Newer components include a comprehensive radiomics toolbox (which has been carefully compared with other open-source toolboxes, such as pyradiomics), and greatly improved contouring tools. Figure 15.2 shows an example of a color wash display of several radiomics features. There are many other relevant open-source informatics tools for radiotherapy research. We maintain a webpage with descriptions and links at http://opensource4rt.info.

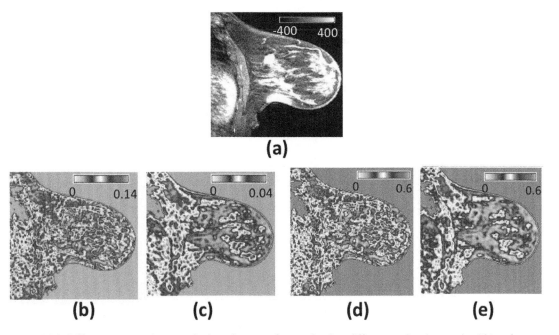

Figure 15.2 Different approaches to calculate the same feature lead to different radiomics results. This, often ignored, aspect is critical for validating radiomics signatures across institutions. (a) T1 post contrast image from a breast cancer patient. (b) Local gray level co-occurrence matrix (GLCM) homogeneity averaged across 2D directional offsets. (c) Local GLCM homogeneity averaged across 3D directional offsets. (d) Local GLCM homogeneity computed by accumulating co-occurrence frequencies from 2D directional offsets into a single co-occurrence matrix. (e) Local GLCM homogeneity computed by accumulating co-occurrence frequencies from 3D directional offsets into a single co-occurrence matrix.

15.10.4 A TOOL TO ENVISION DOSE–RESPONSE CURVES BASED ON PATIENT SPECIFIC TREATMENT PLANS

At our own clinic, we are working toward introducing predictive models for lung, head and neck treatments, and possibly other sites. This requires a useful interface for understanding how prescription changes could impact complication or local control probabilities. Figure 15.3 shows the new Radiotherapy Outcome Estimator (ROE) that we have built as a part of CERR. The concept is to give physicians, physicists, and others a more realistic view of just where the treatment plan sits on the dose–response curves, with simple tools for testing the impact of increasing or decreasing dose or numbers of fractions. ROE is integrated into CERR and is available in the same download. We believe that physicians are more likely to make rational treatment decisions if they can see where they are at on the outcome probability curves, compared with the current lack of such information. ROE could be used to decide between two alternative fractionation schemes, for example. Of course, the value of seeing outcome curves varies for different sites and increases especially for sites with variability of disease location and shape. This makes localized lung cancer a particularly attractive site for this approach, but other treatment indications may still benefit from knowing the rough absolute risk of life-changing complications. Sophisticated models are easy to add via text files (JSON format.) ROE is available as a component of CERR (https://github.com/cerr).

15.11 DATA COHORTS FOR PREDICTIVE MODELING

van Loon et al. (2012) reviewed key requirements for building prospective observational cohorts to evaluate the impact of novel radiotherapy techniques. Planning technique data should include numbers of beams, standard beam arrangements, total dose, and fractionation. Patient cohort baseline characteristics should include age, World Health Organization (WHO) performance score (or equivalent), sex, comorbidity,

Big data in radiation oncology

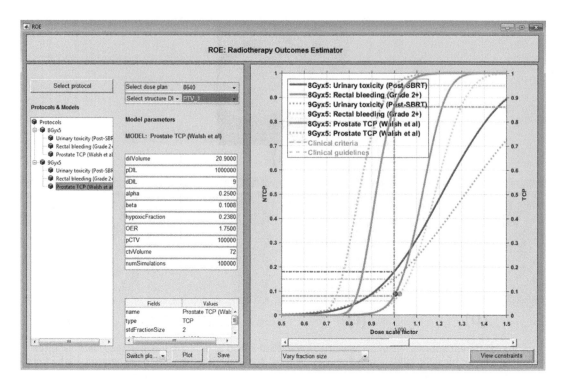

Figure 15.3 The radiotherapy outcomes estimator (ROE), a component of CERR. We developed this open-source tool to demonstrate the application of NTCP and TCP models for given treatment plans, to inform clinical decision making. Uses include selection dose–fractionation regimens. Models are stored in ASCII format as JSON files. The TCP model is for an actual prostate plan. (From Walsh, S. et al., *Med. Phys.*, 43, 734–47, 2016.)

socioeconomic status, quality of life, stage, histological features, and tumor location. Other factors that may confound the analysis should be collected, including: EQD2 (the biologically equivalent dose referred to 2 Gy/weekday fractions), the overall treatment time, and any other treatment the patient received. More generally, the relative dates and fraction doses for all treatments should be stored for further analyses.

Dose distribution specific data should, of course, also be collected, including dose–volume histograms (DVHs) of the relevant organ/tissue regions of interest (ROIs). Reconciling ROI naming by adopting common conventions is critical for data pooling. Recommendations in this area have been made by the AAPM Taskgroup report 263 (Mayo et al. 2017). More usefully, if the entire DICOM-RT file can be exported and converted into CERR format, many graphical and batch analyses can be conveniently performed. CERR provides batch extraction tools that can help to navigate variations in ROI names. Analyses of tumor response including larger tumors will often require a joint analysis with CT and PET imaging.

15.11.1 THE PROBLEM OF DATA POOLING

Pooling data into the same analysis raises special issues. Most importantly, were the endpoints measured or evaluated in a similar and consistent manner across the cohorts? Each contributing patient cohort must itself have significant statistical power to at least understand whether it is comparable to the other cohorts. It has often been found that one or more cohorts simply do not behave like the others, sometimes for unknown reasons (compare Michalski et al. 2010). Do all cohorts have similar rates of complication rates given similar risk factors? On the other hand, variability in treatment factors may be a key reason to pool data. Too much consistency in prescription doses may lead to conclusions that simply do not apply if the prescription or fractionation is changed significantly. In practice, some of these issues are often present. The question then is whether the flaws critically damage the value of the analysis. In any case, transparency in discussing these issues is an absolute requirement.

15.11.2 THE NEED FOR CONSISTENT DATA COLLECTION

Prospective collection of relevant data is the desired method, yielding data that is with a common, consistent understanding of variable definitions. There are, in fact, many data sets that were collected prospectively, and previously analyzed for publication, which could be used to probe new questions, leading to so-called "prospective–retrospective" category of analyses. Completely retrospective data analyses that go back to the original clinical record, rather than research databases, are prone to bias, inconsistency, and incompleteness and are seldom worth the trouble. A more controversial issue is the comparison between cohorts gathered from randomized clinical trials versus nonrandomized observational cohorts.

15.11.3 RANDOMIZED CLINICAL TRIALS (RCTs) VERSUS OBSERVATIONAL COHORTS

There are several crucial differences between randomized clinical trial cohorts and observational cohort studies (Rawlins 2008). Observational cohorts can be of several different kinds: historical controls (new series after a change in treatment, compared with a previous cohort), concurrent cohorts (nonrandomized selection of patients for new treatment versus concurrently accrued comparison cohort), case–control studies (impact of intervention in patients with condition/reaction compared with intervention in patients without condition/reaction), or case series from registries (usually defined by broad criteria and a specific time window).

Case–control studies are often further organized such that cases and controls are *matched*, referring to selection of patients with similar factors that are known to confound the impact of the intervention on the endpoint (e.g., survival). For example, patients could be matched on age, sex, smoking, and comorbidity measures.

There is an inherent tradeoff between RCTs and observational cohorts in terms of validity (here meaning that the data are complete and accurate) and representativeness (here meaning a high proportion of typical patients are considered well represented in the data set). Randomized clinical trials (RCTs) reduce variance in confounding factors, supporting precise statistical measures, whereas observational cohorts can potentially better represent typical populations. True generalizability is nearly impossible if cohorts are limited to RCTs (Rawlins 2008). Nonetheless, the use of observational studies compared with RCT data is controversial. Evidence hierarchies have been proposed that install RCT data at the pinnacle. In particular, RCT weaknesses include

- Strict entrance criteria (in age, stage, and comorbidity status) that limit generalizability of the results.
- Effects that are dependent on technical delivery may not be accounted for, especially in a highly technical, process dependent discipline such as radiation therapy.
- To perform an RCT that accurately identifies a potential change in a late endpoint, which is typically small (e.g., 10% increase in median overall survival), several hundred or even thousands of patients may need to be accrued.
- Time from conception to reporting is often many years for RCTs.
- RCTs are only ethically feasible when there is a true uncertainty or 'equipoise' concerning the question being asked. In such cases, the difference in effects is often expected to be relatively small.

Because radiation oncology advances are often driven by technological improvements (e.g., in-treatment-room cone-beam CT imaging with reduced artifacts), many changes are made to treatments that are not subject to randomized clinical trials. In particular, "drop-in" technical advances that merely improve the performance characteristics of systems do not seem to be attractive candidates for RCTs. Counter examples may exist, however, such as when volumetric CT scanning was introduced, more abnormalities were identified by radiologists of uncertain health importance, but which resulted in many unnecessary follow-up medical procedures.

It has also been argued extensively whether RCTs or observational studies give the more accurate estimate of interventional treatment effects (Benson and Hartz 2000). It has been argued, based on an extensive survey of published comparisons, that when entrance criteria were similar, treatment effect sizes estimated from observational studies are often of a similar size as those derived from RCTs (Concato et al. 2000). Leaving aside the question of superiority of one data source over another, it seems indisputable that a tremendous amount of data is being accrued but not utilized to understand and estimate dose–response curves more thoroughly. Thus, despite the problems of observational cohorts, we agree with the recent *New England Journal of Medicine* editorial from FDA personnel that such data can provide crucial "real-world evidence" as to the actual efficacy and dangers of an intervention, as seen in typical clinics (Resnic and Matheny 2018). Not surprisingly, we are strong advocates of the use of observational cohorts, assuming proper data quality assurance.

As a caveat, we note that, some effects are nearly impossible to authoritatively analyze in combined observational cohorts. For example, we contributed to a paper comparing observational cohort outcomes for prostate patients given radiotherapy versus those treated with surgery alone. Survival was clearly worse for the radiotherapy patients (Kibel et al. 2012). However, radiotherapy patients had worse morbidity scores and higher grade disease on average. Moreover, it was impossible to completely correct for the well-known selection bias that surgeons routinely invoke to reduce the number of patients with post-surgical morbidity or even mortality. In other words, choice of treatment (e.g., RT) was likely associated with a patient with a worse prognosis walking in the door, regardless of treatment.

Inconsistencies in toxicity data reported in the literature are common. In particular, dose–volume thresholds of toxicity risk taken from the literature often vary enormously, as the QUANTEC reports showed in detail. This is due to various reasons, including variations in patient cohort characteristics (e.g., older, sicker cohort rejected for surgery vs. younger), variations in treatment techniques (e.g., 3-dimensional conformal radiation therapy (3D-CRT) vs. image-guided radiation therapy (IGRT), or electronic portal imaging device (EPID) setup vs. volumetric daily IGRT), variations on contouring practice, including the image source of contouring (CT vs. MRI or PET, 4D breathing-cycle CT vs. snapshot CT), variations in prescription dose, and differing analysis methods (Jackson et al. 2010).

15.11.4 POOLING DATA TO IMPROVE NORMAL TISSUE TOXICITY PREDICTION STUDIES

As an example, Thor et al. (2018) recently published a pooled analysis of more than 1000 patients from seven institutions treated with a variety of prescriptions and treatment techniques. Briefly, it was found that the best predictive model for 3D-CRT treatments could not be generalized to IMRT (as many have suspected) and that key dose–volume predictors included a (soft) threshold dose of 55 Gy and, surprisingly, the need to spare some of the rectum even from low doses, was instantiated in the model by the minimum dose.

Importantly, there is a "mirror" effect of tolerance doses derived from clinical studies that reflects prescription doses. It is very common that, if the prescription dose is X, then the most predictive dose–volume threshold of the most critical/radiosensitive structure will also be X, or something close to X. Data pooling is the key technique for reducing this effect, as shown by the results of Thor et al.

15.12 PATIENT-REPORTED OUTCOMES AS A BASIS FOR BIG DATA STUDIES IN RADIATION ONCOLOGY

No discussion of modeling outcomes is complete without emphasizing that patient-reported outcomes measures (PROMs) are a powerful yet underutilized resource for improving our understanding of treatment toxicity (Chera et al. 2014; Atherton et al. 2015; Niska et al. 2017). The power of PROMs derives from their ability to more fully describe the actual symptoms experienced by a patient. Although the use of PROMs has been controversial, and adoption of PROMS has been a slow process, it must be emphasized that it has been shown in multiple studies that providers underestimate the rates of cancer treatment toxicity as well as the level of severity compared with patient-reported outcomes. Fortunately, the tide of physician resistance to PROMs has abated, and it is now common to see PROMS vigorously endorsed. Most so-called objective grading schemes, in contrast, are provider-focused rather than patient-focused. It cannot be emphasized enough that underlying dose–volume treatment characteristics may be related to some atomized symptoms but not others that are usually lumped into the same symptom grade.

In radiotherapy, PROMs can provide key data that is unavailable in any other form. Although it is highly desirable to use the exact same PROM questions in any cohorts to be combined, it is feasible to combine answers from different PROMS. Such questionnaires can be given during radiotherapy, longitudinally over a longer time frame, or even cross-sectionally (i.e., mailed to a cross section of patients treated at different times.) For example, Thor et al. (2015) recently showed how two different cross-sectional PROM data sets, gathered at different time points following RT, could be analyzed together, using the statistical technique of factor analysis. Factor analysis can be used to define underlying latent factors that relate observed outcomes to treatment characteristics (Oh et al. 2016). The hypothesis is that clusters of atomized questions in PROMs can be identified that share an underlying causal. Key advantages of the cross-sectional approach include the ability to ask detailed, atomized questions regarding individual symptoms, as well as the ability to assemble large cohorts of data, and the overall low cost. Key drawbacks include a lack of baseline, or any other longitudinal data points for comparison. We believe that PROMS will play an increasingly important role in understanding radiotherapy outcomes.

15.13 MODELING METHODOLOGY

We briefly summarize standard and emerging methodologies for TCP and NTCP modeling. There are many pitfalls to predictive modeling, and many published results in medicine do not withstand independent validation (Ioannidis and Panagiotou 2011). To improve the methodology of predictive models, Collins et al. (2015) published modeling guidelines titled "Transparent Reporting of a multivariable prediction model for Individual Prognosis Or Diagnosis (TRIPOD): The TRIPOD Statement." They put forward a checklist of 22 items that are important to ensure high quality. In fact, they recommend that a statement of adherence to TRIPOD standards be included in predictive modeling paper submissions. See also Munafò et al. (2017). For radiotherapy predictive modeling, we emphasize the following points.

15.13.1 DATA SPLITTING FOR VALIDATION

Data should be split between training and validation: a typical split is 70%/30%. This serves two important purposes: keeping the researchers honest, and it helps to convince the readership that the results deserve the claimed validity. Data sets that are too small to do this should probably not be used to train models, but should rather be used to test a very small number of predefined models or hypotheses.

15.13.2 DOSE–VOLUME HISTOGRAM PARAMETERS

The morbidity-driving effects of radiotherapy, as summarized by DVH statistics, are typically captured in a few treatment-related parameters: fractional volume receiving higher than x Gy (Vx), minimum dose to the hottest x% (Dx). A less used, but still useful parameter is the mean of the hottest or coldest x% (MOHx/MOCx). Vx parameters can be tracked by selected MOH or MOC parameters (Clark et al. 2008). Despite the fact that we have on occasion used Vx variables in modeling outcomes, we strongly recommend that more robust metrics be used instead. Here robustness refers to the fact that Vx (e.g., V55Gy) seems to contain a "magic" threshold, and if treatments are kept below this threshold, damage will be minimized. In fact, NTCP modeling is not that precise. A much better approach is to use metrics such as Dx or MOHx/MOCx or even possibly combinations of generalized equivalent uniform dose (gEUD) functions that can be chosen to be closely related to DVH parameters (which have the added value of being convex, thereby aiding optimization [Xie 2014]). Most cumulative histograms are estimated based on pretreatment planning CT images. Although this usually does not limit the accuracy of the final outcome model, there are a few categories where intra-treatment changes may be critical. Tumor shrinkage clearly changes delivered dose values in some sites, such as the lung.

15.13.3 CLINICAL VARIABLES

Standard clinical variables (e.g., age, sex, smoking status), as well as any known predictors from the literature, should be included. Cox proportional hazards modeling should be used if there is a long-tailed distribution of event times. If the presence or absence of the event is reliably tallied within the study follow-up time frame, then logistic modeling is perhaps justified. Competing outcomes (risks) must always be considered if they preclude measurement of the endpoint. We often use the least absolute shrinkage operator (LASSO), a modeling process that effectively penalizes sets of counter-balancing terms (that may have weights being either positive or negative), and results in relatively low-dimensional models (Efron and Hastie 2016). Forward step variable selection is less reliable when data size is an issue or the expected model is large.

Big data in radiation oncology

15.13.4 HOW ROBUST ARE THE MODELING RESULTS?

The results of the overall fitting process should be carefully tested on bootstrap resamples of the data. This will give a good sense of the stability and uniqueness of the results. We carefully consider any models that appear frequently in the bootstrap results (say, more than 10% of the time.) We often test interaction terms between frequently selected variables on a second pass of modeling. Another useful cross validation method during curve fitting is the 90/10 method, where 10% of the data are held out, repeated for each 10% in turn. This can be done with shuffling, and repeated many times so that the held-out samples show no bias. Compared with the bootstrap, this approach supplies a bigger slice of data for modeling but has the drawback that the models chosen across validation samples are more alike than they would be for bootstrap samples or for a completely new set of data.

15.13.5 THINGS WERE SIMPLER IN THE PAST

Although simplified modeling of outcomes in the form of using the gEUD function within the Lyman-Kutcher-Burman framework has a long history (Marks et al. 2010), we no longer trust the basic assumption that dose–volume factors can be modeled in this simplistic fashion. In particular, tissue sparing is often important as well as the properties of the hottest dose region. A more agnostic data mining approach is often more predictive, and more insightful.

15.13.6 SPATIAL DOSE PATTERNS FOR NTCP MODELING

Beyond modeling based on the DVH, spatial dose patterns can be predictive. There have been several provocative studies showing that the actual pattern (shape) of dose across a normal structure can be predictive (e.g., Dréan et al. 2016). The methodology often involves mapping of dose distributions to a single reference anatomical coordinate system. The list of patient–doses mapped to a given voxel is then statistically correlated with the outcome in question. This can be particularly useful when the identity of the radiobiologically critical structure is still uncertain. Endpoints that have been studied in this way include rectal bleeding (Dréan et al. 2016), trismus (Beasley et al. 2017), and lung or heart toxicity following lung cancer or lymphoma radiotherapy (Palma et al. 2016). Statistical issues of multiple comparison must be carefully considered in this situation, and it is important to ascertain that highly correlated dose parameters are not simply surrogates for treatment intensity.

15.13.7 IMAGE-BASED PREDICTORS FOR NTCP MODELS

An emerging area of interest is to use image-based predictors, in particular to characterize the susceptibility of a given tissue or organ to manifest damage for a given level of radiotherapy dose. For example, van Dijk et al. (2018) demonstrated that the P90 of the FDG-PET histogram of values for the parotid gland is a significant predictor of xerostomia risk. It seems likely that radiomic predictors will emerge that better capture patient-to-patient tolerance to radiation, especially for organs with known internal heterogeneity, such as kidneys, liver, pancreas, lungs, or brain. Although it is beyond the scope of this chapter, most imaging modalities beyond helical CT require some sort of standardized calibration process to ensure a useful level of generalizability.

15.13.8 THE UNTAPPED VALUE OF LONGITUDINAL IMAGING FOR TCP MODELING

Longitudinal imaging and image changes will undoubtedly offer more clues as to how to interpret tumor heterogeneity. There is no reason for a tumor subvolume that is well vascularized to have the same response to radiotherapy as a poorly vascularized region. Local tumor shrinkage/growth is of interest, as is its correlation with FDG-PET imaging and changes in radiomics features ("delta-radiomics," to use the MAASTRO group's term.)

15.13.9 THE RISE OF THE MACHINES

As more data of a greater variety becomes available for prediction, it is perhaps inevitable that machine learning approaches come to the fore. However, this should only be done alongside more comprehensible (and, hopefully, mechanistic) approaches, that also continue to develop.

15.13.10 SOME COMMON WAYS TO STUMBLE IN TCP AND NTCP MODELING

To summarize modeling pitfalls, we propose "Seven Deadly Sins" of predictive modeling:
- Standard variables not included (e.g., age, stage)
- Cross-validation was not used to characterize the training process
- No validation data was set aside or secured
- Feature extraction was not included in the cross-validation loop (resulting in over confidence)
- Robustness and reproducibility of extracted data were not considered
- No probe of model variability to establish variable importance or redundancy (e.g., using bootstrap)
- Results not compared with previously published models

15.13.11 SOME RULES OF THUMB FROM EXPERIENCE

Given the universality of data limitations in medicine, we make the following suggestions for TCP and NTCP modeling, based on our own experience (note: we define "informative-size" as the smaller number of patients of the two classes in a data set with dichotomized endpoint):
- For small data sets (informative-size less than 20 responders), testing a previously established model may be the only reasonable approach. A related option may be to introduce only a single parameter into the previously published model (e.g., the 50% response dose), and give confidence intervals.
- For large-by-medical standards data sets (with informative-size more than 100), a machine learning or high dimensional modeling approach might be useful, with careful cross-validation and set-aside validation.
- For mid-sized data sets (20< informative-size <100), a judicious use of multivariate modeling using a modest number of input variables (less than 30) could yield useful results. All steps of the modeling process must be contained within the cross-validation loop, including any initial variable pruning/ filtering steps.

We also note two little-recognized but critical technical issues in dose–volume modeling. [We have previously notified the commercial groups involved.]
- Reported DVH metrics may not necessarily represent what they seem to represent. In particular, in the current Pinnacle treatment planning system, exported DVH values are based on the ROI volume that overlaps with the dose calculation grid. If the dose calculation grid does not cover the entire ROI, the values are different from the accepted definition of that DVH parameter. If CERR is used to extract DVHs instead of reading directly from the system, this potential problem does not arise.
- Patient de-identification. Physicists at our institution recently discovered that the Eclipse planning system uses private DICOM tags to store a photograph of the patient (when available). This was not previously appreciated and is not removed by the commonly used open-source anonymizer, dicompyler, at the time of this writing. The CERR anonymizer always removes this private tag and any photograph.

15.14 SUMMARY

The current era of increasing data richness and size represents new opportunities for physicists and other quantitative scientists working in medicine. Speaking now specifically to medical physicists: As individuals who often have extensive skill sets and experience manipulating and analyzing large data sets, and given how well situated we are to understand and interpret clinical data, we should seize the initiative. We are in an excellent position to increase our value to the field of radiotherapy, and more generally to medicine, by supporting multi-institutional data projects. Emphasizing data science as an activity should be an important part of the future of physicists in medicine.

One theme of the chapter is that mechanistic modeling can still guide research in tumor or normal tissue response to radiotherapy, yet mechanisms have to be bound together with rich sources of guiding data. Optimal models are only likely to emerge slowly over an iterative process involving new data sets. As predictive models inevitably become more complicated, it is critical to understand how much of the complication is necessary, and how close the models align with previously established principles. Future advances in TCP modeling will likely combine various sources of data, including clinical data, detailed data from multiple institutions, as well as longitudinal measures of imaging metrics (such as radiomic measures but also simpler measures), and biological assays. Similarly, future advances in NTCP modeling will also likely include similar elements, as well as germline (radio) genomics and, in some cases, functional heterogeneity as determined from imaging studies. Using TCP and NTCP models to drive improved optimized treatment planning (for protons or photons, with or without online MRI imaging) remains a "high risk/high reward" goal, even though past generations of researchers have failed to reach this Promised Land. The only sure bet is that there will be surprises along the way.

ACKNOWLEDGMENTS

We are grateful for many useful discussions, over the years, with Chuck Mayo, Ellen Yorke, Ellen Huang, Issam El Naqa, Mark Kessler, Jung Hun Oh, Margie Hunt, John Humm, Caroline Olsson, Alan Nahum, Jack Fowler, Larry Marks, Randy Ten Haken, Laura Cella, and Soeren Bentzen, among others.

REFERENCES

Ahmed, K. A., P. Chinnaiyan, W. J. Fulp, S. Eschrich, J. F. Torres-Roca, and J. J. Caudell. 2015. "The Radiosensitivity Index Predicts for Overall Survival in Glioblastoma." *Oncotarget* 6 (33): 34414–22.

Appelt, A. L., I. R. Vogelius, K. P. Farr, A. A. Khalil, and S. M. Bentzen. 2014. "Towards Individualized Dose Constraints: Adjusting the QUANTEC Radiation Pneumonitis Model for Clinical Risk Factors." *Acta Oncologica* 53 (5): 605–12.

Atherton, P. J., D. W. Watkins-Bruner, C. Gotay, C. M. Moinpour, D. V. Satele, K. A. Winter, P. L. Schaefer, B. Movsas, and J. A. Sloan. 2015. "The Complementary Nature of Patient-Reported Outcomes and Adverse Event Reporting in Cooperative Group Oncology Clinical Trials: A Pooled Analysis (NCCTG N0591)." *Journal of Pain and Symptom Management* 50 (4): 470–79.

Beasley, W., M. Thor, Alan McWilliam, A. Green, R. Mackay, N. Slevin, C. Olsson et al. 2017. "Image-Based Data Mining for Identifying Regions Exhibiting a Dose-Response Relationship with Radiation-Induced Trismus." *International Journal of Radiation Oncology, Biology, Physics* 99 (2): S165.

Beaton, L. A., C. Ferrarotto, L. Marro, S. Samiee, S. Malone, S. Grimes, K. Malone, and R. C. Wilkins. 2013. "Chromosome Damage and Cell Proliferation Rates in in Vitro Irradiated Whole Blood as Markers of Late Radiation Toxicity after Radiation Therapy to the Prostate." *International Journal of Radiation Oncology, Biology, Physics* 85 (5): 1346–52.

Benadjaoud, M. A., P. Blanchard, B. Schwartz, J. Champoudry, R. Bouaita, D. Lefkopoulos, E. Deutsch, I. Diallo, H. Cardot, and F. de Vathaire. 2014. "Functional Data Analysis in NTCP Modeling: A New Method to Explore the Radiation Dose-Volume Effects." *International Journal of Radiation Oncology, Biology, Physics* 90 (3): 654–63.

Benson, K., and A. J. Hartz. 2000. "A Comparison of Observational Studies and Randomized, Controlled Trials." *American Journal of Ophthalmology* 130 (5): 688.

Birtwhistle, R., and T. Williamson. 2015. "Primary Care Electronic Medical Records: A New Data Source for Research in Canada." *CMAJ: Canadian Medical Association Journal = Journal de l'Association Medicale Canadienne* 187 (4): 239–40.

Brodin, N. P., R. Kabarriti, M. K. Garg, C. Guha, and W. A. Tomé. 2018. "Systematic Review of Normal Tissue Complication Models Relevant to Standard Fractionation Radiation Therapy of the Head and Neck Region Published After the QUANTEC Reports." *International Journal of Radiation Oncology, Biology, Physics* 100 (2): 391–407.

Buettner, F., S. L. Gulliford, S. Webb, M. R. Sydes, D. P. Dearnaley, and M. Partridge. 2012. "The Dose—Response of the Anal Sphincter Region—An Analysis of Data from the MRC RT01 Trial." *Radiotherapy and Oncology: Journal of the European Society for Therapeutic Radiology and Oncology* 103 (3): 347–52.

Calais, J., B. Dubray, L. Nkhali, S. Thureau, C. Lemarignier, R. Modzelewski, I. Gardin, F. Di Fiore, P. Michel, and P. Vera. 2015. "High FDG Uptake Areas on Pre-Radiotherapy PET/CT Identify Preferential Sites of Local Relapse after Chemoradiotherapy for Locally Advanced Oesophageal Cancer." *European Journal of Nuclear Medicine and Molecular Imaging* 42 (6): 858–67.

Carlin, S., H. Zhang, M. Reese, N. N. Ramos, Q. Chen, and S.-A. Ricketts. 2014. "A Comparison of the Imaging Characteristics and Microregional Distribution of 4 Hypoxia PET Tracers." *Journal of Nuclear Medicine: Official Publication, Society of Nuclear Medicine* 55 (3): 515–21.

Cella, L., J. H. Oh, J. O. Deasy, G. Palma, R. Liuzzi, V. D'avino, M. Conson, M. Picardi, M. Salvatore, and R. Pacelli. 2015. "Predicting Radiation-Induced Valvular Heart Damage." *Acta Oncologica* 54 (10): 1796–1804.

Chan, N., M. Koritzinsky, H. Zhao, R. Bindra, P. M. Glazer, S. Powell, A. Belmaaza, B. Wouters, and R. G. Bristow. 2008. "Chronic Hypoxia Decreases Synthesis of Homologous Recombination Proteins to Offset Chemoresistance and Radioresistance." *Cancer Research* 68 (2): 605–14.

Chapman, J. D., and C. J. Gillespie. 2012. "The Power of Radiation Biophysics—Let's Use It." *International Journal of Radiation Oncology*Biology*Physics* 84 (2): 309–11.

Cheng, Q., E. Roelofs, B. L. T. Ramaekers, D. Eekers, J. van Soest, T. Lustberg, T. Hendriks et al. 2016. "Development and Evaluation of an Online Three-Level Proton vs Photon Decision Support Prototype for Head and Neck Cancer—Comparison of Dose, Toxicity and Cost-Effectiveness." *Radiotherapy and Oncology: Journal of the European Society for Therapeutic Radiology and Oncology* 118 (2): 281–85.

Chera, B. S., A. Eisbruch, B. A. Murphy, J. A. Ridge, P. Gavin, B. B. Reeve, D. W. Bruner, and B. Movsas. 2014. "Recommended Patient-Reported Core Set of Symptoms to Measure in Head and Neck Cancer Treatment Trials." *JNCI: Journal of the National Cancer Institute* 106 (7). doi:10.1093/jnci/dju127.

Christianen, M. E. M. C., A. van der Schaaf, H. P. van der Laan, I. M. Verdonck-de Leeuw, P. Doornaert, O. Chouvalova, R. J. H. M. Steenbakkers et al. 2016. "Swallowing Sparing Intensity Modulated Radiotherapy (SW-IMRT) in Head and Neck Cancer: Clinical Validation according to the Model-Based Approach." *Radiotherapy and Oncology: Journal of the European Society for Therapeutic Radiology and Oncology* 118 (2): 298–303.

Clark, V. H., Y. Chen, J. Wilkens, J. R. Alaly, K. Zakaryan, and J. O. Deasy. 2008. "IMRT Treatment Planning for Prostate Cancer Using Prioritized Prescription Optimization and Mean-Tail-Dose Functions." *Linear Algebra and Its Applications* 428 (5–6): 1345–64.

Collins, G. S., J. B. Reitsma, D. G. Altman, and K. G. M. Moons. 2015. "Transparent Reporting of a Multivariable Prediction Model for Individual Prognosis or Diagnosis (TRIPOD): The TRIPOD Statement." *European Journal of Clinical Investigation* 45 (2): 204–14.

Colombié, M., S. Gouard, M. Frindel, A. Vidal, M. Chérel, F. Kraeber-Bodéré, C. Rousseau, and M. Bourgeois. 2015. "Focus on the Controversial Aspects of 64Cu-ATSM in Tumoral Hypoxia Mapping by PET Imaging." *Frontiers in Medicine* 2: 58.

Concato, J., N. Shah, and R. I. Horwitz. 2000. "Randomized, Controlled Trials, Observational Studies, and the Hierarchy of Research Designs." *The New England Journal of Medicine* 342 (25): 1887–92.

Coroller, T. P., P. Grossmann, Y. Hou, E. R. Velazquez, R. T. H. Leijenaar, G. Hermann, P. Lambin, B. Haibe-Kains, R. H. Mak, and H. J. W. L. Aerts. 2015. "CT-Based Radiomic Signature Predicts Distant Metastasis in Lung Adenocarcinoma." *Radiotherapy and Oncology: Journal of the European Society for Therapeutic Radiology and Oncology* 114 (3): 345–50.

Crispin-Ortuzar, M., J. Jeong, A. N. Fontanella, and J. O. Deasy. 2017. "A Radiobiological Model of Radiotherapy Response and Its Correlation with Prognostic Imaging Variables." *Physics in Medicine and Biology* 62 (7): 2658–74.

Dean, J. A., K. H. Wong, L. C. Welsh, A.-B. Jones, U. Schick, K. L. Newbold, S. A. Bhide, K. J. Harrington, C. M. Nutting, and S. L. Gulliford. 2016. "Normal Tissue Complication Probability (NTCP) Modelling Using Spatial Dose Metrics and Machine Learning Methods for Severe Acute Oral Mucositis Resulting from Head and Neck Radiotherapy." *Radiotherapy and Oncology: Journal of the European Society for Therapeutic Radiology and Oncology* 120 (1): 21–27.

Deasy, J. O. 1998. "Inter-Patient and Intra-Tumor Radiosensitivity Heterogeneity." In *Volume & Kinetics in Tumor Control and Normal Tissue Complications*, Paliwal, B., Fowler, J., Herbert, D., Mehta M. (Eds.), pp. 363–80. American Association of Physicists in Medicine, Madison, WI.

Deasy, J. O., C. S. Mayo, and C. G. Orton. 2015. "Treatment Planning Evaluation and Optimization Should Be Biologically and Not Dose/volume Based." *Medical Physics* 42 (6): 2753–56.

Deist, T. M., A. Jochems, J. van Soest, G. Nalbantov, C. Oberije, S. Walsh, M. Eble et al. 2017. "Infrastructure and Distributed Learning Methodology for Privacy-Preserving Multi-Centric Rapid Learning Health Care: euro-CAT." *Clinical and Translational Radiation Oncology* 4: 24–31.

Dréan, G., O. Acosta, J. D. Ospina, A. Fargeas, C. Lafond, G. Corrégé, J.-L. Lagrange et al. 2016. "Identification of a Rectal Subregion Highly Predictive of Rectal Bleeding in Prostate Cancer IMRT." *Radiotherapy and Oncology: Journal of the European Society for Therapeutic Radiology and Oncology* 119 (3): 388–97.

Efron, B., and T. Hastie. 2016. *Computer Age Statistical Inference*. Cambridge University Press, New York, NY.

Emami, B., J. Lyman, A. Brown, L. Coia, M. Goitein, J. E. Munzenrider, B. Shank, L. J. Solin, and M. Wesson. 1991. "Tolerance of Normal Tissue to Therapeutic Irradiation." *International Journal of Radiation Oncology, Biology, Physics* 21 (1): 109–22.

Eschrich, S., H. Zhang, H. Zhao, D. Boulware, J.-H. Lee, G. Bloom, and J. F. Torres-Roca. 2009. "Systems Biology Modeling of the Radiation Sensitivity Network: A Biomarker Discovery Platform." *International Journal of Radiation Oncology, Biology, Physics* 75 (2): 497–505.

Ewing, M., G. A. Funk, A. M. Warren, N. Rapier, M. Reynolds, M. Bennett, C. Mastropieri, and M. L. Foreman. 2015. "Improving National Trauma Data Bank® Coding Data Reliability for Traumatic Injury Using a Prospective Systems Approach." *Health Informatics Journal* 22 (4): 1076–82.

Fenwick, J. D., A. E. Nahum, Z. I. Malik, C. V. Eswar, M. Q. Hatton, V. M. Laurence, J. F. Lester, and D. B. Landau. 2009. "Escalation and Intensification of Radiotherapy for Stage III Non-Small Cell Lung Cancer: Opportunities for Treatment Improvement." *Clinical Oncology* 21 (4): 343–60.

Ferté, C., F. André, and J.-C. Soria. 2010. "Molecular Circuits of Solid Tumors: Prognostic and Predictive Tools for Bedside Use." *Nature Reviews. Clinical Oncology* 7 (7): 367–80.

Fertil, B., and E. P. Malaise. 1985. "Intrinsic Radiosensitivity of Human Cell Lines Is Correlated with Radioresponsiveness of Human Tumors: Analysis of 101 Published Survival Curves." *International Journal of Radiation Oncology, Biology, Physics* 11 (9): 1699–1707.

Finley, C., L. Schneider, and S. Shakeel. 2016. *Approaches to High-Risk, Resource Intensive Cancer Surgical Care in Canada*. Canadian Partnership Against Cancer, Toronto, Canada.

Fowler, J. F. 2006. "Development of Radiobiology for Oncology-a Personal View." *Physics in Medicine and Biology* 51 (13): R263–86.

Fowler, J. F. 2010. "21 Years of Biologically Effective Dose." *The British Journal of Radiology* 83 (991): 554–68.

Fyles, A. W., M. Milosevic, R. Wong, M. C. Kavanagh, M. Pintilie, A. Sun, W. Chapman et al. 1998. "Oxygenation Predicts Radiation Response and Survival in Patients with Cervix Cancer." *Radiotherapy and Oncology: Journal of the European Society for Therapeutic Radiology and Oncology* 48 (2): 149–56.

Garcia-Murillas, I., M. Beanney, M. Epstein, K. Howarth, A. Lawson, S. Hrebien, E. Green, N. Rosenfeld, and N. Turner. 2017. "Abstract 2743: Comparison of Enhanced Tagged-Amplicon Sequencing and Digital PCR for Circulating Tumor DNA Analysis in Advanced Breast Cancer." *Cancer Research* 77 (13 Supplement): 2743–2743.

Goitein, M. 2007. *Radiation Oncology: A Physicist's-Eye View*. Springer Science & Business Media, New York.

Gubin, M. M., X. Zhang, H. Schuster, E. Caron, J. P. Ward, T. Noguchi, Y. Ivanova et al. 2014. "Checkpoint Blockade Cancer Immunotherapy Targets Tumour-Specific Mutant Antigens." *Nature* 515 (7528): 577–81.

Hadley, S. W., M. L. Kessler, D. W. Litzenberg, C. Lee, J. Irrer, X. Chen, E. Acosta et al. 2016. "SafetyNet: Streamlining and Automating QA in Radiotherapy." *Journal of Applied Clinical Medical Physics/American College of Medical Physics* 17 (1): 387–95.

Hill, R. P. 2017. "The Changing Paradigm of Tumour Response to Irradiation." *The British Journal of Radiology* 90 (1069): 20160474.

Hsu, E. R., J. D. Klemm, A. R. Kerlavage, D. Kusnezov, and W. A. Kibbe. 2017. "Cancer Moonshot Data and Technology Team: Enabling a National Learning Healthcare System for Cancer to Unleash the Power of Data." *Clinical Pharmacology and Therapeutics* 101 (5): 613–15.

Huang, Z., K. A. Yuh, S. S. Lo, J. C. Grecula, S. Sammet, C. L. Sammet, G. Jia et al. 2014. "Validation of Optimal DCE-MRI Perfusion Threshold to Classify at-Risk Tumor Imaging Voxels in Heterogeneous Cervical Cancer for Outcome Prediction." *Magnetic Resonance Imaging* 32 (10): 1198–1205.

Institute of Medicine, and Roundtable on Evidence-Based Medicine. 2007. *The Learning Healthcare System: Workshop Summary*. National Academies Press, Washington, DC.

Ioannidis, J. P. A., and O. A. Panagiotou. 2011. "Comparison of Effect Sizes Associated with Biomarkers Reported in Highly Cited Individual Articles and in Subsequent Meta-Analyses." *JAMA: The Journal of the American Medical Association* 305 (21): 2200–2210.

Jackson, A., L. B. Marks, S. M. Bentzen, A. Eisbruch, E. D. Yorke, R. K. T. Haken, L. S. Constine, and J. O. Deasy. 2010. "The Lessons of QUANTEC: Recommendations for Reporting and Gathering Data on Dose-Volume Dependencies of Treatment Outcome." *International Journal of Radiation Oncology, Biology, Physics* 76 (3 Suppl): S155–60.

Jagadish, H. V., J. Gehrke, A. Labrinidis, Y. Papakonstantinou, J. M. Patel, R. Ramakrishnan, and C. Shahabi. 2014. "Big Data and Its Technical Challenges." *Communications of the ACM* 57 (7): 86–94.

Jeong, J., J. H. Oh, J.-J. Sonke, J. Belderbos, J. D. Bradley, A. N. Fontanella, S. S. Rao, and J. O. Deasy. 2017. "Modeling the Cellular Response of Lung Cancer to Radiation Therapy for a Broad Range of Fractionation Schedules." *Clinical Cancer Research: An Official Journal of the American Association for Cancer Research* 23 (18): 5469–79.

Jeong, J., J. S. Setton, N. Y. Lee, J. H. Oh, and J. O. Deasy. 2014. "Estimate of the Impact of FDG-Avidity on the Dose Required for Head and Neck Radiotherapy Local Control." *Radiotherapy and Oncology: Journal of the European Society for Therapeutic Radiology and Oncology* 111 (3): 340–47.

Jeong, J., K. I. Shoghi, and J. O. Deasy. 2013. "Modelling the Interplay between Hypoxia and Proliferation in Radiotherapy Tumour Response." *Physics in Medicine and Biology* 58 (14): 4897–4919.

Jochems, A., T. M. Deist, J. Van Soest, M. Eble, P. Bulens, P. Coucke, W. Dries, P. Lambin, and A. Dekker. 2016. "Distributed Learning: Developing a Predictive Model Based on Data from Multiple Hospitals without Data Leaving the Hospital—A Real Life Proof of Concept." *Radiotherapy and Oncology: Journal of the European Society for Therapeutic Radiology and Oncology* 121 (3): 459–67.

Joiner, M. C., and A. van der Kogel. 2016. *Basic Clinical Radiobiology, Fifth Edition*. CRC Press, London, UK.

Kaanders, J. H., L. A. M. Pop, H. A. M. Marres, I. Bruaset, F. J. A. van den Hoogen, M. A. W. Merkx, and A. J. van der Kogel. 2002. "ARCON: Experience in 215 Patients with Advanced Head-and-Neck Cancer." *International Journal of Radiation Oncology, Biology, Physics* 52 (3): 769–78.

Kibel, A. S., J. P. Ciezki, E. A. Klein, C. A. Reddy, J. D. Lubahn, J. Haslag-Minoff, J. O. Deasy et al. 2012. "Survival among Men with Clinically Localized Prostate Cancer Treated with Radical Prostatectomy or Radiation Therapy in the Prostate Specific Antigen Era." *The Journal of Urology* 187 (4): 1259–65.

Lambin, P., R. G. P. M. van Stiphout, M. H. W. Starmans, E. Rios-Velazquez, G. Nalbantov, H. J. W. L. Aerts, Erik Roelofs et al. 2013. "Predicting Outcomes in Radiation Oncology—Multifactorial Decision Support Systems." *Nature Reviews Clinical Oncology* 10 (1): 27–40.

Lee, S., S. Kerns, H. Ostrer, B. Rosenstein, J. O. Deasy, and J. H. Oh. 2018. "Machine Learning on a Genome-Wide Association Study to Predict Late Genitourinary Toxicity Following Prostate Radiotherapy." *International Journal of Radiation Oncology, Biology, Physics*.

Lee, T.-F., and F.-M. Fang. 2013. "Quantitative Analysis of Normal Tissue Effects in the Clinic (QUANTEC) Guideline Validation Using Quality of Life Questionnaire Datasets for Parotid Gland Constraints to Avoid Causing Xerostomia during Head-and-Neck Radiotherapy." *Radiotherapy and Oncology: Journal of the European Society for Therapeutic Radiology and Oncology* 106 (3): 352–58.

Ling, C. C., J. Humm, S. Larson, H. Amols, Z. Fuks, S. Leibel, and J. A. Koutcher. 2000. "Towards Multidimensional Radiotherapy (MD-CRT): Biological Imaging and Biological Conformality." *International Journal of Radiation Oncology, Biology, Physics* 47 (3): 551–60.

Li, X., E. Kostareli, J. Suffner, N. Garbi, and G. J. Hämmerling. 2010. "Efficient Treg Depletion Induces T-Cell Infiltration and Rejection of Large Tumors." *European Journal of Immunology* 40 (12): 3325–35.

Marks, L. B., E. D. Yorke, A. Jackson, R. K. T. Haken, L. S. Constine, A. Eisbruch, S. M. Bentzen, J. Nam, and J. O. Deasy. 2010. "Use of Normal Tissue Complication Probability Models in the Clinic." *International Journal of Radiation Oncology, Biology, Physics* 76 (3 Suppl): S10–19.

Mayo, C. S., J. O. Deasy, B. S. Chera, J. Freymann, J. S. Kirby, and P. H. Hardenberg. 2016. "How Can We Effect Culture Change Toward Data-Driven Medicine?" *International Journal of Radiation Oncology, Biology, Physics* 95 (3): 916–21.

Mayo, C. S., J. M. Moran, W. Bosch, Y. Xiao, T. McNutt, R. Popple, J. Michalski, et al. 2017. "AAPM TG-263: Standardizing Nomenclatures in Radiation Oncology." *International Journal of Radiation Oncology, Biology, Physics* 100 (4): 1057–66.

Mayr, N. A., W. T. Yuh, J. C. Arnholt, J. C. Ehrhardt, J. I. Sorosky, V. A. Magnotta, K. S. Berbaum et al. 2000. "Pixel Analysis of MR Perfusion Imaging in Predicting Radiation Therapy Outcome in Cervical Cancer." *Journal of Magnetic Resonance Imaging: JMRI* 12 (6): 1027–33.

Mayr, N. A., J. Z. Wang, D. Zhang, J. F. Montebello, J. C. Grecula, S. S. Lo, J. M. Fowler, and W. T. C. Yuh. 2009. "Synergistic Effects of Hemoglobin and Tumor Perfusion on Tumor Control and Survival in Cervical Cancer." *International Journal of Radiation Oncology, Biology, Physics* 74 (5): 1513–21.

Michalski, J. M., H. Gay, A. Jackson, S. L. Tucker, and J. O. Deasy. 2010. "Radiation Dose-Volume Effects in Radiation-Induced Rectal Injury." *International Journal of Radiation Oncology, Biology, Physics* 76 (3 Suppl): S123–29.

Munafò, M. R., B. A. Nosek, D. V. M. Bishop, K. S. Button, C. D. Chambers, N. Percie du Sert, U. Simonsohn, E.-J. Wagenmakers, J. J. Ware, and J. P. A. Ioannidis. 2017. "A Manifesto for Reproducible Science." *Nature Human Behaviour* 1: 0021.

Nahum, A. E., B. Movsas, E. M. Horwitz, C. C. Stobbe, and J. Donald Chapman. 2003. "Incorporating Clinical Measurements of Hypoxia into Tumor Local Control Modeling of Prostate Cancer: Implications for the α/β Ratio." *International Journal of Radiation Oncology*Biology*Physics* 57 (2): 391–401.

Naqa, I. E., J. O. Deasy, Y. Mu, E. Huang, A. J. Hope, P. E. Lindsay, A. Apte, J. Alaly, and J. D. Bradley. 2010. "Datamining Approaches for Modeling Tumor Control Probability." *Acta Oncologica* 49 (8): 1363–73.

Niska, J. R., M. Y. Halyard, A. D. Tan, P. J. Atherton, S. H. Patel, and J. A. Sloan. 2017. "Electronic Patient-Reported Outcomes and Toxicities during Radiotherapy for Head-and-Neck Cancer." *Quality of Life Research: An International Journal of Quality of Life Aspects of Treatment, Care and Rehabilitation* 26 (7): 1721–31.

Oh, J. H., S. Kerns, H. Ostrer, S. N. Powell, B. Rosenstein, and J. O. Deasy. 2017. "Computational Methods Using Genome-Wide Association Studies to Predict Radiotherapy Complications and to Identify Correlative Molecular Processes." *Scientific Reports* 7: 43381.

Oh, J. H., M. Thor, C. Olsson, V. Skokic, R. Jörnsten, D. Alsadius, N. Pettersson, G. Steineck, and J. O. Deasy. 2016. "A Factor Analysis Approach for Clustering Patient Reported Outcomes." *Methods of Information in Medicine* 55 (5): 431–39.

Oorschot, B., S. E. Hovingh, P. D. Moerland, J. P. Medema, L. J. A. Stalpers, H. Vrieling, and N. A. P. Franken. 2014. "Reduced Activity of Double-Strand Break Repair Genes in Prostate Cancer Patients with Late Normal Tissue Radiation Toxicity." *International Journal of Radiation Oncology, Biology, Physics* 88 (3): 664–70.

Overgaard, J. 2007. "Hypoxic Radiosensitization: Adored and Ignored." *Journal of Clinical Oncology: Official Journal of the American Society of Clinical Oncology* 25 (26): 4066–74.

Overgaard, J., J. G. Eriksen, M. Nordsmark, J. Alsner, M. R. Horsman et al. 2005. "Plasma Osteopontin, Hypoxia, and Response to the Hypoxia Sensitiser Nimorazole in Radiotherapy of Head and Neck Cancer: Results from the DAHANCA 5 Randomised Double-Blind Placebo-Controlled Trial." *The Lancet Oncology* 6 (10): 757–64.

Palma, G., S. Monti, V. D'Avino, M. Conson, R. Liuzzi, M. C. Pressello, V. Donato et al. 2016. "A Voxel-Based Approach to Explore Local Dose Differences Associated With Radiation-Induced Lung Damage." *International Journal of Radiation Oncology, Biology, Physics* 96 (1): 127–33.

Patel, S. P., and R. Kurzrock. 2015. "PD-L1 expression as a predictive biomarker in cancer immunotherapy." *Molecular Cancer Therapeutics* 14 (4): 847–56.

Prokopiou, S., E. G. Moros, J. Poleszczuk, J. Caudell, J. F. Torres-Roca, K. Latifi, J. K. Lee, R. Myerson, L. B. Harrison, and H. Enderling. 2015. "A Proliferation Saturation Index to Predict Radiation Response and Personalize Radiotherapy Fractionation." *Radiation Oncology* 10: 159.

Rawlins, M. 2008. "De Testimonio: On the Evidence for Decisions about the Use of Therapeutic Interventions." *Clinical Medicine* 8 (6): 579–88.

Rawlins, M. D. 2011. *Therapeutics, Evidence and Decision-Making.* CRC Press, London, UK.

Resnic, F. S., and M. E. Matheny. 2018. "Medical Devices in the Real World." *The New England Journal of Medicine* 378 (7): 595–97.

Rischin, D., R. J. Hicks, R. Fisher, D. Binns, J. Corry, S. Porceddu, and L. J. Peters. 2006. "Prognostic Significance of [18F]-Misonidazole Positron Emission Tomography—Detected Tumor Hypoxia in Patients with Advanced Head and Neck Cancer Randomly Assigned to Chemoradiation with or without Tirapazamine: A Substudy of Trans-Tasman Radiation Oncology Group Study 98.02." *Journal of Clinical Oncology: Official Journal of the American Society of Clinical Oncology* 24 (13): 2098–2104.

Scott, J. G., A. Berglund, M. J. Schell, I. Mihaylov, W. J. Fulp, B. Yue, E. Welsh et al. 2017. "A Genome-Based Model for Adjusting Radiotherapy Dose (GARD): A Retrospective, Cohort-Based Study." *The Lancet Oncology* 18 (2): 202–11.

Soto, D. E., M. L. Kessler, M. Piert, and A. Eisbruch. 2008. "Correlation between Pretreatment FDG-PET Biological Target Volume and Anatomical Location of Failure after Radiation Therapy for Head and Neck Cancers." *Radiotherapy and Oncology: Journal of the European Society for Therapeutic Radiology and Oncology* 89 (1): 13–18.

Spalding, A. C., K.-W. Jee, K. Vineberg, M. Jablonowski, B. A. Fraass, C. C. Pan, T. S. Lawrence, R. K. Ten Haken, and E. Ben-Josef. 2007. "Potential for Dose-Escalation and Reduction of Risk in Pancreatic Cancer Using IMRT Optimization with Lexicographic Ordering and gEUD-Based Cost Functions." *Medical Physics* 34 (2): 521–29.

Stokes, W. A., M. R. Bronsert, R. A. Meguid, M. G. Blum, B. L. Jones, M. Koshy, D. J. Sher et al. 2018. "Post-Treatment Mortality After Surgery and Stereotactic Body Radiotherapy for Early-Stage Non–Small-Cell Lung Cancer." *Journal of Clinical Orthodontics: JCO* 36 (7): 642–51.

Strom, T., S. E. Hoffe, W. Fulp, J. Frakes, D. Coppola, G. M. Springett, M. P. Malafa et al. 2015. "Radiosensitivity Index Predicts for Survival with Adjuvant Radiation in Resectable Pancreatic Cancer." *Radiotherapy and Oncology: Journal of the European Society for Therapeutic Radiology and Oncology* 117 (1): 159–64.

Taghian, A., H. Suit, F. Pardo, D. Gioioso, K. Tomkinson, W. DuBois, and L. Gerweck. 1992. "In Vitro Intrinsic Radiation Sensitivity of Glioblastoma Multiforme." *International Journal of Radiation Oncology, Biology, Physics* 23 (1): 55–62.

Thames, H. D., Jr, H. R. Withers, L. J. Peters, and G. H. Fletcher. 1982. "Changes in Early and Late Radiation Responses with Altered Dose Fractionation: Implications for Dose-Survival Relationships." *International Journal of Radiation Oncology, Biology, Physics* 8 (2): 219–26.

Thor, M., A. Jackson, M. J. Zelefsky, G. Steineck, A. Karlsdòttir, M. Høyer, M. Liu et al. 2018. "Inter-Institutional Analysis Demonstrates the Importance of Lower than Previously Anticipated Dose Regions to Prevent Late Rectal Bleeding Following Prostate Radiotherapy." *Radiotherapy and Oncology: Journal of the European Society for Therapeutic Radiology and Oncology* 127 (1): 88–95. doi:10.1016/j.radonc.2018.02.020.

Thor, M., C. E. Olsson, J. H. Oh, S. E. Petersen, D. Alsadius, L. Bentzen, N. Pettersson et al. 2015. "Relationships between Dose to the Gastro-Intestinal Tract and Patient-Reported Symptom Domains after Radiotherapy for Localized Prostate Cancer." *Acta Oncologica* 54 (9): 1326–34.

Titz, B., and R. Jeraj. 2008. "An Imaging-Based Tumour Growth and Treatment Response Model: Investigating the Effect of Tumour Oxygenation on Radiation Therapy Response." *Physics in Medicine and Biology* 53 (17): 4471–88.

Torres-Roca, J. F., W. J. Fulp, J. J. Caudell, N. Servant, M. A. Bollet, M. van de Vijver, A. O. Naghavi, E. E. Harris, and S. A. Eschrich. 2015. "Integration of a Radiosensitivity Molecular Signature Into the Assessment of Local Recurrence Risk in Breast Cancer." *International Journal of Radiation Oncology, Biology, Physics* 93 (3): 631–38.

Torres-Roca, J. F., N. Erho, I. Vergara, E. Davicioni, R. B. Jenkins, R. B. Den, A. P. Dicker, and S. A. Eschrich. 2014. "A Molecular Signature of Radiosensitivity (RSI) Is an RT-Specific Biomarker in Prostate Cancer." *International Journal of Radiation Oncology, Biology, Physics* 90 (1): S157.

Trani, D., A. Yaromina, L. Dubois, M. Granzier, S. G. J. A. Peeters, R. Biemans, G. Nalbantov et al. 2015. "Preclinical Assessment of Efficacy of Radiation Dose Painting Based on Intratumoral FDG-PET Uptake." *Clinical Cancer Research: An Official Journal of the American Association for Cancer Research* 21 (24): 5511–18.

Uzan, J., A. E. Nahum, and I. Syndikus. 2016. "Prostate Dose-Painting Radiotherapy and Radiobiological Guided Optimisation Enhances the Therapeutic Ratio." *Clinical Oncology* 28 (3): 165–70.

Vallières, M., E. Kay-Rivest, L. J. Perrin, X. Liem, C. Furstoss, H. J. W. L. Aerts, N. Khaouam et al. 2017. "Radiomics Strategies for Risk Assessment of Tumour Failure in Head-and-Neck Cancer." *Scientific Reports* 7 (1): 10117.

Vallières, M., C. R. Freeman, S. R. Skamene, and I. El Naqa. 2015. "A Radiomics Model from Joint FDG-PET and MRI Texture Features for the Prediction of Lung Metastases in Soft-Tissue Sarcomas of the Extremities." *Physics in Medicine and Biology* 60 (14): 5471–96.

van Dijk, L. V., W. Noordzij, C. L. Brouwer, R. Boellaard, J. G. M. Burgerhof, J. A. Langendijk, N. M. Sijtsema, and R. J. Steenbakkers. 2018. "18F-FDG PET Image Biomarkers Improve Prediction of Late Radiation-Induced Xerostomia." *Radiotherapy and Oncology: Journal of the European Society for Therapeutic Radiology and Oncology* 126 (1): 89–95.

van Loon, J. van, J. Grutters, and F. Macbeth. 2012. "Evaluation of Novel Radiotherapy Technologies: What Evidence Is Needed to Assess Their Clinical and Cost Effectiveness, and How Should We Get It?" *The Lancet Oncology* 13 (4): e169–77.

Vogelius, I. R., and S. M. Bentzen. 2017. "Dose Response and Fractionation Sensitivity of Prostate Cancer after External Beam Radiotherapy: A Meta-Analysis of Randomized Trials." *International Journal of Radiation Oncology, Biology, Physics* 100 (4): 858–865.

Wack, L. J., D. Mönnich, W. Van Elmpt, C. M. L. Zegers, E. G. C. Troost, D. Zips, and D. Thorwarth. 2015. "Comparison of [18F]-FMISO, [18F]-FAZA and [18F]-HX4 for PET Imaging of Hypoxia—A Simulation Study." *Acta Oncologica* 54 (9): 1370–77.

Walsh, S., E. Roelofs, P. Kuess, P. Lambin, B. Jones, D. Georg, and F. Verhaegen. 2016. "A Validated Tumor Control Probability Model Based on a Meta-Analysis of Low, Intermediate, and High-Risk Prostate Cancer Patients Treated by Photon, Proton, or Carbon-Ion Radiotherapy." *Medical Physics* 43 (2): 734–47.

West, C. M., S. E. Davidson, S. A. Roberts, and R. D. Hunter. 1993. "Intrinsic Radiosensitivity and Prediction of Patient Response to Radiotherapy for Carcinoma of the Cervix." *British Journal of Cancer* 68 (4): 819–23.

Wijmenga, C., and A. Zhernakova. 2018. "The Importance of Cohort Studies in the Post-GWAS Era." *Nature Genetics*. doi:10.1038/s41588-018-0066-3.

Williamson, J. F., S. K. Das, M. S. Goodsitt, and J. O. Deasy. 2017. "Introducing the Medical Physics Dataset Article." *Medical Physics* 44 (2): 349–50.

Xie, Y. 2014. "Applications of Nonlinear Optimization." Washington University in St. Louis, Dissertation. doi:10.17605/OSF.IO/EUMZJ.

16 Multi-parameterized models for early cancer detection and prevention

Gregory R. Hart, David A. Roffman, Ying Liang, Bradley J. Nartowt, Wazir Muhammad, and Jun Deng

Contents

16.1 Introduction 265
16.2 A Big Data Approach to Early Cancer Detection and Prevention 266
16.3 A Multi-parameterized Artificial Neural Network Model for Cancer Prediction 267
    16.3.1 What Is an Artificial Neural Network? 267
    16.3.2 Training an Artificial Neural Network 269
    16.3.3 Measuring Accuracy 270
    16.3.4 Predicting Cancer with an ANN 271
        16.3.4.1 Case Study: Lung Cancer 271
        16.3.4.2 Results for All Cancers 273
16.4 Other Machine Learning Algorithms for Big Data Mining and Cancer Prediction 277
    16.4.1 Prequel to Machine Learning: Conditional Probability Analysis 277
    16.4.2 Multivariable Logistic Regression Applied to Lung Cancer Prediction 277
    16.4.3 Support Vector Machine Applied to Lung Cancer Prediction 278
    16.4.4 Decision Trees and Random Forests Applied to Lung Cancer Prediction 279
16.5 Potential Applications of the Multi-parameterized Neural Network Model 280
    16.5.1 Personalized Medicine 280
    16.5.2 Mobile Health App and Platform for Patient-Reported Outcome 281
    16.5.3 Health Insurance Industry 281
16.6 Conclusion and Outlook 281
References 281

16.1 INTRODUCTION

Men have a 39.66% chance of acquiring cancer during their lifetimes and a 22.03% chance of dying from it. For women, these values are 37.65% and 18.76%, respectively (Howlader et al. 2017). Cancer is the second most common cause of death in the United States and worldwide (Centers for Disease Control and Prevention 2017a; World Health Organization 2017). It is responsible for 22.0% of the deaths in the United States annually. In addition to this high mortality, it is estimated that in 2010 the economic cost of cancer was $1.16 trillion and has been increasing every year since then (World Health Organization 2017).

Although lung, colon, and pancreatic cancer kill the most people, the three cancers with the highest incidence in the United States are lung, breast, and prostate cancer (American Cancer Society 2017). Breast and prostate cancer have 90% and 99% 5-year survival rates, respectively, whereas the rate for lung cancer is 18%, and for pancreatic cancer 8%. This large discrepancy in the survival rate is mainly due to early detection. Breast and prostate cancer both have well-established screening guidelines, so these cancers are often detected at an early (local) stage when they are easier to treat. There is no recommended screening for pancreatic cancer: Normally it is not discovered until a very late stage. The American Cancer Society reports relative 5-year survival rates by stage at diagnosis for 18 cancer types (American Cancer Society 2017). The average overall 5-year survival rate for these 18 cancers is 61.3%, whereas the average rate for cancers detected at an early (local) stage is 78.2%. Using these numbers as a rough estimate and, without any improvements to treatment, early detection can improve survival by nearly one-third. Therefore, early detection is essential.

At the time of writing (March 2018), the American Cancer Society's article on "Guidelines for the Early Detection of Cancer" only discusses six types of cancer (The American Cancer Society Medical and Editorial Content Team 2017); however, there is ongoing research in early detection methods for numerous cancers (e.g., see Macdonald et al. 2017, 198–213; Oeffinger et al. 2015, 1599; Nardi-Agmon and Peled 2017, 31–38; Nolen et al. 2015, 111–119). These screening methods vary greatly from observing moles to CT scans to biopsies. However, many are intrusive, expensive, and/or have non-negligible risks (American Cancer Society 2017; Bach et al. 2012, 2429; Wolf et al. 2010, 70–98; O'Connor and Hatabu 2012, 698–708). A simple nonintrusive way of determining disease risk would allow health care providers and individuals to better weigh the costs and benefits of screenings on an individual level. Better stratification of cancer risk would decrease the burden—both to the individual and the health care system—resulting from high false-positive rates, overdiagnosis, and follow-up tests.

In this chapter, we discuss how the methods of machine learning can be used to develop such a nonintrusive tool for risk stratification and screening recommendations. In Section 16.2, we will discuss the "big data" approach and the data that we use. In Section 16.3, we will describe artificial neural networks (ANNs), apply that description to one cancer, and discuss its application to a wider range of cancers. In Section 16.4, we will discuss traditional statistical and other machine learning methods in comparison with our ANN. In Section 16.5, we discussion our vision for how this approach can be used to benefit individuals, health care providers, and the health care system.

16.2 A BIG DATA APPROACH TO EARLY CANCER DETECTION AND PREVENTION

A quick literature review will reveal studies that often look at only one risk factor at a time for tumor-specific cancers. However, in order to attain a more complete understanding of cancer risk, many of the risk factors must be considered at once. Studying the synergetic effects of multiple risk factors requires big data because the frequency of certain combinations of parameters will be missed in smaller samples. For example, we know smoking increases lung cancer risk, but how does lung cancer risk vary if also considering chronic obstructive pulmonary disease (COPD), age, sex, body mass index (BMI), race, and asthma? Using combinations of easy-to-understand clinical parameters in consort with big data to predict cancer would allow patients to receive the necessary screening at the early stages of the disease, thereby reducing mortality rates and financial costs.

The use of machine learning in the prediction and prognosis of cancer is not new (e.g., Cruz and Wishart 2006, 59–77; Kourou et al. 2015, 8–17; Bibault, Giraud, and Burgun 2016, 110–117). However, most of those studies relied on detailed clinical data such as blood tests, imaging, biopsy, and/or DNA sequences. Although these types of data produce good results and should be used when available, the difficulty of gathering such data often leads to relatively small data sets. Furthermore, the intrusiveness and expensive of gathering such data both necessitates and limits the cheap, nonintrusive type of model we are proposing.

To overcome these difficulties, we focus on using data that is routinely gathered and stored in electronic health records (EHRs) such as age, BMI, and existing diseases. The direct use of EHR data is complicated by legal and privacy issues as well as issues in extracting the data accurately. For the purposes of this chapter, we turn to similar data already in the public domain. Specifically, we use the National Health Interview Survey (NHIS) sample adult data sets (Centers for Disease Control and Prevention 2017b) from 1997 to 2016, except for 2004, which does not contain some of the basic information such as sex and age. This combined data set contains information on more than 588,000 individuals, more than 47,100 of which have had at least one type of cancer.

Table 16.1 **The parameters that can be used in our models after selecting a specific cancer type**

PARAMETER	INPUT TYPE	INPUT RANGE	DETAILS
Sex	Binary	0 or 1	0 is a man, and 1 is a woman.
Age	Continuous	0.2118–1	Age range is 18–85, with >85 is treated as 85.
BMI	Continuous	0 to 1	BMI >99.95 is treated as 99.95.
Smoker	Binary	0 or 1	Never-smokers are 0, and current and former smokers are 1.
Emphysema	Binary	0 or 1	No emphysema is 0, and emphysema is 1.
Asthma	Binary	0 or 1	No asthma is 0, and asthma is 1.
Diabetes	Binary	0 or 1	Nondiabetics and prediabetics are 0, and diabetics are 1.
Strokes	Binary	0 or 1	No strokes is 0, and having had a stroke is 1.
Hypertension	Binary	0 or 1	No recording of hypertension is 0, and having single measurement of it is 1.
Heart disease score	Continuous	0 to 1	Coronary heart disease, angina, heart attacks, and other heart complications each contribute 0.25 to the score.
Race	Continuous	0.0083–1	Each race is assigned a value equal to its fractional percentage in the sample plus the fractional percentage of each less common race being added to the race of interest.
Hispanic ethnicity	Binary	0 or 1	No Hispanic ethnicity is 0, and having Hispanic ethnicity is 1.
Vigorous exercise	Continuous	0 to 1	Number of times per week vigorous exercise is performed, with the >28 being treated as 28. All years' criteria were ≥10 min, with the exception of the first half of 1997, which was 20 min.

From this data set, we extracted 13 parameters (Table 16.1) and used them to predict each of the 30 types of cancer recorded in the survey. Some of the parameters we used, such as BMI, can change over time and all of them need to be observed before cancer is readily detectable in order to be useful as predictors. In an effort to balance this time dependence with maximizing the number of samples, we discarded all samples for which the diagnosis of cancer occurred more than 4 years before the survey was taken. Then, we used only data for those who answered all the questions about the parameters we were using.

It should be noted that, aside from smoking, we were not focusing on known indicators or risk factors such as ultraviolet light exposure, family history, or human papillomavirus (HPV) infection. Some of this information has been gathered for a subset of the data set we are using (years 2000, 2005, 2010, and 2015), but using it would reduce our data set to one-fifth its current size. The careful observer will also note that other improvements could have been made here, such as using more specific data like pack-years smoked instead of just asking, "Have you ever smoked?" All of these things could improve the predictive power of our models, but the point of the models presented here is to show that even a relatively simple model can perform well.

16.3 A MULTI-PARAMETERIZED ARTIFICIAL NEURAL NETWORK MODEL FOR CANCER PREDICTION

16.3.1 WHAT IS AN ARTIFICIAL NEURAL NETWORK?

An ANN is a machine learning method that relies on optimization theory in which inputs are fed-forward to produce predictions and then the error can be backpropagated to train the model. (See the simple tutorial in Stanford University [2015]; the Bishop [2006] textbook is also good.) An ANN is a collection of artificial neurons or nodes, each of which is a computational unit. Each neuron takes a number of inputs, computes a weighted sum of these inputs, and then uses this sum in an activation function to produce its output:

$$f\left(\sum_{i=1}^{n} W_i x_i + b\right),$$

where $\{x_i\}$ are the inputs, $\{W_i\}$ are the weights, $f(\ )$ is the activation function, and b is known as the bias and is analogous to the intercept term in linear regression. Neurons are organized into layers. The neurons in a layer receive their inputs from the neurons of the previous layer and pass their outputs to the neurons in the next layer (Figure 16.1). There are three types of layers: input, hidden, and output. The input layer has a neuron for each input parameter and may scale the data, but it generally leaves the data untouched and passes it to the first hidden layer. There can be one or more hidden layers. The number of hidden layers and the number of neurons in each hidden layer are chosen to balance accuracy and overfitting. The last hidden layer's output is passed to the output layer. As the name implies, the output layer generates the output (predictions) that we are interested in. A neural network is trained through the selection of appropriate weights and bias terms. The network makes predictions and learns from its mistakes via a backpropagation algorithm (see Section 16.3.2).

For this chapter we used a single neural network architecture and common set of inputs to train 30 neural networks, 1 for each type of cancer reported in the NHIS data. We use 13 inputs, two hidden layers with 12 neurons each, and a single neuron in the output layer indicating whether a person has that type of cancer or not. We chose two hidden layers because most bounded polynomial functions can be fit with a two-layered neural network (Andoni et al. 2014). Meaning, if we assume that cancer risk is a function of our input parameters, we can write it as a polynomial using a Taylor series expansion. Therefore, in principle, a two-layered ANN should prove adequate.

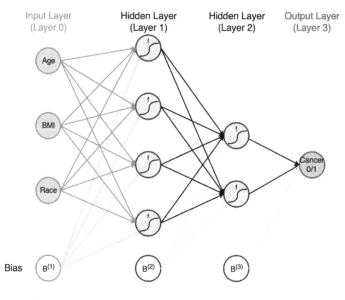

Figure 16.1 A schematic of an ANN. Each circle is a neuron. Each line is a weight (lighter lines are the biases). The number of weights (and biases) grows rapidly with the number of neurons. Here we have 10 neurons (plus 3 bias neurons) and 22 weights (plus 7 biases).

16.3.2 TRAINING AN ARTIFICIAL NEURAL NETWORK

The standard method for the training of an ANN is via backpropagation of errors with a gradient-based optimizer, such as gradient descent. Backpropagation allows for the efficient evaluation of the partial derivatives in the gradient and allows us to measure how much each node was "responsible" for the error in the output. In this section, we present an example of this algorithm for our neural network architecture using gradient descent, a sum-of-squares error function, and sigmoidal activation function.

Let the matrix X contain all the training examples' inputs, with each row being a type of input (BMI, age, and so on; 13 total) and each column being a sample (information for a single patient, N total). The variable Y will be a row vector with each entry containing the output of an individual sample (cancer status of a single patient, length N). Let the bias terms $B^{(n)}$ (added to the inputs of the nth layer) be column vectors of 1's with length equal to the number of neurons in the next layer. For our architecture the length of $B^{(1)}$ is 12, $B^{(2)}$ is 12, and $B^{(3)}$ is 1.

The weights connecting the neurons in layer $n-1$ to those in layer n is a matrix, $W^{(n)}$. There is a plethora of literature on the topic of weight initialization, but we used a simple selection of uniform random numbers between –1 and 1. The number of rows in each weight matrix is equal to the number of neurons in layer n and the number of columns is equal to the number of neurons in layer $n-1$. Because the bias terms are analogous to intercepts, they do not affect the dimensions of the $W^{(n)}$. For our architecture the dimensions of $W^{(1)}$ are 12×13, $W^{(2)}$ are 12×12, and $W^{(3)}$ are 1×12. The number of neurons to select in each hidden layer has no hard set of rules; the reader must learn based on trial and error (i.e., the results of your simulation). We are using 12 for each hidden layer.

With the initialization outlined, we will outline the training algorithm on a general ANN with two hidden layers. The number of inputs is N_i. The first hidden layer has N_1 neurons and the second hidden layer has N_2 neurons. The number of outputs is N_o. We assume N samples in the training data. The training is broken into two main parts: feed-forward and backpropagation. The feed-forward process gives the models prediction based on the current weights and biases, and the backpropagation provides an efficient means of calculating the gradient of the error with respect to the weights and biases. This information is used to update the weights and biases and the process repeats until converged or the error is below a desired threshold. The feed-forward algorithm proceeds as follows:

1. Calculate the (summed) input matrix for the first hidden layer: $Z^{(1)} = W^{(1)} * X + B^{(1)}$.

2. Use $Z^{(1)}$ to calculate the activation matrix for the first hidden layer: $a^{(1)} = \dfrac{1}{1 + \exp\left(-Z^{(1)}\right)}$.

3. Use $a^{(1)}$ to calculate input matrix for the second hidden layer: $Z^{(2)} = W^{(2)} * a^{(1)} + B^{(2)}$.

4. Use $Z^{(2)}$ to calculate the activation matrix for the second hidden layer: $a^{(2)} = \dfrac{1}{1 + \exp\left(-Z^{(2)}\right)}$.

5. Use $a^{(2)}$ to calculate input matrix for the output (third) layer: $Z^{(3)} = W^{(3)} * a^{(2)} + B^{(3)}$.

6. Use $Z^{(3)}$ to calculate the output (activation) matrix for the output layer: $a^{(3)} = \dfrac{1}{1 + \exp\left(-Z^{(3)}\right)} = \overline{Y}$.

With the ANN's predictions $\overline{Y}$, and the correct values Y, we can calculate the cost, $E = \dfrac{1}{N} \sum_{n=1}^{N} \dfrac{1}{2}\left(Y - \overline{Y}\right)^2$. We update the weights and biases based on the gradient of the cost,

$$W_{ij}^{(n)} = W_{ij}^{(n)} - \varepsilon \frac{\partial}{\partial W_{ij}^{(l)}} E\left(W, b\right), \ b_j^{(n)} = b_j^{(n)} - \varepsilon \frac{\partial}{\partial b_j^{(l)}} E\left(W, b\right).$$

Where ε is the learning rate and controls the size of step in every iteration of the algorithm. If ε is too small, the algorithm is slow to converge and can get caught in the shallowest of local minimums. If it is too big, the algorithm is unstable and may never converge.

Big data in radiation oncology

The backpropagation algorithm gives an efficient way to evaluate these partial derivatives. It uses the chain rule to calculate an error term, $\delta_{in}^{(l)}$, for each neuron, i, for each sample, n, for layer l, based on a weighted average of the error terms of the neurons that use its output for their input. Thus, we are propagating the error backward through the network. Before laying out the algorithm, we remind the reader that these details depend on both the form of the cost function and the activation function. The algorithm proceeds as follows:

1. Calculate the error term matrix for the output (third) layer:

$$\delta_{in}^{(3)} = \frac{\partial}{\partial Z_{in}^{(3)}} \frac{1}{2} \left(Y_{in} - \bar{Y}_{in} \right)^2 = -\left(Y_{in} - a_{in}^{(3)} \right) * \left(a_{in}^{(3)} * \left(1 - a_{in}^{(3)} \right) \right).$$

2. Use $\delta^{(3)}$ to calculate the error term matrix for the second hidden layer:

$$\delta_{in}^{(2)} = \left(\sum_{j=1}^{N_o} W_{ji}^{(2)} \delta_{jn}^{(3)} \right) * \left(a_{in}^{(2)} * \left(1 - a_{in}^{(2)} \right) \right).$$

3. Use $\delta^{(2)}$ to calculate the error term matrix for the first hidden layer: $\delta_i^{(1)} = \left(\sum_{j=1}^{N_2} W_{ji}^{(1)} \delta_{jn}^{(2)} \right) * \left(a_{in}^{(1)} * \left(1 - a_{in}^{(1)} \right) \right).$

4. Calculate the partial derivatives for the weights: $dW^{(3)} = \delta^{(3)} * \left(a^{(2)} \right)^T$, $dW^{(2)} = \delta^{(2)} * \left(a^{(2)} \right)^T$, and $dW^{(1)} = \delta^{(1)} * X^T$.

5. Calculate the partial derivatives for the biases: $B_i^{(3)} = \sum_{j=1}^{N} \delta_{ij}^{(3)}$, $B_i^{(2)} = \sum_{j=1}^{N} \delta_{ij}^{(2)}$, and $B_i^{(1)} = \sum_{j=1}^{N} \delta_{ij}^{(1)}$.

6. Update weights and biases, normalizing for the number of samples: $W^{(n)} = W^{(n)} - \epsilon * dW^{(n)} / N$ and $B^{(n)} = B^{(n)} - \epsilon * dB / N$.

This simple gradient descent algorithm works; there are, however, several ways to speed up its convergence. There are three modifications we employed, which we briefly describe here and leave up to the reader to implement in the language of their choice. The first change was to check that the cost function decreased with each iteration, and if it increased, we selected the weights and biases from the previous iteration and decreased ϵ. If the cost function increased, the network would be losing performance, likely because ϵ was too large or we were nearing a minimum. The second change was either to use the momentum approach or to simply increase the learning rate each time the cost function decreased in the previous iteration. This allowed the algorithm to converge faster if it was already moving in the "right" direction. Modification three was to scramble the training data after each iteration. While not necessary, this may help in certain circumstances.

We used the conventional validation method, randomly selecting 70% of the data to make up the training set and the remaining 30% for the validation set. In creating these two data sets, we divided the cancer and noncancer cases separately to guarantee that the ratio of cancer to noncancer patients was the same in both sets. Because the cancer cases make up a small fraction of the overall data set, we did this to guarantee that both sets got a significant number of cancer cases. After training the ANN on the training set, various measures of accuracy could be calculated (see Section 16.3.3) on the training set and the validation set (which the model had never seen), then we compared them to test the model's generalizability or if it overfit the training data.

16.3.3 MEASURING ACCURACY

We wanted our model to be accurate (to make good predictions) and generalized (to predict well on new data sets). To evaluate these, we first used the ANN to make predictions for both the training and validation sets. With these predictions and the known states for these data sets we could calculate the true-positive rate or *sensitivity*—the fraction of those who have cancer that the model correctly identifies as having cancer—and the true negative rate or *specificity*—the fraction of those who do not have cancer that the model correctly identifies as not having cancer:

$$\text{Sensitivity} = \frac{\text{\# true positives}}{\text{\# positive conditions}}, \text{Specificity} = \frac{\text{\# true negatives}}{\text{\# negative conditions}}.$$

If both values are high for the training data set, the model is accurate; and if these values for the validation set are similar to the values for the training set, the model is generalized. In calculating sensitivity and specificity, it is assumed that a binary prediction is made—either someone has cancer or they do not—and ideally our model would give a binary prediction, a 0 or 1. After all, the data it was trained on had binary states. Nevertheless, in practice it returns a number in the range of [0,1].

Because of this, we must choose a threshold, anything above the threshold is a 1 and anything below it is a 0. Instead of calculating sensitivity and specificity as single numbers, we calculate them as functions of this threshold. When varying the threshold, there is a trade-off between sensitivity and specificity, so depending on the desired application, the threshold can be chosen to have high sensitivity, high specificity, or we could try to balance them.

Having the sensitivity and specificity as a function of the threshold allows us to use another measure of accuracy. We can plot the false-positive rate or 1–specificity against sensitivity in what is known as the receiver operating characteristic curve (ROC). This plot effectively captures the trade-off between sensitivity and specificity and the area under the curve (AUC) is used as a measure of predictive power. An AUC of 0.5 is a worthless test (equivalent to random guessing), and higher AUC values mean a more significant test, with 1.0 being a perfect test.

In addition to calculating these different measures of accuracy, we can also calculate 95% confidence intervals (CIs) based on the sample size to indicate their significance.

16.3.4 PREDICTING CANCER WITH AN ANN

We are now ready to apply our ANN to the problem of cancer prediction. First, we apply our model to a single cancer type, Lung, and look in depth at how it performs. Then we apply it to all 30 cancer types in our data set.

16.3.4.1 Case study: Lung cancer

For an in-depth look at how our ANN performs, we focus on lung cancer. Lung cancer was chosen for this case study for several reasons. First, it is the most common kind of cancer in the United States and is killing the greatest number of people (American Cancer Society 2017). Next, early detection can greatly benefit lung cancer patients. Currently, the overall 5-year survival rate is 18%; however, with early detection, it is 55% (American Cancer Society 2017). Last, the current screening guidelines are relatively risky, expensive, and time consuming. A low-dose CT scan (LDCT) is recommended for those over 55 years of age and with at least 30 pack-years smoked. The screening has a high false-positive rate, leading to follow-up imaging, tests, and/or biopsies. Those not in this high-risk group would increase their cancer risk by being screened (American Cancer Society 2017; Bach et al. 2012, 2429; O'Connor and Hatabu 2012, 698–708). An ANN trained to predict cancer risk can help refine who is screened.

The ANN presented in the preceding Sections 16.3.2 2013 16.3.3 was trained on 341,893 noncancer samples and 454 cancer samples. The validation set had 146,526 noncancer samples and 195 cancer samples. For the training data, the sensitivity was 75.3% (95% CI: 71.4%, 79.3%), specificity was 85.2% (95% CI: 85.1%, 85.3%), and AUC was 0.86 (95% CI: 0.85, 0.88). For the validation data, the sensitivity was 73.3% (95% CI: 67.1%, 79.5%), specificity was 85.2% (95% CI: 85.0%, 85.4%), and AUC was 0.86 (95% CI: 0.84, 0.89). As discussed above, these values are for a specific threshold value; a plot of the sensitivity and specificity as a function of the threshold can be seen in Figure 16.2. The ROC curve is shown in Figure 16.3.

Although our model's sensitivity is not as high as that of LDCT, its specificity is higher. A high specificity amounts to a low false-positive rate, meaning if the model predicts someone has cancer they mostly likely do. Therefore, our model can be used to identify those that will benefit most from LDCT. We formalize this by using the model to stratify individuals into different risk categories. To do this we do not use a threshold value, but instead normalize the output of the model to produce a percentage of risk. Using this risk, we divide the population into three categories: low risk (green), medium risk (yellow), and

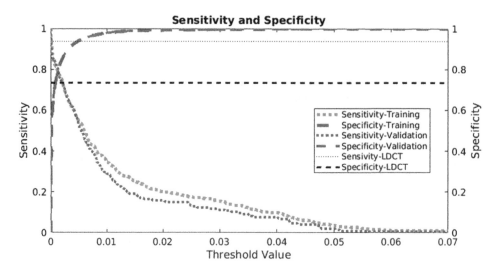

Figure 16.2 Note the trade-off between sensitivity and specificity as a function of the threshold value. The difference between the training and validation performance is small, indicating that the model generalizes well. We have marked for reference the sensitivity and specificity for the LDCT screening which are 93.8% (95% CI: 90.6%, 96.3%) and 73.4% (95% CI: 72.8%, 73.9%), respectively.

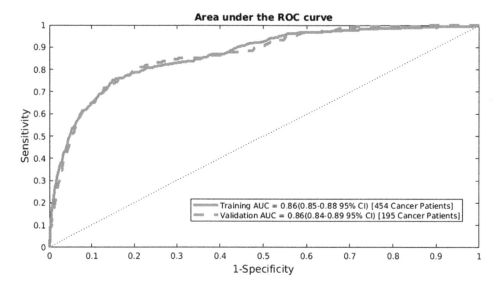

Figure 16.3 An ROC plot for our ANN's training and validation data sets with the corresponding 95% confidence intervals. Again, the performance between the training and validation is very similar indicating good generalizability. However, the validation set has less than half the data the training set has and so the confidence interval is a little larger.

high risk (red). The boundaries are chosen based on the training data. The boundaries are chosen so that no more than 1% of the cancer cases are classified as low risk and 1% of the noncancer cases are classified as high risk (see Figure 16.4).

For these three risk categories, we suggest that those of high risk should immediately be screened, those of medium risk should continue the following the standard guidelines, and those of low risk do not need to be screened as frequently. The results of this stratification are shown in Table 16.2, where we can see that over 15% of the individuals with cancer are marked as high risk. Almost 15% of those without cancer are marked as low risk, leaving about 85% of both groups as medium risk. Because the current screening guidelines are based on age and amount smoked, many of these will not be recommended for screening.

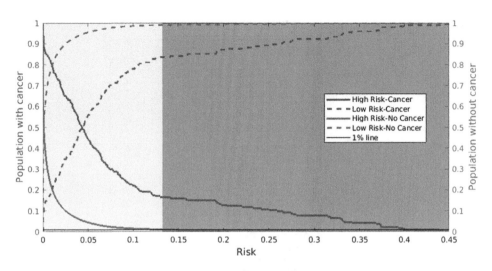

Figure 16.4 Cumulative distribution function for high risk (solid line) and low risk (dashed line) population without cancer (orange) and population with cancer (blue) in the validation data set. Allowing for a 1% misclassification rate in the training data (black line), we can divide individual cancer risk into three categories: high (red), medium (yellow), and low (green, which is too narrow to see on the left of this figure).

Table 16.2 **The results for stratifying the validation data set**

	LOW RISK		MEDIUM RISK		HIGH RISK	
	NUMBER	PERCENT	NUMBER	PERCENT	NUMBER	PERCENT
Cancer	0	0.0	163	83.6	32	16.4
No cancer	19,323	13.2	125,737	85.8	1,465	1.0

We would like to remind the reader that the simple model presented here relied solely on whether someone was a current smoker or had ever smoked, and it was still able to catch 15% of the people with cancer. If we used more detailed information such as the amount smoked, we could further shrink the medium risk category and increase the number of those put in the low- and high-risk categories.

16.3.4.2 Results for all cancers

Having explored lung cancer in some detail, we now discuss training the ANN on all 30 types of cancer reported in the NHIS data and give the results here. It should be noted that not all of the "cancer types" reported in the NHIS data are clinically useful or combined the way they are clinically. For example, cancer of the colon and cancer of the rectum are reported separately instead of as a single colorectal cancer. If someone had skin cancer but could not remember whether it was melanoma or not, it was marked as skin cancer of unknown kind (Skin-DKkind). There is also another category for cancers that did not fit in the other 29 types. Although these are not particularly useful in the clinic we include them anyway, because our goal currently is to show wide applicability of the ANN in predicting cancer. Remember, we are not currently creating exact models for clinical use, we are showing their viability and will build up to fully relevant models. In evaluating the ANN for each cancer type, we were careful to use only the appropriate subpopulation for sex-specific cancers, such as cervical or prostate cancer, to avoid overestimating our accuracy. In doing this, we evaluated breast cancer separately for men and women, and thus had 31 cancer types.

Let us begin by reviewing the AUCs for the different cancer types (Figure 16.5). The mean AUC for the validation sets is 0.732. It is generally accepted that to be clinically relevant we need an AUC of 0.8 or higher. Nine of our 31 models have an AUC above 0.8. Twelve more have AUCs between 0.7 and 0.8, with 2 very close to 0.8. Ten have AUCs below 0.7.

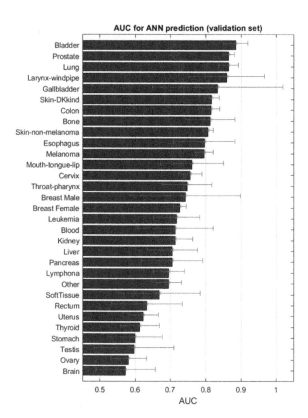

Figure 16.5 Training the ANN on each of the 30 cancers in the NHIS data. Here we show the AUCs from the data sets with 95% confidence intervals. Sex-specific cancers were only evaluated on the data from the pertinent sex.

Our simple ANN can predict one-third of the cancer types with the accuracy needed to be clinical useful. This is remarkable considering that it was built without any consideration of the cancer type and using just readily available information. If we were to go back and use more detailed information such as amount smoked, add more variables such as alcohol use or family history of cancer, and/or add cancer type specific risk factors such as tanning or radiation exposure, we could further strengthen our predictive power. This would likely allow us to push the AUCs of the middle third of cancer types into the clinically relevant range.

Furthermore, some of the cancer types are very rare in the data set; three of those in the least accurate third have less than 100 cancer cases each and another has only 123 cases. Regardless of our method, it is hard to create a robust predictive model with so little data.

Acknowledging that only about one-third of our ANNs are accurate enough to be clinically useful, we can still show how the stratification of risk (see Section 16.3.4.1) works on all the cancer types (Table 16.3). Averaging the percentage of cancer cases marked as high risk we get 5.3%; for noncancer cases marked as low risk we get 11.9%, leaving an average of 92.9% and 87.1% of the cancer and noncancer cases, respectively, marked as medium risk. Because most of these cancers do not have standard screenings, being able to identify even 5% with cancer as high risk can help individuals investigate options to detect cancers that they might not otherwise think about. In addition, for those cancers with standard screenings, identifying 54% (prostate) or 20% (colon) of the population without cancer as low risk and needing less frequent screenings or cheaper, but less accurate, screenings would ease the expense and burden of screenings and allow more focus on those at risk.

Table 16.3 **The results for stratifying the validation data set for each cancer type**

		LOW RISK		MEDIUM RISK		HIGH RISK	
		NUMBER	PERCENT	NUMBER	PERCENT	NUMBER	PERCENT
Bladder	Cancer	2	1.8	94	84.5	14	12.7
	No cancer	46,342	31.6	98,682	67.4	1,502	1.0
Prostate	Cancer	5	0.9	507	94.2	26	4.8
	No cancer	36,183	54.4	29,587	44.5	692	1.0
Lung	Cancer	0	0.0	163	83.6	32	16.4
	No cancer	19,323	13.2	125,737	85.8	1,465	1.0
Larynx-windpipe	Cancer	0	0.0	12	92.3	1	7.69
	No cancer	258	0.2	144,838	98.9	1,430	1.0
Gallbladder	Cancer	1	20.0	4	80.0	0	0.0
	No cancer	98,029	66.9	47,083	32.1	1,414	1.0
Skin-DKkind	Cancer	5	1.3	372	95.1	14	3.6
	No cancer	31,833	21.7	113,212	77.3	1,480	1.0
Colon	Cancer	1	0.3	299	95.8	12	3.9
	No cancer	30,477	20.8	114,541	78.2	1,508	1.0
Bone	Cancer	0	0.0	37	100.0	0	0.0
	No cancer	3,147	2.2	141,908	96.9	1,471	1.0
Skin-non-melanoma	Cancer	11	1.4	751	91.9	55	6.7
	No cancer	34,213	23.4	110,784	75.6	1,528	1.0
Esophagus	Cancer	0	0.0	24	92.3	2	7.7
	No cancer	2,682	1.8	142,391	97.2	1,453	1.0
Melanoma	Cancer	1	0.4	259	91.5	23	8.1
	No cancer	12,435	8.5	132,672	90.6	1,419	1.0
Mouth-tongue-lip	Cancer	0	0.0	24	88.9	3	11.1
	No cancer	4,654	3.2	140,366	95.8	1,506	1.0
Cervix	Cancer	5	2.5	178	89.9	15	7.6
	No cancer	6,643	8.3	72,910	90.7	855	1.1
Throat-pharynx	Cancer	2	4.4	38	84.4	5	11.1
	No cancer	11,965	8.2	133,094	90.8	1,466	1.0
Breast (male)	Cancer	0	0.0	8	88.9	1	1.0
	No cancer	4,848	7.3	60,677	91.7	658	1.0
Breast (female)	Cancer	6	0.9	668	95.8	23	3.3
	No cancer	9,963	12.4	69,574	86.6	806	1.0
Leukemia	Cancer	1	1.9	50	92.6	3	5.6
	No cancer	2,313	1.6	142,739	97.4	1,474	1.0

(*Continued*)

Table 16.3 (*Continued*) **The results for stratifying the validation data set for each cancer type**

		LOW RISK		MEDIUM RISK		HIGH RISK	
		NUMBER	PERCENT	NUMBER	PERCENT	NUMBER	PERCENT
Blood	**Cancer**	0	0.0	20	100.0	0	0.0
	No cancer	22,511	15.4	122,615	83.7	1,400	1.0
Kidney	**Cancer**	5	5.4	85	91.4	3	3.2
	No cancer	10,107	6.9	134,876	92.1	1,543	1.1
Liver	**Cancer**	0	0.0	45	97.8	1	2.2
	No cancer	11,154	7.6	133,900	91.4	1,472	1.0
Pancreas	**Cancer**	0	0.0	30	96.8	1	3.2
	No cancer	1,549	1.1	143,546	98.0	1,431	1.0
Lymphoma	**Cancer**	0	0.0	120	97.6	3	2.4
	No cancer	2,339	1.6	142,739	97.4	1,448	1.0
Other	**Cancer**	2	1.1	170	93.9	9	5.0
	No cancer	11,333	7.7	133,764	91.3	1,429	1.0
Soft tissue	**Cancer**	0	0.0	18	100.0	0	0.0
	No cancer	22,065	15.1	123,011	84.0	1,449	1.0
Rectum	**Cancer**	0	0.0	24	100.0	0	0.0
	No cancer	3,297	2.3	141,859	96.8	1,370	0.9
Uterus	**Cancer**	2	1.5	123	91.8	9	6.7
	No cancer	3,531	4.4	75,984	94.	790	1.0
Thyroid	**Cancer**	1	1.3	75	93.8	4	5.0
	No cancer	5,770	3.9	139,330	95.1	1,425	1.0
Stomach	**Cancer**	0	0.0	37	92.5	3	7.5
	No cancer	614	0.4	144,336	98.5	1,576	1.1
Testis	**Cancer**	1	5.3	17	89.5	1	5.3
	No cancer	7,807	11.8	57,672	87.2	663	1.0
Ovary	**Cancer**	6	6.5	86	92.5	1	1.1
	No cancer	3,677	4.6	75,349	94.4	799	1.0
Brain	**Cancer**	0	0.0	34	100.0	0	0.0
	No cancer	678	0.5	144,322	98.5	1,536	1.0

Note: The number of cancer cases for several of types of cancer is not large enough to draw meaningful conclusions from this.

16.4 OTHER MACHINE LEARNING ALGORITHMS FOR BIG DATA MINING AND CANCER PREDICTION

We will now briefly introduce several other methods that could be applied to this model and compare them with the ANN results.

16.4.1 PREQUEL TO MACHINE LEARNING: CONDITIONAL PROBABILITY ANALYSIS

The first approach we tried was for overall cancer risk prediction (all tumor types lumped into one) using the definition of conditional probability. We specified a set of parameters and then calculated the fraction of people in that subset who had cancer. Although intuitively easy to understand, this approach suffers from small sample sizes (specify too many parameters and the number of people matching those criteria will quickly reach zero) and, hence, large confidence intervals. We only used this to compute probabilities, with the results being shown in Figure 16.6.

A modification of this method is to instead divide the sample into cancer and noncancer samples and *then* compute the probability of having a set of parameters given cancer or no cancer. Bayes' Rule can then be applied to flip this to be the probability of acquiring cancer given a set of parameters. However, this method has the same flaws as the previous method.

16.4.2 MULTIVARIABLE LOGISTIC REGRESSION APPLIED TO LUNG CANCER PREDICTION

Multivariable logistic regression is designed to fit a function of n variables with $n + 1$ coefficients, with the extra value being an intercept term and equivalent to an ANN with all the hidden layers removed. Two pros of this approach when compared with a multilayer ANN are a faster run time for the code and its simplicity. However, there is a significant con: In general, because there are fewer coefficients/weights to be fit, the results will typically be inferior to those of a multilayer ANN.

We now apply the multivariable logistic regression to this simple lung cancer (please note that the training and validation sets were exactly the same as those used in the ANN). Figure 16.7 shows the sensitivity and specificity as a function of threshold, whereas Figure 16.8 is an ROC curve. We see that its performance is very similar to the ANN's in terms of both accuracy and generalizability.

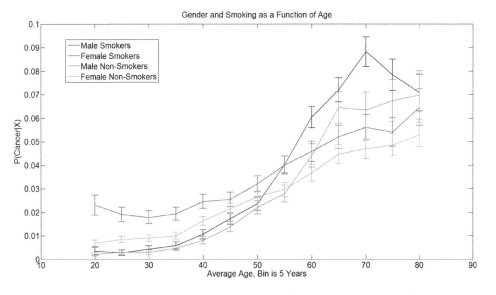

Figure 16.6 The probability of acquiring cancer given sex, smoking status, and age (age is on the x-axis). Notice how age has been discretized to 5 years and the 95% confidence intervals are large. If we add diabetes as a parameter, the size of the 95% confidence interval becomes roughly the same order of magnitude as the probability. This means the results are not useful.

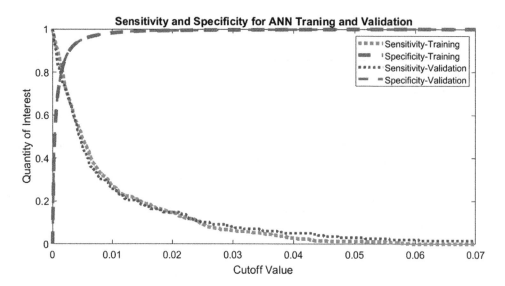

Figure 16.7 Notice how the sensitivity falls off a little faster than with the ANN, but otherwise this is very similar to the ANN results. The sensitivity (76.4%, 95% CI: 70.1%, 82.4%) is slightly higher than the ANN's, but the specificity (81.6%, 95% CI: 81.4%, 81.8%) is slightly lower.

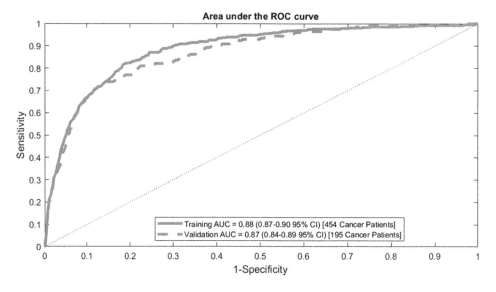

Figure 16.8 The validation AUC is slightly higher than the ANN, but their 95% confidence intervals almost completely overlap.

16.4.3 SUPPORT VECTOR MACHINE APPLIED TO LUNG CANCER PREDICTION

A support vector machine (SVM) is conceptually very simple. Each sample is plotted in the space of the input parameters. Then the algorithm searches for a (hyper)surface that divides the data based on the output value, in our case: cancer or no cancer. Predictions are made by plotting the new points in the input space and "seeing" which side of the surface they fall on (Bishop 2006).

Applying SVM to the case of lung cancer, we see that it fits the training data well but does not generalize to the validation data very well (Figures 16.9 and 16.10). We also note that the trade-off between sensitivity and specificity happens much faster.

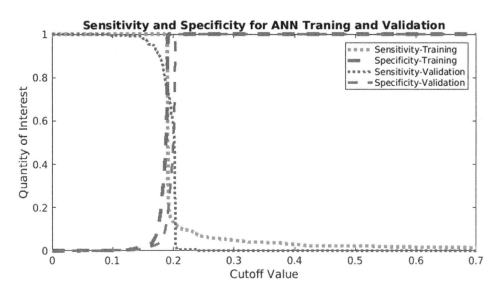

Figure 16.9 Notice how the sensitivity falls off much faster than with the ANN and that the gap in specificity and sensitivity between training and testing is bigger. The sensitivity (74.9% [95% CI: 68.8%, 81.0%]) is slightly higher than the ANN's, but the specificity (16.9% [95% CI: 16.7%, 17.1%]) is much lower.

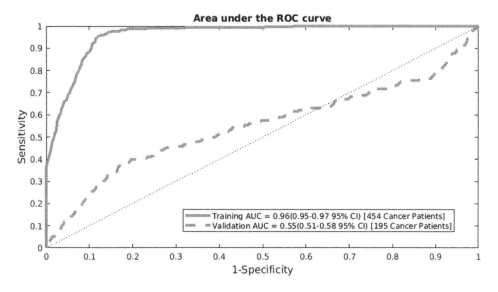

Figure 16.10 The training AUC is much higher than the ANN, but the validation AUC is much lower than the ANN and their 95% CI do not even overlap.

16.4.4 DECISION TREES AND RANDOM FORESTS APPLIED TO LUNG CANCER PREDICTION

Decision trees look at each parameter individually and then sort the inputs of each parameter into a group. Then, subgroups are formed by looking at an additional parameter in each group, with the process continuing onward. Based on the outcomes at the end of each of these decisions, predictions can be made. A random forest is a modification that makes the decision tree algorithm more robust (Quinlan 1986, 81–106; James et al. 2013).

Applying the random forest model to the case of lung cancer, we see that it fits the training data very well but does not generalize to the validation data as well (Figures 16.11 and 16.12). We also note that the sensitivity drops off very quickly and that there is a big discrepancy between the training and validation sensitivity.

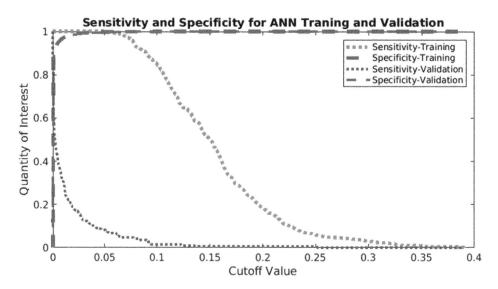

Figure 16.11 Notice how the sensitivity falls off much faster than with the ANN for the validation set and that the gap in sensitivity between training and testing is bigger. The sensitivity (8.2% [95% CI: 4.4%, 81.0%]) is much lower than the ANN's, but the specificity (99.6% [95% CI: 99.5%, 99.6%]) is much higher.

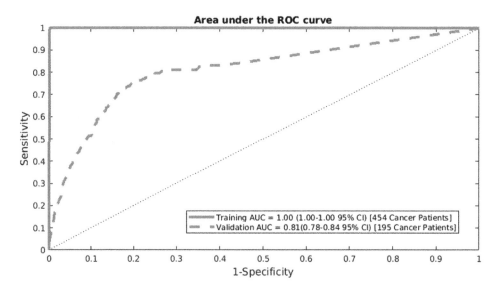

Figure 16.12 The training AUC is much higher than the ANN's, but the validation AUC is lower than the ANN's.

16.5 POTENTIAL APPLICATIONS OF THE MULTI-PARAMETERIZED NEURAL NETWORK MODEL

16.5.1 PERSONALIZED MEDICINE

Instead of using generalized screening guidelines, multi-parameterized models would provide each individual patient with a yes or no answer for screening based on his or her own clinical parameters or a risk value. By taking this approach as an early warning system, cancer mortality would drop, as also would the financial costs of health care. This would also reduce unnecessary screening and tests.

As of right now, personalized medicine is currently based off genetic or molecular information of patients to predict how they will respond to treatment (Hamburg and Collins 2010, 301–304). According to Hamburg, as of 2010 only 10% of FDA approved drugs have pharmacogenomic information (i.e., noting how genetic information impacts drug performance) on the labels. For example, this has been applied to breast cancer to predict that Trastuzumab will better treat those who overexpress human epidermal growth factor receptor type 2 (HER2).

Despite the rapidly decreasing cost of genetic sequencing, very few people bother with the test. As of 2014, the cost of sequencing was $2,500 and fewer than 1 million people had participated (Topol 2014, 241–253). A brief check of the medical records of 192 cancer patients at our institution revealed that less than 5% had any sort of genetic testing done (as of 2016). This will prove to be an impediment to progress but will gradually be overcome.

16.5.2 MOBILE HEALTH APP AND PLATFORM FOR PATIENT-REPORTED OUTCOME

In addition to clinicians taking advantage of the power of our models, the best way to ensure that screening is done for those at higher risk is to put the results of our models into the hands of patients via smartphones. Ideally, the users would answer a series of simple questions in dialog boxes, with the output telling the user what cancer (if any) to get screened for. If the patient does decide to get screened, they can then send feedback through the app to indicate the test results. This information can be used to further refine the models; however, this would be optional and may not produce enough data to make a difference.

16.5.3 HEALTH INSURANCE INDUSTRY

The major implication for the health insurance companies is that early screening will result in fewer deaths and lower costs in treatment (i.e., insurance profits will increase). Higher-grade cancers require more aggressive treatment and, hence, will be more expensive. In addition, the operation of these models would be essentially free and can limit the number of unnecessary and often expensive screenings.

16.6 CONCLUSION AND OUTLOOK

Machine learning approaches for cancer prediction are very promising, with robust results already being shown for one-third of the cancers reported in the NHIS data. Using commonly known clinical parameters makes it easy for patients to answer a survey to provide accurate data for personalized medicine. This would lead to a better screening process and decreased mortality due to detecting the disease in the early stages.

There are however, complications, one of which is the acquisition of better and larger data sets as a starting point. Although the NHIS data set is vast, we had to compromise between the quality and quantity of data with the 4-year restriction between cancer diagnosis and data collection (a 1-year restriction is ideal, but would shrink even further). It is also missing family history of cancer for many of the years, and this parameter provides vital genetic information for cancer prediction. Furthermore, many of the cancer types appear infrequently in the data. The best source of data would be EMRs to ensure the parameters are recorded at the time of and leading up to cancer diagnosis.

Machine learning is a cutting edge technique; however, it will take time to integrate this as a universally accepted method of cancer prediction. If the reader has a way of accessing a better data set, we encourage the use of that data set to improve upon what we have presented here.

REFERENCES

American Cancer Society. 2017. Cancer Facts and Figures 2017.
Andoni, A., R. Panigrahy, G. Valiant, and L. Zhang. 2014. "Learning Polynomials with Neural Networks." *Proceedings of Machine Learning Research.*
Bach, P. B., J. N. Mirkin, T. K. Oliver, C. G. Azzoli, D. A. Berry, O. W. Brawley, T. Byers et al. 2012. "Cost-Effectiveness of Computed Tomography Lung Cancer Screening." *JAMA* 307 (22): 2429.
Bibault, J.-E., P. Giraud, and A. Burgun. 2016. "Big Data and Machine Learning in Radiation Oncology: State of the Art and Future Prospects." *Cancer Letters* 382: 110–117.
Bishop, C. M. 2006. *Pattern Recognition and Machine Learning.* Springer.
Centers for Disease Control and Prevention. 2017a. "Deaths and Mortaility." National Center for Health Statistics, https://www.cdc.gov/nchs/fastats/deaths.htm.

Centers for Disease Control and Prevention. 2017b. "NHIS - Data, Questionnaires and Related Documentation," https://www.cdc.gov/nchs/nhis/data-questionnaires-documentation.htm.

Hamburg, M. A. and F. S. Collins. 2010. "The Path to Personalized Medicine." *New England Journal of Medicine* 363 (4): 301–304.

Howlader, N., A. M. Noone, M. Krapcho, D. Miller, K. Bishop, C. L. Kosary, M. Yu, et al. 2017. *SEER Cancer Statistics Review, 1975–2014.* Bethesda, MD: National Cancer Institute.

James, G., D. Witten, T. Hastie, and R. Tibshirani. 2013. *An Introduction to Statistical Learning : With Applications in R.* 1st ed. New York: Springer-Verlag.

Macdonald, I. K., C. B. Parsy-Kowalska, and C. J. Chapman. 2017. "Autoantibodies: Opportunities for Early Cancer Detection." *Trends in Cancer* 3 (3): 198–213.

Nardi-Agmon, I. and N. Peled. 2017. "Exhaled Breath Analysis for the Early Detection of Lung Cancer: Recent Developments and Future Prospects." *Lung Cancer (Auckland, N.Z.)* 8: 31–38.

Nolen, B. M., A. Lomakin, A. Marrangoni, L. Velikokhatnaya, D. Prosser, and A. E. Lokshin. 2015. "Urinary Protein Biomarkers in the Early Detection of Lung Cancer." *Cancer Prevention Research (Philadelphia, PA)* 8 (2): 111–119.

O'Connor, G. T. and H. Hatabu. 2012. "Lung Cancer Screening, Radiation, Risks, Benefits, and Uncertainty." *JAMA* 307 (22): 698–708.

Oeffinger, K. C., E. T. H. Fontham, R. Etzioni, A. Herzig, J. S. Michaelson, Y.-C. Tina Shih, L. C. Walter et al. 2015. "Breast Cancer Screening for Women at Average Risk." *JAMA* 314 (15): 1599.

Quinlan, J. R. 1986. "Induction of Decision Trees." *Machine Learning* 1 (1): 81–106.

Stanford University. "Multi-Layer Neural Network." UFLDL Tutorial, http://ufldl.stanford.edu/tutorial/supervised/MultiLayerNeuralNetworks/.

The American Cancer Society medical and editorial content team. 2017. "Cancer Screening Guidelines | Detecting Cancer Early." ACS website, https://www.cancer.org/healthy/find-cancer-early/cancer-screening-guidelines/american-cancer-society-guidelines-for-the-early-detection-of-cancer.html.

Topol, E. J. 2014. "Individualized Medicine from Prewomb to Tomb." *Cell* 157 (1): 241–253.

Wolf, A. M. D., R. C. Wender, R. B. Etzioni, I. M. Thompson, A. V. D'Amico, R. J. Volk, D. D. Brooks et al. 2010. "American Cancer Society Guideline for the Early Detection of Prostate Cancer: Update 2010." *CA: A Cancer Journal for Clinicians* 60 (2): 70–98.

World Health Organization (WHO). 2017. "WHO Cancer Fact Sheet." World Health Organization, http://www.who.int/mediacentre/factsheets/fs297/en/.

Index

Note: Page numbers in italic and bold refer to figures and tables respectively.

ACID (atomicity, consistency, isolation, durability) properties 28
Adaboost (adaptive boosting) 232
Affordable Care Act (ACA) 194
aggregation layer, role 66
agnostic feature 224
AI (artificial intelligence) 114
Akaike information criteria (AIC) 46
AllegroGraph 36
AMA (American Medical Association) 14
Amazon Artifact 72
Amazon EC2 67
Amazon machine images (AMIs) 67
American Association of Physicists in Medicine (AAPM) Task Group 263 (TG-263) 18
American Board of Medical Specialties (AMBS) 197
American Medical Association (AMA) 14
AMIs (Amazon machine images) 67
artificial intelligence (AI) 114
artificial neural network (ANN) 7–8, 232; accuracy, measuring 270–1; cancer prediction with 271–6; definition 267–8; layers 268; for overall survival prediction 8; schematic for 268; training of 269–70
ASP.NET platform 68
Association of Surgeons of the Netherlands (ASN) 157
associative arrays 30
as treated plan sums (ATPS) 16–17
atomicity, consistency, isolation, durability (ACID) properties 28
auto-extracted features via DL 104–5
automatic dose accumulation 135–6
automation, treatment planning: DVH 101; iso-dose distributions 102; prior knowledge 100; VMAT/IMRT plan optimization 101
autonomic computing systems 63
Azure platform 68

backpropagation algorithm 269–70
Bayesian analysis 83
Bayesian information criterion (BIC) 46

Bayesian network (BN) approach 48; to error detection 114–16, 115, 116; for LC prediction 53; with nested CV 50; novel 49–51; pre-treatment full 52; results 51–2
Bayes net model 115
BD2K (Big Data to Knowledge) 76, 150
Benjamini-Hochberg method 233
Berkeley Data Analysis Stack (BDAS) 36
bias 250
BIC (Bayesian information criterion) 46
big data 2, 243; for biomedicine 79–80; to cancer detection/prevention 266–7; and CER 145–51, 146; collection 168–9; and dark data to smart data 4; definition 145; distributed processing, software platforms for 68–70; ethics and 8–9; exchange 173–6; four Vs 2; in health care informatics and analytics 13–14; medical data specificities 2–3; and medical informatics 185; national and international collaborative initiatives 6; in radiation oncology process 166–8; systems security 38–9; technologies 34–8; in translational radiation oncology 24–6, 25; into translational research platforms 39
Big Data to Knowledge (BD2K) 76, 150
Bigtable concept 35
biologically guided radiotherapy planning 250
biomedical sciences, machine learning in: biomarker discovery 212–13; genetic interactions 211–12, 212; protein function prediction 210–11, 211
bit-depth resampling 223
block learning methods 45
Blue Cloud 68
BN approach *see* Bayesian network (BN) approach
boosted trees–Cox (BT-Cox) model 234
boosted trees–Weibull (BT-Weibull) model 234
boosting model algorithms 84, 232
bootstrapping 47
brachytherapy sources 133
BSON format 35
BT-Cox (boosted trees–Cox) model 234
BT-Weibull (boosted trees–Weibull) model 234
business associate 72

CAD (computer-aided diagnosis) 219–20
cancer 265–6; ANN for 271–6; conditional probability
        analysis 277; detection/prevention, big data
        to 266–7; prediction parameters for **267**;
        prognosis/prediction, application in 42–3;
        risk monitoring and control 139
Cancer Imaging Archive 222
CancerLinQ 76
cancer registries (CRs) 155; architecture 165–6, *165*;
        barriers 164–6; basic data elements 161, **162**;
        centralized model 170; data quality for 164;
        data sources 157–61, **158–61**; hospital–based
        155, 156; optional data elements 161–3,
        162–3; population-based **155**, 156; reporting
        information 164
CAT (computer assisted theragnostics) 175–6, *176*
CDS (clinical decision support) *188*
Centers for Disease Control and Prevention (CDC) 14
centralized data pooling architecture 169–70
CER *see* comparative effectiveness research (CER)
chaotic/haphazard phenomenon 42
classic meta-analysis technique 88
classification models 229, 231
clinical challenges: EHR 183; justice and fair distribution
        196; overview 182–3; physician referral
        challenges 183
clinical decision support (CDS) *188*
clinical informatics, law of 244
Cloud computing 62–3; commercial providers 67–8;
        datacenter for *66*; economics 66–7; for
        handling health care data *62*; medical
        information storage 73–4; overview *63*;
        physical implementation 66; security and
        privacy 70–2; service provision model *64*,
        64–6
Cloud computing in radiation oncology 72; challenges
        76; clinical trials 75–6; information systems in
        72–3; medical computation in 74–5; medical
        information storage 73–4
CloudWatch service 67
CNN (convolutional neural network) 104
colorectal lesion, CT image *225*
column family databases 31
common terminology criteria for adverse events
        (CTCAE) 20
comparative effectiveness research (CER) 146;
        endometrial cancer 149; in health care 147–8;
        NIH 150; prostate cancer 148
Computational Environment for Radiological Research
        (CERR) software system 250
computational technique 91–2
computer-aided diagnosis (CAD) 219–20
computer assisted theragnostics (CAT) 175–6, *176*
conditional probability tables 115
conditioning–integration strategy 84–6, **85**
confounding effects 250

contouring 4
contrasted analysis strategy 90
convolutional neural network (CNN) 104
Cox proportional hazards model 229, 255
cross-validation (CV) technique 46–7
CRs *see* cancer registries (CRs)
CTCAE (common terminology criteria for adverse
        events) 20
cultural challenges 182–3
culture metrics 117
CV (cross-validation) technique 46–7

dashboards 34
data access 6–7
data-driven decision making 4
data exchange: administrator barrier 176–7; centralized
        versus distributed learning architecture 174–6;
        ethical barrier 177; political barrier 177;
        send data out 173–4; send questions in 174;
        technical barrier 177
data farming 112
data granularity 2–3
data interoperability: ontology 172–3; semantic web
        171–2
data loss 71–2
data management 34, *82*
data modeling requirements **29**, 30; column family
        databases 31; distributed NoSQL systems *31*;
        document databases 31; graph databases 32;
        key–value stores 30; NewSQL databases 32
datanodes 69–70
data pooling architecture *170*; centralized model 169–70;
        decentralized model 170; hybrid model 170–1;
        to improve tissue toxicity prediction studies 254;
        problem 252
data processing platforms 36
data relevance 2
data security risks 70–1
data volume 6, **7**
death indexes **160**
decentralized data pooling model 170
decision trees (DTs) 7, 279
decomposition–integration strategy 86, *86*
deep learning (DL) approaches 8, 234; auto-extracted
        features via 104–5; machine learning
        algorithms 45
DeepMind Technologies 4
deformable image registration (DIR) 75, 134
demilitarized zone (DMZ) 38
derivative features 224–5
DFS (distributed file systems) 30
diagnosis and staging information 15–16
Digital Imaging and Communication in Medicine
        (DICOM) 169; library 127–8; MapReduce
        processing paradigm *33*
dimension reduction approaches 83

DIR (deformable image registration) 75, 134
discriminant *versus* generative models 44
distributed file systems (DFS) 30
distributed learning architecture 175–6
distributed NoSQL systems *31*
divide and conquer algorithm 91
DL approaches *see* deep learning (DL) approaches
DMZ (demilitarized zone) 38
document databases 31
domain model validation 228
Donebedian, A. 113
dose calculation 74
dose–distribution models 243
dose painting approach 248
dose report and morbidity analysis 136
dose–volume histogram (DVH) 105; clinical *versus* autopiloted plans *101*; curves 17, *101*; metrics 18, *19*; parameters 255; prediction 103–5
dosimetric variation-controlled model (DVCM) 102
dosimetry optimization 4
DTs (decision trees) 7, 279
Dutch Surgical Colorectal Audit (DSCA) 157
DVCM (dosimetric variation-controlled model) 102
DVH *see* dose–volume histogram (DVH)
DVH prediction: auto-extracted features 104–5; handcrafted features 103–4
dynamic machine learning 44–5

elasticity 63
electronic health records (EHRs) 20, 244, 266; advantages and disadvantages *159*; data 196; genome sequence 196; informed consent 196; and patient information *184*; physician referral challenges and 183; text fields 16; time demands 183
electronic medical record (EMR) 146, 244
electronic systems, practical constraints in 19–20
Elekta's cancer care systems 186
embedded feature selection methods 231
EMR (electronic medical record) 146, 244
endometrial cancer 149
ETLs (extraction, transformation, and loading) systems 13–14
EUROCAT project 244
exploratory model 227–8
extended Markov blanket (MB) approach 49, *51*
extraction, transformation, and loading (ETLs) systems 13–14
ex vivo irradiation 249

FAIR (findable, accessible, interoperable, and reusable) principles 3, 154
false discoveries 232–3
Fast Healthcare Interoperability Resources (FHIR) 165
feature engineering 225–7

feature extraction via convolutional auto-encoders 106
feature selection, radiomics 229–30; methods 230–1; redundancy tests 230; robustness tests 230
feed-forward process 269
FHIR (Fast Healthcare Interoperability Resources) 165
filter feature selection methods 231
findable, accessible, interoperable, and reusable (FAIR) principles 3, 154
first-order intensity histogram features 224
fitting process 45
Foundational Model of Anatomy (FMA) 5, 19
four Vs of data 24
Fowler equation 246
Fowler, J. 246
functional-group–based analysis 86–7

GARD (genomic-adjusted radiation dose) 149, 204
gene expression model 149
gene signatures 202–3
genomewide association studies (GWAS) 205–6, 208, 249
genomic-adjusted radiation dose (GARD) 149, 204
genomics: decipher 210; gene signatures 202–3; hypoxia signatures 204; PORTOS 210; RadiotypeDx 203; RSI 203–4; uses 201–2
geographic model validation 227–8
Google App Engine 64, 67
Google File System (GFS) 34
GPU-based MC dose engine 127–9; brachytherapy sources 133; DICOM library 128; GPU basics 129–30; kV photons, calculations 133; LINACs 131; PBT 132–3; phase–space file 131–2; PSL 131; special procedures, calculations 133–4; thread divergence 130–1
gradient boosting algorithms 232
gradient descent algorithm 269–70
graph databases 32
graph theory 32
gray-level co-occurrence matrix feature 224
gray-level resampling 223
Grid computing model 63
GWAS (genomewide association studies) 205–6, 208, 249

Hadoop distributed file system (HDFS) 34, 69
Hadoop software framework 34–5, 69–70
handcrafted features 103–4
Haralick features 224
hardware layer 66
HBase tables 35
HDFS (Hadoop distributed file system) 34, 69
health care clouds, issues in 72
health care informatics systems 13–14
health care system 147–8, 194–5
health insurance industry 281

Health Insurance Portability and Accountability Act of 1996 (HIPAA) 72; accessibility 190–2; overview 189; patient data storage 190–2, *191*; PHI *see* protected health information (PHI)

Health Level-7 (HL7) 165, 169

health literacy 194–5

HealthMyne industry 186

heterogeneity model 89

high-intensity focused ultrasound (HIFU) 148

high risk cancer 272

HIPAA *see* Health Insurance Portability and Accountability Act of 1996 (HIPAA)

Hive Query Language (HiveQL) 36

Hive software 36–7

HL7 (Health Level-7) 165, 169

Holm-Bonferroni method 233

homogeneity model 89

hospital–based CR **155**, 156

human performance metrics 117

hybrid architecture model 170–1

hybrid Cloud model 65

hypothesis and data-driven research workflows 26–8, *27*

hypothesis testing 47

hypoxia: PET measurements 248; signatures 204

IaaS (infrastructure as a service) 64

ICISS (Integrated Cancer Information and Surveillance System) 150

IFPV (iso-dose feature-preserving voxelization) 106–7

ILS (incident learning systems) 117

Image Biomarker Standardization Initiative 224

imaging: acquisition 221–2; features 2; preprocessing 223

IMRT (intensity-modulated radiotherapy) 4

incident learning systems (ILS) 117

information systems 72–3

information theoretic approaches 46

infrastructure as a service (IaaS) 64

in-memory databases 30, 35

Integrated Cancer Information and Surveillance System (ICISS) 150

integrative analysis methods 88–9, *89*, **90**

intensity-modulated radiotherapy (IMRT) 4

internal–external cross-validation approach 228

Internet technologies 71

interstitial pneumonia 225–6, *226*

inverse planning: class solution to 99; clinical domain knowledge 97–8, *98*

iso-dose feature-preserving voxelization (IFPV) 106–7

isoeffect fractionation regimes 246

Italian Association of Cancer Registries (AIRTUM) 157

Jeong model 247

JobTracker task 70

key–value stores technology 30

k-fold CV technique 46

k-nearest neighbors (*k*-NN) 7

knowledge-based planning 103; DVH prediction 103–5; IFPV 106–7; voxel-wise 3D dose prediction 105–6

Kohane, I. 2

Laplacian of Gaussians (LoG) filters 224

LASSO (least absolute shrinkage and selection operator) 209, 231–2

LDCT (low-dose CT scan) 271

learning health system model 5, 147

least absolute shrinkage and selection operator (LASSO) 209, 231–2

leave-one-out CV (LOO-CV) 46–7

linear accelerator (LINAC) 131

linked data *see* semantic web

Lloyd's algorithm 49

LoG (Laplacian of Gaussians) filters 224

Logical Observation Identifiers Names and Codes (LOINC) 14–15

logistic regression 7, 229, 232

longitudinal imaging, TCP modeling 256

low-dose CT scan (LDCT) 271

low-level API 32

low risk cancer 272

"LQ+time" equation 246

lung cancer 265–6, 271–4

lung cancer prediction: DTs and RF 279–80; multivariable logistic regression 277–8; SVM 278–9

lung tumor *220*

machine learning methods 14, 41–2; algorithms validation 45–8; approaches 43–5; art of learning 42; biomedical sciences 210–13; cancer prognosis/prediction, application in 42–3; classification *42*; dynamic 44–5; and pattern recognition *193*; PHI 192–3; predictive modeling 207–8; radiogenomics using 207–10; in radiotherapy big data 48–57; static 44–5; supervised learning 207; systems 192–3, *193*; unsupervised learning 207

manually segmented ROIs 222–3

Map function 33

MapReduce framework 32–3, *33*, 68, 68–9

marginal screening-based approach 82–3

MC *see* Monte Carlo (MC)

Memorial Sloan Kettering Cancer Center (MSKCC) 185–6

metaheuristic search method 49

mitigation via network level security 38

mobile health app 281

model building, radiomics 227; algorithms 231–2; exploratory/validation 227–8; false discovery, controlling 232–3; feature selection 229–31; prediction/interpretability 229; prognostic/predictive 229
model validation 227–8
molecular features 2
MongoDB databases 35
Monte Carlo (MC): dose calculations 129; kV photons, treatment planning *131*; MV photons, treatment planning *132*; simulation 75, 132
MSKCC (Memorial Sloan Kettering Cancer Center) 185–6
multidimensional data 81
multi-parameterized neural network model 280–1
multi-patient atlas-based dose-prediction approach 106
multiple independent data sets 88
multiple resampling 47
multi-tenancy configurations 65, *65*, 71
multivariable logistic regression 277–8

naïve Bayes (NB) classifiers 7
namenode 69–70
national and international collaborative initiatives 6
National Death Index (NDI) **160**
National Institute of Standards and Technology (NIST) 62
National Institutes of Health (NIH) 76, 150
National Network of Libraries in Medicine (NNLM) 194
National Radiation Oncology Registry (NROR) 156–7
natural language processing (NLP) 3
NB (naïve Bayes) classifiers 7
NDI (National Death Index) **160**
neighborhood gray-tone difference matrix feature 224
neurons, ANN 268
NewSQL databases 32
NIH (National Institutes of Health) 76, 150
90/10 method 256
NIST (National Institute of Standards and Technology) 62
NLP (natural language processing) 3
NNLM (National Network of Libraries in Medicine) 194
non-small-cell lung cancer (NSCLC) 48, **48**
normal tissue complication probability (NTCP) model 243, 248–9; bias and confounding factors 250; image-based predictors 256; methodologies 255–7; spatial dose patterns for 256
normal tissue tolerances 137, **138**, 140
not only SQL (NoSQL) databases 32, 35; AllegroGraph 36; attack vectors 38, **39**; default security 38; HBase 35; MongoDB 35
novel BN building approach 49–51

NROR (National Radiation Oncology Registry) 156–7
NTCP model *see* normal tissue complication probability (NTCP) model

objective function *versus* iteration index *100*
observational cohort 253–4
OLAP (online analytical processing) 28
OLTP (online (real-time) transaction processing) 28
omics data 81
omics measurements *84*
Oncospace database 76
online analytical processing (OLAP) 28
online (real-time) transaction processing (OLTP) 28
ontology 5–6, *6*, 172–3
optimized radiotherapy 242
optional data elements 161–3, **162–3**
organ dose 125; *see also* personal organ dose archive (PODA); distributions in patient *125*; kVCBCT-contributed mean doses *127*; MC 126; OAR 125; out-of-field doses 124–5; WPI, PO-IMRT, and IGRT **126**
outcomes studies, quality and safety data 113
oxygen enhancement ratio 246

PaaS (platform as a service) 64
PACSs (picture archiving and communications systems) 74, 168
patient-centered care 149
patient cohort homogeneity 221
patient health literacy 194–5
patient-reported outcomes (PROs) 17, 149–50
patient-reported outcomes measures (PROMs) 254–5
penalization technique 83, 90
penalized model 209
permutation test 47, 233
personalized medicine 280–1
personalized radiation oncology 4–5
personal organ dose archive (PODA) 127–9, *128*; automatic dose accumulation 135–6; clinical applications 139–40; database *135*, 135–6; DIR 134; dose report and morbidity analysis 136; MC dose calculations 129; normal tissue tolerances 137, **138**; warnings/intervention 137
PET measurements 248
phase–space file 131
PHI *see* protected health information (PHI)
Physician Quality Reporting System 195
picture archiving and communications systems (PACSs) 74, 168
Pig Latin query language 36–7, 70
platform as a service (PaaS) 64
PODA *see* personal organ dose archive (PODA)

PODA, clinical applications: cancer risk monitoring and control 139; normal tissue tolerance benchmark 140; organ protection 139; personal health data archive 140; radiation treatment intervention 139
population-based CR **155**, 156
portal implementations 37–8
portal requirements **29**, 33–4
post-operative radiation therapy outcomes score (PORTOS) 203
precision medicine, predictive analytics models for 196–7
preconditioning random forest regression 209
predicting disease progression/treatment response 4–5
predictive model/modeling 207–8, 250; clinical variables 255; data cohorts for 251–4; data splitting for validation 255; dose–volume histogram parameters 255; Seven Deadly Sins 257
predictive studies 229
prior clinical knowledge/data, roles *98*
private Cloud model 65
processing requirements **29**, 32; low-level API 32; MapReduce 32–3, *33*; query languages 33
process metrics, quality and safety data 114–16
process standardizations and cultural shifts 15; ATPS 16–17; diagnosis and staging 15–16; PROs 17
prognostic studies 229
progressive analysis approach 91–2
PROMs (patient-reported outcomes measures) 254–5
PROs (patient-reported outcomes) 17, 149–50
prospective–retrospective analyses 253
prostate cancer 148–50
protected health information (PHI) 72; data visualization 192–3; innovations in 189–90; machine learning 192–3
protein function prediction 210–11, **211**
proton beam therapy (PBT) 132–3
pseudo-inverse method 45
public Cloud model 65

qualification process 4
quality and safety data *113*; data standards/exchange/ interoperability 112; outcomes 113; process metrics 114–16; quality gap 111–12; structure metrics 117–18; validation of 118
Quantitative Analysis of Normal Tissue Effects in the Clinic (QUANTEC) 124, 137, 243
quantitative feature 223; agnostic 224; derived 224–5; engineered 225–7
quantitative health care 244–5
quantitative imaging 220
query languages 33

Radiation Oncology Ontology (ROO) 5–6, 172
radiation treatment intervention 139
radiobiological assays 247–8

radiogenomics 220, 249; biological data analysis 209–10; large-scale projects 206–7; projects in 205–6; using machine learning 207–10
radiogenomics consortium (RGC) 204–5
radiomics 219; acquisition, image 221–2; analysis stages 220, *221*; future work 234–5; model building 227–33; preprocessing, image 223; quantitative feature extraction *see* quantitative feature; *versus* quantitative imaging 220; results 233–4; ROI 221; segmentation, ROI 222–3; software packages 224, 230
radiophenotype 243
radiosensitivity index (RSI) 203–4, 248
radiotherapy: data crisis in 243–4; data types **169**; low model generalizability and 245; obsolete 242; optimized 242
Radiotherapy Outcome Estimator (ROE) 251, *252*
Radiotherapy Toxicity to Reduce Side-effects and Improve Quality-of-Life in Cancer Survivors (REQUITE) project 206–7
RadiotypeDx 203
RadLex 5
random forest (RF) 231–2; approach 53–6; ensemble method 229; lung cancer prediction 279–80; model 208–9
randomized clinical trial (RCT) cohorts 253–4
randomized controlled trial 146–7
random sampling 42
Rapid Learning Health Care (RLHC) 154
RCT (randomized clinical trial) cohorts 253–4
RDDs (resilient distributed data sets) 70
RDF (Resource Description Framework) 32, 171–2
Real-time Recursive Learning methods 45
receiver operating characteristic (ROC) curve 50, 271, *272*; during-treatment full BN *54*; methodology 118; pre-treatment full BN *50*; RF model *55*; two-step BN approach *54*
recursive Bayesian method 45
Reduce function 33
redundancy tests 230
regions of interest (ROIs), segmentation 221–3
Registro Tumori Ospedaliero (RTO) 157
regression models 229
regularization technique 45–6
Remote Dictionary Server (Redis) 35
repeatability 230
repeated *k*-fold CV 47
reproducibility 230
REQUITE (Radiotherapy Toxicity to Reduce Side-effects and Improve Quality-of-Life in Cancer Survivors) project 206–7
resilient distributed data sets (RDDs) 70
Resource Description Framework (RDF) 32, 171–2
RF *see* random forest (RF)
RGC (radiogenomics consortium) 204–5
RLHC (Rapid Learning Health Care) 154

robustness tests 230
ROC curve *see* receiver operating characteristic (ROC) curve
ROE (Radiotherapy Outcome Estimator) 251, *252*
ROO (Radiation Oncology Ontology) 5–6, 172
RSI (radiosensitivity index) 203–4, 248
RTO (Registro Tumori Ospedaliero) 157

SaaS (software as a service) 64–5
safety profile assessment 118
Samuel, A. 41
Sawzall programming language 70
scalability 63
SEER (Surveillance, Epidemiology and End Results) program 3, 156
semantic web 32, 171–2
sensitivity, measuring accuracy 270–1
sequential forward/backward selection 231
shape features 224
shared decision making 154
significance testing 47–8
simple resampling 46
single nucleotide polymorphisms (SNPs) 49, 206, 208
skilled community approach 18–19, *19*
small-data analytics 80
smearing 223
SNPs (single nucleotide polymorphisms) 49, 206, 208
software as a service (SaaS) 64–5
solidity, engineered feature 226–7, *227*
Spark software 36–7, 70
spatial dose patterns 256
specificity, measuring accuracy 270–1
standardizations development principles 18; electronic systems, practical constraints in 19–20; extensibility 20–1; skilled community approach 18–19, *19*; templates/automation in workflow 20
static machine learning 44–5
statistical methods 80–1; additive models, joint analysis on 83–4; classic meta-analysis 88; computation problem 91–2; conditioning–integration strategy 84–6; data structures 89–90; decomposition–integration strategy 86; functional-group–based analysis 86–7; integrative analysis 88–9, *89*; marginal screening-based approach 82–3; multidimensional data 81–7; multiple independent data sets 88–91; TCGA BRCA data *82*
statistical resampling approaches 46–7
storage system requirements 29, **29**; DFS 30; in-memory databases 30
structured data 3
structure metrics, quality and safety data: culture metrics 117; human performance metrics 117; ILS 117; safety profile assessment 118
supervised learning 207

supervised *versus* unsupervised methods 43, *44*
support vector machines (SVMs) 7, 232, 278–9
Surveillance, Epidemiology and End Results (SEER) program 3, 156
survival analysis 229
sustainable big data research platforms 28; data modeling requirements **29**, 30–2; portal requirements **29**, 33–4; processing requirements **29**, 32–3; storage system requirements **29**, 29–30
SVMs (support vector machines) 7, 232, 278–9

Tableau software 193
TCGA BRCA data *82*
TCP model *see* tumor control probability (TCP) model
temporal model validation 228
10:1 rule 228
tenant 65–6
texture features 224
three-dimensional conformal radiation therapy (3D-CRT) 123–4
tissue toxicity/tumor control, model for 242–3
topology, Bayes net model 115
transactional/analytic workloads, attributes 28, **28**
transactional systems, exchange unit standardization in 14–15
tranSMART data warehouses 37–8
treatment evaluation techniques 4
treatment planning 97; automated determination 100–3; convergence and computational efficiency 99; inverse planning 97–9; knowledge-based planning 103–7
triplestores 32
TRIPOD Statement 255
tumor control probability (TCP) model 243; bias and confounding factors 250; imaging to refinement of 248; isoeffect fractionation regimes 246; longitudinal imaging for 256; mechanistic TCP models 246; methodologies 255–7; radiobiological assays 247–8
tumor response–evolution models 246–7

unstructured data 3
unsupervised learning 207

variable selection methods 83–4
virtualization 62–3
virtual machine (VM) 62
VMAT/IMRT plan optimization *101*
volume *versus* real-time response 34; Hadoop and HDFS 34–5; in-memory databases 35
voxel-wise 3D dose prediction 105–6

waterfall model 26, *26*
Watson oncology 185; MSKCC 185–6; for radiation oncology 186
wrapper feature selection methods 230–1

Printed and bound by CPI Group (UK) Ltd, Croydon, CR0 4YY

24/10/2024

01778295-0013